Nanobiomaterials

Perspectives for Medical Applications in the Diagnosis and Treatment of Diseases

Edited by

Jorddy Neves Cruz

Institute of Biological Sciences, Federal University of Pará, Belém, Pará, Brazil

Published by **Materials Research Forum LLC**
Millersville, PA 17551, USA

Published as part of the book series
Materials Research Foundations
Volume 145 (2023)
ISSN 2471-8890 (Print)
ISSN 2471-8904 (Online)

Print ISBN 978-1-64490-236-3
eBook ISBN 978-1-64490-237-0

Distributed worldwide by

Materials Research Forum LLC
105 Springdale Lane
Millersville, PA 17551
USA
https://www.mrforum.com

Manufactured in the United States of America
10 9 8 7 6 5 4 3 2 1

Table of Contents

Preface

Nanotechnology is a rapidly evolving field that has already had a significant impact on a wide range of industries, including medicine, environmental cleanup, and materials science. The fundamental aspects of nanotechnology, as well as its biomedical and technological applications, are the focus of this book.

We begin with an overview of the synthesis of nanomaterials, including both top-down and bottom-up approaches. We also delve into the principles of green chemistry and their application to the synthesis of nanomaterials.

One important application of nanotechnology is the synthesis and characterization of silver nanoparticles with antimicrobial action. In this book, we explore the processes of synthesis and characterization of these nanoparticles, as well as the biology, mechanisms, and future prospects of their antimicrobial action.

Nanomaterials have also been shown to have potential in bone regeneration and wound treatment. We examine the use of nanomaterials in these applications, as well as the in silico methods for evaluating the mode of interaction of nanoparticles with molecular targets.

Finally, we discuss the use of nanotechnology in cancer diagnosis and environmental cleanup, as well as the ethical considerations of nanomedicine.

This book aims to provide a comprehensive overview of the diverse and rapidly expanding field of nanotechnology, highlighting both its current and future potential. The work reported in the following twelve chapters is as follows:

Chapter 1: In this chapter, the synthesis of nanomaterials and nanocomposites is discussed, including the various synthetic routes for nanoparticle formation and the properties that depend on these routes. Nanomaterials are substances with at least one dimension smaller than 100 nanometers that show special properties due to their size. They can be tailored to meet specific needs by controlling their size, shape, synthesis conditions, and functionalization, and can be classified based on various factors. The field of nanoscience was opened up by Richard Feynman's famous talk in 1959, and recent technological advancements have allowed for closer study of nanomaterials.

Chapter 2: In this chapter, the characteristics of nanomaterials are highlighted, including their exceptional magnetic, electrical, mechanical, optical, and catalytic properties. The properties of nanomaterials can be tailored to meet specific needs by carefully controlling their size, shape, synthesis conditions, and functionalization. The chapter also outlines and defines the terminologies associated with nanomaterials and covers a range of nanomaterial synthesis techniques, including top-down and bottom-up methods. Nanomaterials are materials with at least one dimension between 1 and 100 nanometers that have high surface area and differ significantly from their bulk counterparts.

Chapter 3: Nanomaterials have had numerous industrial breakthroughs in the 21st century, but their synthesis often involves hazardous chemicals from non-renewable sources. One alternative is the synthesis of nanomaterials using the principles of green chemistry, which prioritize environmental friendliness, cost-effectiveness, and biocompatibility. These principles use natural resources for nanomaterial synthesis and are on their way from the laboratory to commercial application. In this chapter, the principles of green chemistry for nanomaterial synthesis are presented, along with recent advances and challenges in the field.

Chapter 4: In summary, nanoparticles synthesized using green methods have a wide range of potential uses in the medicinal and environmental fields. These methods utilize safe organic resources like plants and microorganisms to produce nanoparticles that can be used for various applications such as catalysis, sensing, electronics, photonics, and medicine. The production of plant-mediated nanoparticles and current uses of materials such as gold, silver, copper, palladium, platinum, zinc oxide, and titanium dioxide are discussed, as well as the principles of green chemistry which aim to minimize waste and maintain efficacy.

Chapter 5: Nanomaterials have shown potential as effective antibacterial agents due to their special properties and ability to effectively treat infectious diseases, including those that are antibiotic-resistant. In this chapter, the characteristics of microorganisms and how they differ among strains are discussed, as well as the toxicity mechanisms and effectiveness of nanomaterials in treating different bacteria. The potential of nanomaterials to combat drug resistance is also examined, along with predictions for their future in science. The search for novel therapies has been driven by the increasing resistance of microbes to conventional antibiotics.

Chapter 6: Silver nanoparticles are synthesized through various methods such as chemical, physical, and biology, with the biology method being the safest, most eco-friendly, and environmentally sustainable option. The size and shape of the nanoparticles are important factors in their ability to enter and affect bacteria cells. Silver nanoparticles can disrupt the cell membrane and enter cells, where they can interact with proteins, enzymes, DNA, ribosomes, and mitochondria, leading to their inactivation and the death of the cells. The obtained nanoparticles are characterized using techniques such as TEM, SEM, XRD, and UV.

Chapter 7: The use of nanomaterials in wound treatment has gained significant attention in recent years due to their unique properties, such as high surface area, chemical stability, biocompatibility, and the ability to deliver drugs and growth factors effectively. In this chapter, the various types of nanomaterials used in wound treatment, their mechanisms of action, and their potential advantages and disadvantages are discussed. These include nanofiber dressings, nanoparticles, nanocapsules, and nanogels, among others. The chapter also covers the current state of research on nanomaterials for wound treatment, including clinical trials and ongoing studies. It concludes with a discussion of the future prospects and challenges of using nanomaterials in wound treatment.

Chapter 8: Nanomaterials are being explored as a means of facilitating wound healing due to their antibacterial, antifungal, and anti-inflammatory properties. These materials, including metal-based nanomaterials and organic-inorganic and organic-organic nanocomposites, have the potential to revolutionize treatment and reduce the economic burden of wound care. Acute and chronic wounds are distinguished by the healing process, and proper wound care is crucial to prevent infection and protect the internal environment from exposure to harmful microbes.

Chapter 9: Molecular docking is a useful tool for structure-based drug design, which involves predicting the optimal spatial arrangement between a ligand and its target and calculating the associated complex free energy. While molecular docking techniques are widely used, it can be challenging to find accurate scoring systems. This chapter discusses the current methods for molecular docking and their applications in the field of nanostructures.

Chapter 10: Early detection of cancer is critical for successful treatment, but traditional diagnostic methods often have limited sensitivity and specificity. Nanotechnology offers new materials and contrast agents that may allow for earlier and more accurate diagnosis of cancer and the ability to track cancer treatment progress. This chapter discusses recent developments in the use of nanotechnology for cancer diagnosis and the challenges that must be overcome for these approaches to be widely adopted in clinical settings. Nanotechnology has the potential to provide highly sensitive and specific measurements of extracellular cancer biomarkers and to detect cancer cells.

Chapter 11: Nanotechnology, or the use of particles or materials at the nanoscale (1 to 100 nanometers), has been explored as a means of capturing and degrading environmental pollutants. The reactivity of nanotechnologies depends on the surface area-to-volume ratio of the nanoparticles used, and the selection of specific nanoparticles for a given pollution problem is crucial. This chapter discusses the various types of nanotechnologies, including polymeric-, semiconductor-, ceramic-, metal-, and carbon-based nanoparticles, and their applications in the remediation of different types of environmental pollutants, including heavy metals, organic and inorganic substances, chemical herbicides, volatile and aromatic compounds, and more. The impact of these nanoparticles on living organisms, including plants, animals, and humans, is also considered.

Chapter 12: Nanomedicine is the application of nanomaterials and nanotechnologies in healthcare, and has enabled the development of mRNA COVID-19 vaccines and other technologies such as digital twins, organ-on-chip systems, and wearables. While nanomedicine has the potential to improve patient outcomes, it also raises concerns about nanosafety and ethical impacts, including issues related to freedom, equality, data protection, and biosecurity. Researchers in the field of nanomedicine should strive to follow the principles of responsible research and innovation in collaboration with governments and other stakeholders, with a focus on inclusivity, anticipation, openness, and responsiveness.

Materials Research Foundations 145 (2023) 1-18 https://doi.org/10.21741/9781644902370-1

Chapter 1

Nanotechnology: Fundamental Aspects and Biomedical and Technological Applications

Indrani Das Sarma[1,*], Debashis Bhowmick[2], Pawan Bhilkar[3], Rohit Sharma[4] and Ratiram G. Chaudhary[3,*]

[1]Jhulelal Institute of Technology, Nagpur, India

[2]Smt. Rajkamal Baburao Tidke Mahavidyalaya, Mouda, Nagpur, India

[3]Post Graduate Department of Chemistry, Seth Kesarimal Porwal College of Arts and Science and Commerce, Kamptee-441001, India

[4]Department of Rasa Shastra and Bhaishajya Kalpana, Faculty of Ayurveda, Institute of Medical Science, Banaras Hindu University, Varanasi, India

Indrani Das Sarma, dassarmaindrani@gmail.com and Ratiram G. Chaudhary, chaudhary_rati@yahoo.com

Abstract

Richard Feynman's famous talk titled '*There's Plenty of Room at the Bottom: An Invitation to Enter a New Field of Physics*' at Caltech's annual American Physical Society meeting in 1959 highlighted the importance of nanoscience and opened a whole new vista of the nanoworld. In the International System of Units nanoscience, the prefix nano means one-billionth or 10^{-9}. So 1 nanometer (nm) = 10^{-9} m. Nanomaterials are defined as a set of substances whose at least one dimension is approximately less than 100 nm and show special properties due to minuscule size. For instance, variable colors shown by gold and silver nanoparticles, change in the oxidation state of nanosized aluminum crystals, and so on. Nanomaterials already exist in nature. However, the recent advancements in microscopy and technology have empowered scientists and technologists to witness phenomena occurring naturally at the nanoscale dimension. These phenomena are mostly based on quantum mechanical interactions and expanded surface area. The nanomaterials can be classified in various modes: dimension, composition, morphology, geometry, etc. This chapter intend to overview the classification of nanomaterials and their applications in the different fields.

Keywords

Nanomaterials, Quantum Dots, Nanomaterials in Biomedicals, Nanomaterials in Agriculture

Materials Research Foundations 145 (2023) 1-18 https://doi.org/10.21741/9781644902370-1

Contents

1. Introduction to nanoscience and nanomaterial

'*There's Plenty of Room at the Bottom: An Invitation to Enter a New Field of Physics*' at the annual American Physical Society meeting at Caltech in 1959 by Sir Richard Feynman highlighted the importance of nanoscience and opened a whole new vista of the nanoworld. The nanoscale as against bulk is depicted in Fig.1. The nanoscience in the International System of Units, the prefix nano means one-billionth or 10^{-9}. So 1 nanometer (nm) = 10^{-9} m.

Materials Research Forum LLC
https://doi.org/10.21741/9781644902370-1

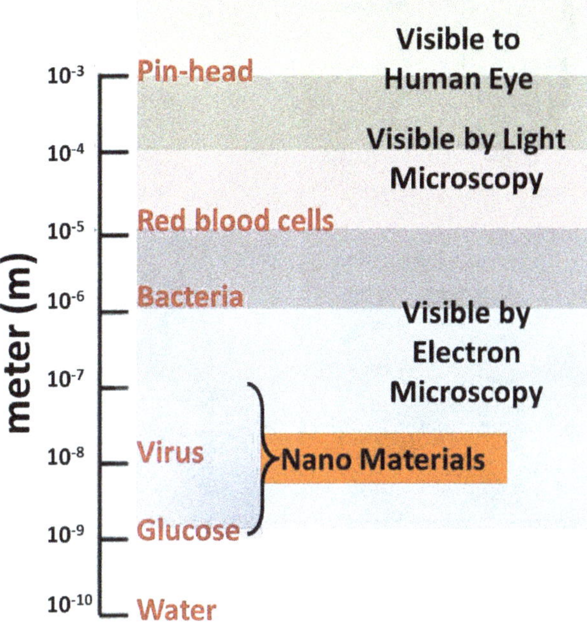

Figure 1. The dimension of nanoscale as compared to bulk.

Nanomaterials are a set of substances with at least one dimension of approximately 100 nm and show special properties due to their minuscule size. For instance, variable colors shown by gold and silver nanoparticles, change in the oxidation rate of nanosized aluminum crystals, and so on.

Since ancient times, the unusual and intriguing properties of materials at minuscule scales have been observed. However, the reasons for this anomaly were not fully understood. The secrets of ancient *nano* production are passed from generation to generation by word of mouth and/or practice. The utility of cotton, jute, wool, etc., as fabrics has been well-known since immemorial, as evidenced by archaeological findings. They cultivated them and processed them into fibers and fabrics. The network of pores with a size range of 1-20 nanometers (typical nanoporous materials) provides natural fabrics with beneficial properties [1].

Materials Research Foundations 145 (2023) 1-18
https://doi.org/10.21741/9781644902370-1

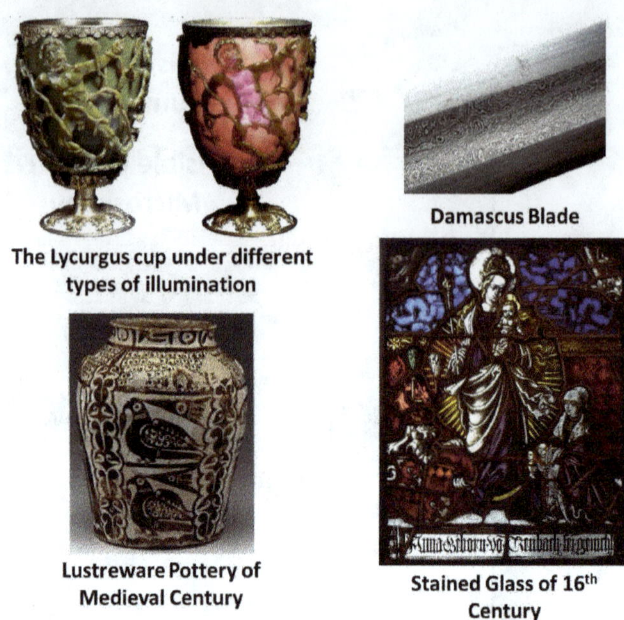

Figure 2 Ancient artifacts across the world depicting the use of nanomaterials.

The various properties of metals and their oxides in nano-dimensions have been used since time immemorial, either knowingly or unknowingly. The presence of awe-inspiring ancient artifacts across the world, as shown in Fig. 2 is a testament to the wonders of nanomaterials. The famous *Lycurgus cup* made up of dichroic glass, gives evidence that the art of making gold and silver nanoparticles was known to Roman civilization during the 4th century C.E. The Lusterware Pottery by Abbasid Potters in the 8th century C.E. stands as testimony to the usage of nanoparticles in production of the ceramics. These medieval technologists knowingly or unknowingly used the plasmonic effect of the nanometal particles. The famous *Damascus* blades, known for their remarkable hardness, sharpness, and lightweight, were popular for their military use. This is a high-carbon content steel originating in Southern India. Microscopic analysis of the material showed the presence of carbon nanotubes in the steel [2].

Similarly, using various otherwise toxic heavy metals like mercury and lead as life savings *Bhasmas* (herbs-metallic medicinal preparations) was prevalent in India and well-documented in Ayurveda since time immemorial [3,4]. The therapeutic action of the otherwise toxic metals may be attributed to their particle size, which has to be in the nano-dimension region. The method documented shows the probability of a top-down approach in its preparation. Thus, nanomaterials profoundly impacted the ancient and medieval world in various industries comprising glass, ceramics, metal forging, medicine, etc.

Nanomaterials are a significant part of volcanic ash, dust, oceanic spray, viruses, etc. Microclusters of polyaromatic hydrocarbons have been noticed in various galaxies, including our Milky Way [2]. However, the recent advancements in microscopy, spectroscopy and instrumentations have empowered scientists to comprehend the natural phenomena of matter at the nanoscale.

2. Properties of nanomaterials

The nanomaterials have different mechanical, chemical, electrical, optical, and magnetic properties by their size as compared to their bulk counterparts. These unusual properties are primarily due to quantum effects and expanded surface area. The properties of nanomaterials are discussed following way.

1. The particles in the nano or quantum realm have widely different properties from that of bulk. For instance, gold and silver particles in the nano realm exhibiting particle size-dependent color are well-known. The electronic motion is confined at the nano dimension, influencing their properties. Almost all the properties, such as melting point, fluorescence, electrical conductivity, magnetic permeability, and chemical reactivity, are influenced at this realm.

2. The concept of tunability of the nanomaterials due to the quantum effects. So, by changing the particle size, the material's properties can be finely tuned.

3. The nanomaterial applications for flash memory devices and scanning tunneling microscopy (S.T.M.) are due to the nanoscale's tunneling effect.

4. The expanded surface areas of nanomaterials compared to their bulk counterparts make them an ideal candidate for water treatment. The large surface area increases the contact area for adsorption or catalytic activities.

5. Nanomaterials' chemical, mechanical, optical, and physical properties vary differently from those of bulk materials. This has been evidenced since medieval times, as given in the previous section. Also, the non-linear optical properties of nanostructured semiconductors are well-known.

6. The properties of materials, such as elastic modulus, fatigue resistance, hardness, etc., vary for nanomaterial compared to the bulk. By virtue of their smaller dimensions, these materials are prone to practically no or lesser imperfections like dislocations, defects, impurity precipitates, etc. This leads to better material with a lesser chance of failure during operation. Also, the defects or impurities trapped in the nanomaterials, if any, would be highly energetic and would find their way to the surface to release the energy under a suitable purification process like annealing. This eventually results in perfect structure within the nanomaterial[5]. The improvised mechanical properties have higher

Materials Research Foundations 145 (2023) 1-18
https://doi.org/10.21741/9781644902370-1

applicability in nanoscale, composites, and macroscale. A plethora of super hard nanocomposites of borides, carbides, nitrides, and so on have been reported [6]. They have promising potential as hard protective coatings. The nanometer size is instrumental in many novel properties which are otherwise absent in bulk materials.

7. Significant changes in properties like resistance to scratch, hardness, fracture toughness, and elastic modulus are seen as a function of size.

3. Classification of nanomaterials

Various aspects such as dimension, chemical composition, morphology, geometry, and so on form the basis of nanomaterials' classification. This chapter only briefly describes type based on the measurement and chemical composition.

3.1 Dimensions (QDs, one dimension, two & three dimensional)

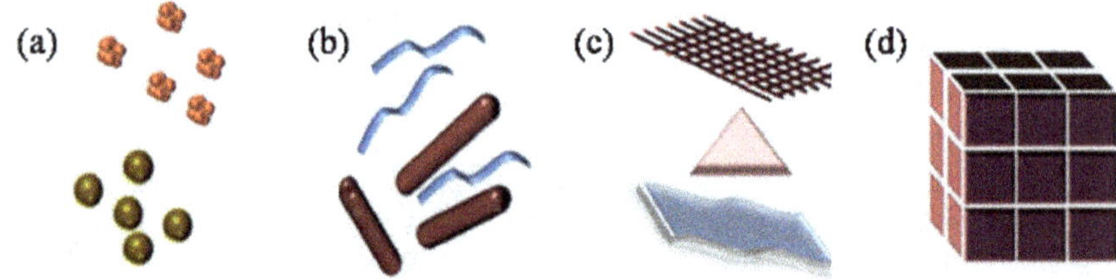

Figure 3. Classification based on dimensions: (a) 0-D, (b) 1-D, (c) 2-D and (d) 3-D.

The nanomaterials are classified based on dimensions, as shown in Fig 3.

(a) Zero dimensional (0-D): All three dimensions, i.e., length, breadth, and height, are in the nanometers. They represent point-like entities on a geometrical scale, hence are 0-D. For instance, fullerenes, nanoparticles, and quantum dots.

(b) One-dimensional (1-D): They have one dimension outside the nanoscale. They represent rod and fiber-like entities on a geometrical scale, hence are 1-D. For instance, carbon nanotubes, nanofibers, nanorods, nanowires, etc.

(c) Two-dimensional (2-D): They have two dimensions outside the nanoscale. They represent layer-like entities in geometrical scale, hence are 2-D. For instance, nanosheets, nanofilms, and nanolayers.

(d) Three-dimensional (3-D): They are composite nanomaterials. The overall structure is not confined to the nanometer. However, they are arrays of nanowires/ nanotubes or dispersions of nanoparticles.

3.2 Chemical composition

Nanomaterials are classified on the basis of chemical composition, as shown in Fig. 4

(a) Organic nanoparticles: They are made up of organic compounds like proteins, carbohydrates, lipids, polymers, and so on [6,7] . For instance, dendrimers, micelles and protein complexes and so on. They have wide range of application in drug delivery and therapeutics.

(b) Carbon-based nanoparticles: They are made from carbon atoms. For instance, fullerenes, carbon black nanoparticles, carbon nanotubes and carbon quantum dots. They have varied applications in drug delivery, energy storage, bioimaging, photovoltaics and environmental sensing and so on [8,9].

(c) Inorganic nanoparticles: They are metal, ceramic and semiconductor nanoparticles. Metal nanoparticles can be sub-classified as mono-, bi- or poly- metallic nanoparticles. They have characteristic optical and electrical properties which could be attributed to surface plasmon resonance [10]. Ceramic nanoparticles are carbonates, carbides, sulfides and oxides of metals or metalloids. They may be amorphous, polycrystalline, dense, porous or hollow in structure [10]. Apart from biomedical applications, they find applications in catalysis, dye degradation and optoelectronics [11].

Semiconductor nanoparticles have properties between metals and non-metals. They have wide bandgaps which can be tuned. The tuning of band gap significantly affects its properties [10]. They are instrumental in photocatalysis, optics, electronics, etc.

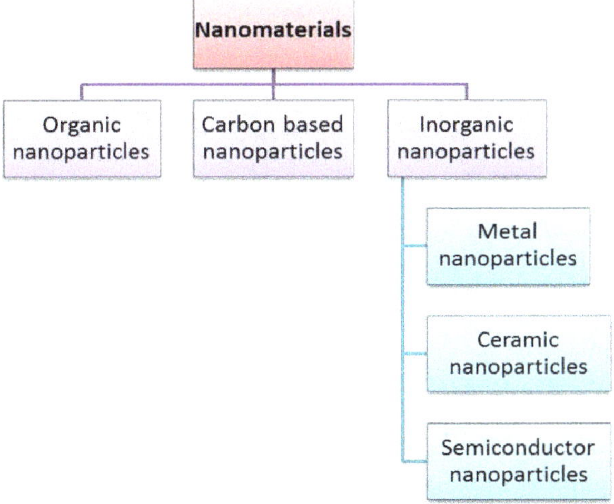

Figure 4 Classification of nanomaterials based on chemical composition.

Materials Research Foundations 145 (2023) 1-18 https://doi.org/10.21741/9781644902370-1

4. Applications of nanomaterials in different sectors

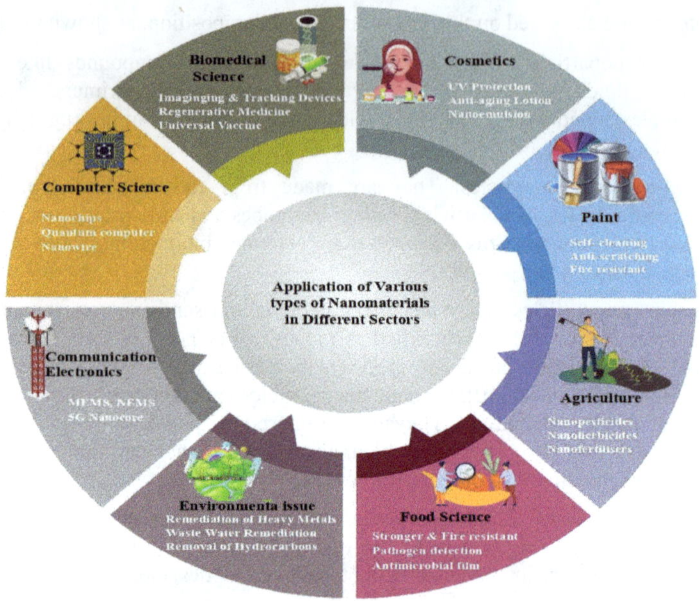

Figure 5. Multiple applications of nanomaterials in various sectors.

Nanomaterials offer gigantic possibilities for various fields of contemporary science and industry. The recent research developments in nanomaterial science is delivering a wide array of benefits in multiple disciplines in both expected and unexpected ways, viz. biomedical science, cosmetics, food industry, agriculture, environmental science, electronics and communication, and computer science, among many others (Fig 5).

4.1 Nanomaterials in computer science and communication electronics

Nanotechnology enables the designing and manufacturing electrical devices and parts, enabling the development of more trustworthy, rapid, and portable computers.

Both top-down and bottom-up strategies are used in nanomaterial science. These, for example, are used to create nanotubes and nanolayers, which are then used to create computer nanochips and nano electrical components [12]. Carbon nanotube computers have processors made of carbon nanotubes (CNTs). Compared to silicon chips, CNT exhibits functional similarities to silicon transistors and the potential to develop energy-efficient and speedier electronic gadgets [13]. Qubits serve as the foundation for quantum computers (quantum Bits). The atomic spin state is controlled via magnetic fields. By

Materials Research Foundations 145 (2023) 1-18
https://doi.org/10.21741/9781644902370-1

altering the magnetic field on a subatomic scale, massive amounts of data may be sent and stored at an unprecedented velocity [14,15].

Devices with deep memory are being created using magnetic nanowires formed of an alloy of iron and nickels. The fabrication process for low-cost, highly dependable, high-density memory devices has been created by I.B.M. researchers [16]. In terms of the radio solution, scaling and nanotechnology enable the construction of systems with multiple nanoscale resonators, such as NEMS devices, which could be employed for GHz signal processing applications. Future wireless communication systems will benefit greatly from the features of microelectromechanical systems (MEMS) and nanoelectromechanical systems (NEMS). This is particularly crucial for wireless communications systems with high data rates [17]. The development of Transparent Resistive Random Access Memory (T.R.R.A.M.) enables high-capacity storage and transparency in mobile devices, opening up new possibilities for data storage [18]. Nano Equipment (N.E.), the 5G Nano core, is included with mobile devices. Utilizing this system, computation and communication are constantly available and prepared to serve the customer intelligently [19].

4.2 Nanomaterials in biomedical science

Especially through helping in the detection of disorders, diagnostics, and as a targeted and effective treatment strategy, nanoscience is offering a wide range of roles in diagnostics, therapeutics, and disease prevention [20]. Applications include surgery, cancer diagnosis and treatment, cellular imaging, implants, tissue engineering, and delivering devices for drugs, proteins, genes, and radionuclides [21–23]. Various nanomedicines are being developed pharmaceutically to manage various health ailments [24,25]. Compelling evidence reports the immense potential of Au nanoparticles in managing cancer and several other disorders, and they have found commercial use as probes for detecting certain nucleic acid sequences. Quantum dots have also demonstrated significant potential as intracellular imaging and tracking agents [26–28]. Different drug-encapsulating nanocarriers, including liposomes, polymer nanoparticles, and produced nanocrystals of drugs, can facilitate sustained and targeted drug administration to combat infectious diseases. Therapies of local infections of the female genitourinary system, lungs, and skin have been tested using these methods. Additionally, being investigated for systemic drug delivery are injectable nanocarriers [29]. Many different types of NPs have been created, including metals or metal oxides, liposomes, polymers, fullerene, nanoemulsion, solid-lipid, and polylactide-co-glycoside (P.L.G.A.) NPs. These various types are used in the diagnostics and treatment of various health disorders, particularly the more serious conditions like cancer metastasis, brain cancers, and neurodegeneration [30,31].

Liposomes, polymeric micelles, dendrimers, polymeric nanoparticles, carbon nanotubes, and mesoporous silicon constitute a family of nanomaterials that are frequently utilised in the treatment of cardiovascular disorders [32]. Nanoscience is emerging as a beneficial application in bone tissue and neural tissue engineering in regenerative medical therapies [33]. New crystal-mineral materials can be developed that resemble the bone tissue structure, and they also have valuable applications in dentistry [34].

Researchers in the field of nanomedicine is exploring how nanotechnology might enhance vaccinations, such as by enabling needle-free administration of vaccines [35]. Additionally, scientists aim to construct a universal vaccine scaffold that would cover more forms of the flu and need fewer resources to develop each year [36]. Compelling evidence shows that bio-nano materials have immense potential against several microbes and vectors involved in pyrexia and related disorders [37–39]. Recent reports substantiate nanomedicines' beneficial roles in treating neurological or neurodegenerative diseases [40–43]. Comprehensive information recently discussed nanoparticles' valuable role in COVID-19 management and vaccine development [44,45].

4.3 Nanomaterials in cosmetics and paints

Nanomaterials are being used in cosmetology to develop various products with several benefits, such as improved U.V. protection, deeper skin penetration, long-lasting effects, heightened color, superior finish, etc. Micellar nanoparticles are also a top-rated cosmetic product used in markets of different geographical regions. Makeup removers, facial cleansers, anti-aging lotion, sunscreen, aqua-based cosmetic preparations, and oil-in-water nanoemulsions are highly effective formulations in the cosmetic industry [46].

Nanomaterials used for various cosmetic applications include inorganic nanoparticles like titanium dioxide (TiO_2), zinc oxide (ZnO), silica (SiO_2), carbon black, nano-organic particles like tris-biphenyl triazine, nanohydroxyapatite, gold and silver nanoparticles, and buckyballs (buckminsterfullerene/C60) [47]. Surfaces with bactericidal coatings are becoming more and more necessary for use in homes and hospitals. Antimicrobial paints with incorporated silver nanoparticles are created [48]. In addition, engineered TiO_2, Ag, and SiO_2 nanoparticles offer well-proven U.V. protection, self-cleaning, and air purification properties. Additionally, they serve as anti-scratch coatings and fire retardants [49].

4.4 Nanomaterials in agriculture and food science

The agrotechnological revolution, which has the immediate ability to transform the entire agricultural system while ensuring food security, has been made possible in part by nanotechnology. Numerous nanoparticles, including multiwalled carbon nanotubes (MWCNTs), silica NPs, TiO_2 NPs, cerium NPs, and cerium oxide NPs, as well as nanoparticle-based preparations, such as nano-sized pesticidal, herbicidal, fungicidal products, fertilizers, sensors, etc are being studied for plant growth, stress tolerance, and soil improvement [50,51].

In order to give consumers food that is safe and free from contamination and to ensure that the food has enhanced functional capabilities that are acceptable to consumers, the potential for using nanoparticles in the food business is being researched. Food manufacture, processing, and packaging are all covered by the use of nanoscience. Food quality, food safety, and the health advantages that food provides are all greatly improved by nanomaterials. When used in food processing, storage, and consumption, nanoencapsulation prevents moisture, heat, chemical, or biological deterioration and manages interactions between bioactive ingredients and the food matrix. They also contain the release of bioactive ingredients and improve availability at a predetermined rate and time [52,53].

Techniques for nanoencapsulation increase food quality and flavor. Numerous techniques like nanocomposite, nano-emulsification, and nanostructured have been used for the effective and precise distribution of nutrients like protein and antioxidants [54]. The use of "smart" and "active" food packaging made of nanocomposites has several benefits over traditional packaging techniques, including improved packing material with increased strength, barrier features, antimicrobial films, and nano-sensing for microbial detection and consumer notification of the state of quality or safety of food [55]. Utilizing innocuous nanoscale fillers like chitin or chitosan, silica (SiO_2) nanoparticles, clay and silicate nanoplatelets, or chitin or chitosan results in a polymer matrix lighter, stronger, more fire resistant, and has better thermal properties [56].

4.5 Nanomaterials in environmental challenges

Nanotechnology may interact with the environment in three ways: first, through biological congruency; second, through geochemical and biological cycle-based transformation processes; and third, through the use of technological means to address environmental issues [57]. Silica nanoparticles are intended to adsorb mercury, cadmium, and lead from combustion gases [58–60]. To capture and store CO_2 emissions from thermal power plants' smokestacks and tailpipes, a metal-organic framework of silica with nanopores, magnesium oxide nanocrystals, and porous nanocubes of chromium terephthalate has been incorporated [61,62]. Alumina (Al_2O_3) nanoparticles, Silica (SiO_2) nanoparticles, Titania (TiO_2) nanoparticles, graphene-based nanostructures, zerovalent iron nanomaterials, Cu or ZnO nanoparticles, and many other nanomaterials are used in water remediation [63,64]. Transforming or detoxifying the pesticides has been recently reported following the use of metal nanoparticles of metals, bimetals, metal oxides, and carbon-nanotubes ions [65,66]. Heavy metals, fluorides, chlorophenols, dyes, hydrocarbons, and radionuclides have all been removed from the hydrosphere using nano-adsorbents [67,68]. Derived from nanomaterials, nano fertilizers (Nfs) have increased nutrient penetration potential and improved nutrient use efficiency due to large surface area and smaller particle size. Nanofertilizers come in various forms, including nanocarriers, nanoscale additives, and fertilizers with nanoparticle coatings [69].

Conclusions

This chapter provides an overview of classification and application of nanomaterials. The particles produced have good size distribution and hence suitable properties. The nanomaterials are promising tools of the future as it seems that this is the scale at which nature works. All the reactions in a living organism or in nature occur at this dimension. The diffusion of ions from soil to plant tissues, oxygen across the alveoli to hemoglobin, or harnessing energy catalyzed by enzymes of the class ATPases occur at the nanoscale. Thus, it seems that nature has mastered this domain. With the upgradation of technology and miniaturization of devices, the nano dimension is an interesting domain to explore. In conclusion, the nanotechnology is a powerful technique can mitigates several issues like environmental challenges, electronic communication, biomedical sciences, agriculture-based and various industrials issues.

References

[1] S. Bayda, M. Adeel, T. Tuccinardi, M. Cordani, F. Rizzolio, The history of nanoscience and nanotechnology: From chemical-physical applications to nanomedicine, Molecules. 25 (2020) 112. https://doi.org/10.3390/molecules25010112

[2] Munir Nayfeh, Nanoeffects in Ancient Technology and Art and in Space, in: Fundam. Appl. Nano Silicon Plasmon. Fullerines, Elsevier, 2018: pp. 497–518. https://doi.org/10.1016/b978-0-323-48057-4.00016-5

[3] D. Pal, C.K. Sahu, A. Haldar, Bhasma: The ancient Indian nanomedicine, J. Adv. Pharm. Technol. Res. 5 (2014) 4–12. https://doi.org/10.4103/2231-4040.126980

[4] S. Nagarajan, S. Krishnaswamy, B. Pemiah, K. Rajan, U. Krishnan, S. Sethuraman, Scientific insights in the preparation and characterisation of a lead-based naga bhasma, Indian J. Pharm. Sci. 76 (2014) 38–45

[5] G. Cao, Nanostructures and Nanomaterials, PUBLISHED BY IMPERIAL COLLEGE PRESS AND DISTRIBUTED BY WORLD SCIENTIFIC PUBLISHING CO., 2004. https://doi.org/10.1142/p305

[6] S. Veprek, A.S. Argon, Mechanical properties of superhard nanocomposites, Surf. Coatings Technol. 146–147 (2001) 175–182. https://doi.org/10.1016/S0257-8972(01)01467-0

[7] K. Pan, Q. Zhong, Organic Nanoparticles in Foods: Fabrication, Characterization, and Utilization, Annu. Rev. Food Sci. Technol. 7 (2016) 245–266. https://doi.org/10.1146/annurev-food-041715-033215

[8] W.K. Oh, H. Yoon, J. Jang, Size control of magnetic carbon nanoparticles for drug delivery, Biomaterials. 31 (2010) 1342–1348. https://doi.org/10.1016/j.biomaterials.2009.10.018

[9] S. Chandra, P. Das, S. Bag, D. Laha, P. Pramanik, Synthesis, functionalization and bioimaging applications of highly fluorescent carbon nanoparticles, nanoscale. 3 (2011) 1533–1540. https://doi.org/10.1039/c0nr00735h

[10] I. Khan, K. Saeed, I. Khan, Nanoparticles: Properties, applications and toxicities, Arab. J. Chem. 12 (2019) 908–931. https://doi.org/10.1016/j.arabjc.2017.05.011

[11] P.B. Chouke, T. Shrirame, A.K. Potbhare, A. Mondal, A.R. Chaudhary, S. Mondal, S.R. Thakare, E. Nepovimova, M. Valis, K. Kuca, R. Sharma, R.G. Chaudhary, Bioinspired metal/metal oxide nanoparticles: A road map to potential

applications, Mater. Today Adv. 16(2022) 100314. https://doi.org/10.1016/j.mtadv.2022.10031

[12] P. Iqbal, J.A. Preece, P.M. Mendes, Nanotechnology: The "Top-Down" and "Bottom-Up" Approaches, in: Supramol. Chem., John Wiley & Sons, Ltd, Chichester, UK, 2012. https://doi.org/10.1002/9780470661345.smc195

[13] R. Maheswaran, B.P. Shanmugavel, A Critical Review of the Role of Carbon Nanotubes in the Progress of Next-Generation Electronic Applications, J. Electron. Mater. 51 (2022) 2786–2800. https://doi.org/10.1007/s11664-022-09516-8

[14] G. A.I. Quantum, Hartree-Fock on a superconducting qubit quantum computer, science (80-.). 369 (2020) 1084–1089. https://doi.org/10.1126/science.abb9811

[15] N.C. da R. Galucio, D. de A. Moysés, J.R.S. Pina, P.S.B. Marinho, P.C. Gomes Júnior, J.N. Cruz, V.V. Vale, A.S. Khayat, A.M. do R. Marinho, Antiproliferative, genotoxic activities and quantification of extracts and cucurbitacin B obtained from Luffa operculata (L.) Cogn, Arab. J. Chem. 15 (2022) 103589. https://doi.org/10.1016/j.arabjc.2021.103589

[16] L. Sun, Y. Hao, C.L. Chien, P.C. Searson, P.C. Searson, Tuning the properties of magnetic nanowires, I.B.M. J. Res. Dev. 49 (2005) 79–102. https://doi.org/10.1147/rd.491.0079

[17] D. Dubuc, K. Grenier, L. Rabbia, A. Tackac, M. Saadaoui, P. Pons, P. Caudrillier, O. Pascal, H. Aubert, H. Baudrand, J. Tao, P. Combes, J. Graffeuil, R. Plana, MEMS and NEMS technologies for wireless communications, in: 2002 23rd Int. Conf. Microelectron. M.I.E.L. 2002 - Proc., IEEE, 2002: pp. 91–98. https://doi.org/10.1109/MIEL.2002.1003153

[18] K.C. Liu, W.H. Tzeng, K.M. Chang, Y.C. Chan, C.C. Kuo, C.W. Cheng, Transparent resistive random access memory (T.R.R.A.M.) based on Gd 2O3 film and its resistive switching characteristics, in: I.N.E.C. 2010 - 2010 3rd Int. Nanoelectron. Conf. Proc., IEEE, 2010: pp. 898–899. https://doi.org/10.1109/INEC.2010.5425137

[19] S. Sharma, H.I. Rasool, V. Palanisamy, C. Mathisen, M. Schmidt, D.T. Wong, J.K. Gimzewski, Nano core-A Review on 5G Mobile Communications, A.C.S. Nano. 4 (2010) 1921–1926.

[20] D.B. Buxton, S.C. Lee, S.A. Wickline, M. Ferrari, Recommendations of the National Heart, Lung, and Blood Institute Nanotechnology Working Group, Circulation. 108 (2003) 2737–2742. https://doi.org/10.1161/01.CIR.0000096493.93058.E8

[21] W. Ma, Y. Zhan, Y. Zhang, C. Mao, X. Xie, Y. Lin, The biological applications of D.N.A. nanomaterials: current challenges and future directions, Signal Transduct. Target. Ther. 6 (2021) 351. https://doi.org/10.1038/s41392-021-00727-9

[22] H. Chopra, A.K. Mishra, I. Singh, Y.K. Mohanta, R. Sharma, T. Bin Emran, S. Bibi, Nano-chitosan: A novel material for glioblastoma treatment, Int. J. Surg. 104 (2022) 106713. https://doi.org/10.1016/j.ijsu.2022.106713

[23] C.B.R. Santos, K.L.B. Santos, J.N. Cruz, F.H.A. Leite, R.S. Borges, C.A. Taft, J.M. Campos, C.H.T.P. Silva, Molecular modeling approaches of selective adenosine receptor type 2A agonists as potential anti-inflammatory drugs, J. Biomol. Struct. Dyn. 39 (2021) 3115–3127. https://doi.org/10.1080/07391102.2020.1761878

[24] R. Sharma, P. Bedarkar, D. Timalsina, A. Chaudhary, P.K. Prajapati, Bhavana, an Ayurvedic Pharmaceutical Method and a Versatile Drug Delivery Platform to Prepare Potentiated Micro-Nano-Sized Drugs: Core Concept and Its Current Relevance, Bioinorg. Chem. Appl. 2022 (2022) 1–15. https://doi.org/10.1155/2022/1685393

[25] R. Sharma, P.K. Prajapati, Nanotechnology in medicine: Leads from Ayurveda, J. Pharm. Bioallied Sci. 8 (2016) 80–81. https://doi.org/10.4103/0975-7406.171730

[26] M. Dahan, S. Lévi, C. Luccardini, P. Rostaing, B. Riveau, A. Triller, Diffusion Dynamics of Glycine Receptors Revealed by Single-Quantum Dot Tracking, Science (80-.). 302 (2003) 442–445. https://doi.org/10.1126/science.1088525

[27] F.S. Alves, J. de A. Rodrigues Do Rego, M.L. Da Costa, L.F. Lobato Da Silva, R.A. Da Costa, J.N. Cruz, D.D.S.B. Brasil, Spectroscopic methods and in silico analyses using density functional theory to characterize and identify piperine alkaloid crystals isolated from pepper (Piper Nigrum L.), J. Biomol. Struct. Dyn. 38 (2020) 2792–2799. https://doi.org/10.1080/07391102.2019.1639547

[28] X. Gao, Y. Xing, L.W.K. Chung, S. Nie, Quantum Dot Nanotechnology for Prostate Cancer Research, in: Prostate Cancer, Humana Press, Totowa, NJ, 2007: pp. 231–244. https://doi.org/10.1007/978-1-59745-224-3_13

[29] A.R. Kirtane, M. Verma, P. Karandikar, J. Furin, R. Langer, G. Traverso, Nanotechnology approaches for global infectious diseases, Nat. Nanotechnol. 16 (2021) 369–384. https://doi.org/10.1038/s41565-021-00866-8

[30] E.E. Ngowi, Y.Z. Wang, L. Qian, Y.A.S.H. Helmy, B. Anyomi, T. Li, M. Zheng, E.S. Jiang, S.F. Duan, J.S. Wei, D.D. Wu, X.Y. Ji, The Application of Nanotechnology for the Diagnosis and Treatment of Brain Diseases and Disorders, Front. Bioeng. Biotechnol. 9 (2021). https://doi.org/10.3389/fbioe.2021.629832

[31] A.R.J.A. de M. Lima, A.S. Siqueira, M.L.S. Möller, R.C. de Souza, J.N. Cruz, A.R.J.A. de M. Lima, R.C. da Silva, D.C.F. Aguiar, J.L. da S.G.V. Junior, E.C. Gonçalves, In silico improvement of the cyanobacterial lectin microvirin and mannose interaction, J. Biomol. Struct. Dyn. (2020). https://doi.org/10.1080/07391102.2020.1821782

[32] Y. Deng, X. Zhang, H. Shen, Q. He, Z. Wu, W. Liao, M. Yuan, Application of the Nano-Drug Delivery System in Treatment of Cardiovascular Diseases, Front. Bioeng. Biotechnol. 7 (2020). https://doi.org/10.3389/fbioe.2019.00489

[33] M. Fathi-Achachelouei, H. Knopf-Marques, C.E. Ribeiro da Silva, J. Barthès, E. Bat, A. Tezcaner, N.E. Vrana, Use of Nanoparticles in Tissue Engineering and Regenerative Medicine, Front. Bioeng. Biotechnol. 7 (2019). https://doi.org/10.3389/fbioe.2019.00113

[34] S.A. Saunders, Current practicality of nanotechnology in dentistry. Part 1: Focus on nanocomposite restoratives and biomimetics, Clin. Cosmet. Investig. Dent. 1 (2009) 47–61. https://doi.org/10.2147/cciden.s7722

[35] B. Mangla, S. Javed, M.H. Sultan, W. Ahsan, G. Aggarwal, K. Kohli, Nanocarriers-Assisted Needle-Free Vaccine Delivery Through Oral and Intranasal Transmucosal Routes: A Novel Therapeutic Conduit, Front. Pharmacol. 12 (2022). https://doi.org/10.3389/fphar.2021.757761

[36] Y. Wang, L. Deng, S.M. Kang, B.Z. Wang, Universal influenza vaccines: from viruses to nanoparticles, Expert Rev. Vaccines. 17 (2018) 967–976. https://doi.org/10.1080/14760584.2018.1541408

[37] P. Chaudhary, R. Sharma, S. Rawat, P. Janmeda, Antipyretic Medicinal Plants, Phytocompounds, and Green Nanoparticles: An Updated Review, Curr. Pharm. Biotechnol. 24 (2022) 23–49. https://doi.org/10.2174/1389201023666220330005020

[38] S. Bawazeer, A. Rauf, T. Bin Emran, A.S.M. Aljohani, F.A. Alhumaydhi, Z. Khan, L. Ahmad, H.A. Hemeg, N. Muhammad, R. Sharma, A. Maalik, I. Khan, Biogenic Synthesis of Silver Nanoparticles Using Rhazya stricta Extracts and Evaluation of Its Biological Activities, J. Nanomater. 2022 (2022) 1–11. https://doi.org/10.1155/2022/7365931

[39] P. Kumari, K. Raina, S. Thakur, R. Sharma, N. Cruz-Martins, P. Kumar, K. Barman, S. Sharma, D. Kumar, P.K. Prajapati, R. Sharma, A. Chaudhary, Ethnobotany, Phytochemistry and Pharmacology of Palash (Butea monosperma (Lam.) Taub.): a Systematic Review, Curr. Pharmacol. Reports. 8 (2022) 188–204. https://doi.org/10.1007/s40495-022-00286-9

[40] M.M. Rhaman, M.R. Islam, S. Akash, M. Mim, M. Noor alam, E. Nepovimova, M. Valis, K. Kuca, R. Sharma, Exploring the role of nanomedicines for the therapeutic approach of central nervous system dysfunction: At a glance, Front. Cell Dev. Biol. 10 (2022). https://doi.org/10.3389/fcell.2022.989471

[41] R. Sharma, N. Garg, D. Verma, P. Rathi, V. Sharma, K. Kuca, P.K. Prajapati, Indian medicinal plants as drug leads in neurodegenerative disorders, in: Nutraceuticals Brain Heal. Beyond, Elsevier, 2020: pp. 31–45. https://doi.org/10.1016/B978-0-12-820593-8.00004-5

[42] R. Sharma, A. Kabra, M.M. Rao, P.K. Prajapati, Herbal and Holistic Solutions for Neurodegenerative and Depressive Disorders: Leads from Ayurveda, Curr. Pharm. Des. 24 (2018) 2597–2608. https://doi.org/10.2174/1381612824666180821165741

[43] R. Sharma, K. Kuca, E. Nepovimova, A. Kabra, M.M. Rao, P.K. Prajapati, Traditional Ayurvedic and herbal remedies for Alzheimer's disease: from bench to bedside, Expert Rev. Neurother. 19 (2019) 359–374. https://doi.org/10.1080/14737175.2019.1596803

[44] A. Rauf, T. Abu-Izneid, A.A. Khalil, N. Hafeez, A. Olatunde, M. Rahman, P. Semwal, Y.S. Al-Awthan, O.S. Bahattab, I.N. Khan, M.A. Khan, R. Sharma, Nanoparticles in clinical trials of COVID-19: An update, Int. J. Surg. 104 (2022) 106818. https://doi.org/10.1016/j.ijsu.2022.106818

[45] C.M.A. Rego, A.F. Francisco, C.N. Boeno, M. V. Paloschi, J.A. Lopes, M.D.S. Silva, H.M. Santana, S.N. Serrath, J.E. Rodrigues, C.T.L. Lemos, R.S.S. Dutra, J.N. da Cruz, C.B.R. dos Santos, S. da S. Setúbal, M.R.M. Fontes, A.M. Soares, W.L. Pires, J.P. Zuliani, Inflammasome NLRP3 activation induced by Convulxin, a C-type lectin-like isolated from Crotalus durissus terrificus snake venom, Sci. Rep. 12 (2022) 1–17. https://doi.org/10.1038/s41598-022-08735-7

[46] Z.A.A. Aziz, H. Mohd-Nasir, A. Ahmad, S.H. Siti, W.L. Peng, S.C. Chuo, A. Khatoon, K. Umar, A.A. Yaqoob, M.N. Mohamad Ibrahim, Role of Nanotechnology for Design and Development of Cosmeceutical: Application in Makeup and Skin Care, Front. Chem. 7 (2019). https://doi.org/10.3389/fchem.2019.00739

[47] V. Gupta, S. Mohapatra, H. Mishra, U. Farooq, K. Kumar, M.J. Ansari, M.F. Aldawsari, A.S. Alalaiwe, M.A. Mirza, Z. Iqbal, Nanotechnology in Cosmetics and Cosmeceuticals—A Review of Latest Advancements, Gels. 8 (2022) 173. https://doi.org/10.3390/gels8030173

[48] G. Moxham, Protective paints, Nat. Nanotechnol. (2008). https://doi.org/10.1038/nnano.2008.27

[49] S. Smulders, K. Luyts, G. Brabants, K. Van Landuyt, C. Kirschhock, E. Smolders, L. Golanski, J. Vanoirbeek, P.H.M. Hoet, Toxicity of nanoparticles embedded in paints compared with pristine nanoparticles in mice, Toxicol. Sci. 141 (2014) 132–140. https://doi.org/10.1093/toxsci/kfu112

[50] N.B. Singh, R.G. Chaudhary,M.F. Desimone, A. Agrawal, S. K. Shukla, Green synthesized nanomaterials for safe technology in sustainable agriculture, Curr. Pharm. Biotechnol. 24 (2023) 61-85. https://doi.org/10.2174/1389201023666220608113924

[51] L. Zhao, L. Lu, A. Wang, H. Zhang, M. Huang, H. Wu, B. Xing, Z. Wang, R. Ji, Nano-Biotechnology in Agriculture: Use of Nanomaterials to Promote Plant Growth and Stress Tolerance, J. Agric. Food Chem. 68 (2020) 1935–1947. https://doi.org/10.1021/acs.jafc.9b06615

[52] J. Ubbink, J. Krüger, Physical approaches for the delivery of active ingredients in foods, Trends Food Sci. Technol. 17 (2006) 244–254. https://doi.org/10.1016/j.tifs.2006.01.007

[53] Nitin Yadav, Anil K. Singh,Talha Bin Emran

,Ratiram G. Chaudhary, Rohit Sharma

,Swati Sharma, and Kalyan Barman, Salicylic acid treatment reduces lipid peroxidation and chlorophyll degradation and preserves quality attributes of pointed gourd fruit, J. Food Quality, (2022) 2090562. https://doi.org/10.1155/2022/2090562

[54] T. Singh, S. Shukla, P. Kumar, V. Wahla, V.K. Bajpai, Application of nanotechnology in food science: Perception and overview, Front. Microbiol. 8 (2017). https://doi.org/10.3389/fmicb.2017.01501

[55] S.D.F. Mihindukulasuriya, L.T. Lim, Nanotechnology development in food packaging: A review, Trends Food Sci. Technol. 40 (2014) 149–167. https://doi.org/10.1016/j.tifs.2014.09.009

[56] S.H. Othman, Bio-nanocomposite Materials for Food Packaging Applications: Types of Biopolymer and Nano-sized Filler, Agric. Agric. Sci. Procedia. 2 (2014) 296–303. https://doi.org/10.1016/j.aaspro.2014.11.042

[57] Y. Jiang, X. Quan, G. Jiang, X. Li, Current Prospective on Environmental Nanotechnology Research in China, Environ. Sci. Technol. 53 (2019) 4001–4002. https://doi.org/10.1021/acs.est.9b01489

[58] M.H. Lee, K. Cho, A.P. Shah, P. Biswas, Nanostructured sorbents for capture of cadmium species in combustion environments, Environ. Sci. Technol. 39 (2005) 8481–8489. https://doi.org/10.1021/es0506713

[59] E. Pitoniak, C.Y. Wu, D.W. Mazyck, K.W. Powers, W. Sigmund, Adsorption enhancement mechanisms of silica-titania nanocomposites for elemental mercury vapor removal, Environ. Sci. Technol. 39 (2005) 1269–1274. https://doi.org/10.1021/es049202b

[60] P. Biswas, M.R. Zachariah, In situ immobilization of lead species in combustion environments by injection of gas phase silica sorbent precursors, Environ. Sci. Technol. 31 (1997) 2455–2463. https://doi.org/10.1021/es9700663

[61] A.R. Millward, O.M. Yaghi, Metal-organic frameworks with exceptionally high capacity for storage of carbon dioxide at room temperature, J. Am. Chem. Soc. 127 (2005) 17998–17999. https://doi.org/10.1021/ja0570032

[62] C. Férey, C. Mellot-Draznieks, C. Serre, F. Millange, J. Dutour, S. Surblé, I. Margiolaki, Chemistry: A chromium terephthalate-based solid with unusually large pore volumes and surface area, science (80-.). 309 (2005) 2040–2042. https://doi.org/10.1126/science.1116275

[63] A. Kaur, Nanoscavengers for the Waste Water Remediation, in: D.G.K. Prof. Rajeev Kumar, Dr. Raman Kumar (Ed.), New Front. Nanomater. Environ. Sci., Springer Singapore, Singapore, 2021: pp. 73–89. https://doi.org/10.1007/978-981-15-9239-3_4

[64] R.G. Chaudhary, V. Sonkusare, G. Bhusari, A. Mondal, A.K. Potbhare, R. Sharma, H.D. Juneja, A.A. Abdala, Preparation of mesoporous ThO2 nanoparticles: influence of calcination on morphology and visible-Light-driven photocatalytic degradation of indigo carmine and methylene blue, Enviornmental Research, 222 (2023) 115363. https://doi.org/10.1016/j.envres.2023.115363

[65] W. Chen, L. Duan, D. Zhu, Adsorption of polar and nonpolar organic chemicals to carbon nanotubes, Environ. Sci. Technol. 41 (2007) 8295–8300. https://doi.org/10.1021/es071230h

[66] A. Khalid, P. Ahmad, A. Khan, S. Muhammad, M.U. Khandaker, M.M. Alam, M. Asim, I.U. Din, R.G. Chaudhary, D. Kumar, R. Sharma, M.R.I. Faruque, T. Bin Emran, Effect of Cu Doping on ZnO Nanoparticles as a Photocatalyst for the Removal of Organic Wastewater, Bioinorg. Chem. Appl. 2022 (2022) 1–12. https://doi.org/10.1155/2022/9459886

[67] J.M. Schnorr, T.M. Swager, Emerging applications of carbon nanotubes, Chem. Mater. 23 (2011) 646–657. https://doi.org/10.1021/cm102406h

[68] X. Ren, C. Chen, M. Nagatsu, X. Wang, Carbon nanotubes as adsorbents in environmental pollution management: A review, Chem. Eng. J. 170 (2011) 395–410. https://doi.org/10.1016/j.cej.2010.08.045

[69] L.R. Khot, S. Sankaran, J.M. Maja, R. Ehsani, E.W. Schuster, Applications of nanomaterials in agricultural production and crop protection: A review, Crop Prot. 35 (2012) 64–70. https://doi.org/10.1016/j.cropro.2012.01.007

Materials Research Foundations 145 (2023) 19-53 https://doi.org/10.21741/9781644902370-2

Chapter 2

An Overview of the Synthesis of Nanomaterials

Arati Gavali[1], Jolina Rodrigues[1], Navinchandra Shimpi, Purav Badani[1,*]

[1] Department of Chemistry, University of Mumbai, Santacruz (East), Mumbai, India

*pmbadani@chem.mu.ac.in

Abstract

Nanomaterials have distinguished themselves as an outstanding class of materials with at least one dimension falling between 1 and 100 nm. The logical design of nanoparticles allows exceptionally high surface area. It is possible to create nanomaterials with exceptional magnetic, electrical, mechanical, optical, and catalytic capabilities that differ significantly from their bulk counterparts. By carefully regulating the size, shape, synthesis conditions, and appropriate functionalization, the properties of nanomaterials can be tailored to meet specific needs. This chapter highlights the particular characteristics of nanomaterials. We specifically outline and define terminologies associated with nanomaterials. The discussion covers a range of nanomaterial synthesis techniques, including top-down and bottom-up methods.

Keywords

Nanomaterials, Synthesis, Top-Down and Bottom-Up, Electrospinning, Catalysis

Contents

Materials Research Foundations 145 (2023) 19-53 https://doi.org/10.21741/9781644902370-2

1. Introduction

Nanotechnology is the modification and/or synthesis of materials/substances at the atomic or molecular scale. The Greek word "nanos", which implies "dwarf," is the source of the phrase "nanoscale." The nanoscale is typically measured in nanometres, or billionths of a meter. Materials constructed at this scale frequently exhibit unique physical and chemical properties because of quantum mechanical effects. [1] The complete array of engineering disciplines and basic science (including physics, chemistry, biology, and materials science) are heavily incorporated into nanotechnology. Both the science and the technology of this new subject are often referred to collectively as "nanotechnology." In its most basic sense, Nanoscience is the study of chemical, physical and biological characteristics at the atomic and near-atomic scales. Since the turn of the century, researchers have been studying nanotechnology. Nanotechnology uses a controlled modification of physical and chemical properties to produce materials and functional systems with distinctive traits. [2]

American physicist Richard Feynman, a Nobel Prize laureate, coined the nanotechnology theory in 1959. "There's Plenty of Room at the Bottom" was the title of a lecture that Feynman delivered at the California Institute of Technology in the annual conference of the American Physical Society. In this lecture, Feynman raised the question, "*Why can't we write the full 24 volumes of the Encyclopedia Britannica on the tip of a pin*?" and sketched out a scope of utilizing technologies to build smaller machines down to the molecular level. The concept of nanotechnology revealed that Feynman's hypotheses have been valid. Several ground-breaking advancements have been achieved in the field of nanotechnology, and for this motive, he is acknowledged as the "Father of Modern Nanotechnology."[3] In 1974, Norio Taniguchi was the foremost to employ the term "nanotechnology," which mainly consists of the processing of separation, consolidation, and deformation of materials by one atom or one molecule. [4]

Nanoparticles (NPs) are a diverse class of materials encompassing particulate substances with at least one dimension of less than 100 nm. NPs are the building blocks of nanotechnology. [5] They can exhibit quantum phenomena because of their extremely small scale, large specific surface area per unit of volume, high atom concentration in the surface, and near-surface layers. A straightforward extrapolation of the properties of bulk materials cannot predict the consequent unique properties of nanoparticles. NPs can be synthesized from diverse chemical components, including organic substances, metals, metal oxides, semiconductors, polymers, carbon compounds, and biological materials. They also have various morphological forms, including spheres, cylinders, discs, platelets, hollow spheres, and tubes. Nanomaterials can be categorized as i) zero-dimensional (0D), ii) one-dimensional (1D), iii) two-dimensional (2D), or iv) three-dimensional (3D), depending on their shape. [6] Due to the nanoscale dimension of NPs, they are highly reactive and/or physically aggregative. Therefore, the surface of the generated nanoparticles must typically be changed to passivate and stabilize them. [7–10] The fundamental building blocks for many applications of nanotechnology include NPs. Several synthetic methods for synthesizing NPs use procedures in the gas, liquid, or solid phase. After Feynman sketched a fascinating field of research, several scientists found interest in this field and pioneered two main synthesis approaches. Synthesis of nanomaterials constitutes two subgroups, namely the top-down approach and bottom-up approach, that vary in quality, cost, and speed. [11,12]

2. Synthesis of nanoparticles

Nanostructure materials have gained much attention due to their dimensions, size, and shape. Nanostructure materials are made up of atoms and molecules. NPs have different physical, chemical, electrical and magnetic properties compared to their higher-dimensional counterparts. Numerous techniques exist for creating nanostructured materials with controlled shape, size, orientation, dimensionality, and structure. A material's performance is determined by its qualities. The kinetics and thermodynamics of the synthesis influence the molecular and atomic structure, composition, morphology, defects, and interfaces, which in turn control the characteristics of the material. There are primarily two methods for creating nanomaterials. (Fig. 1) Top-down approach is a method for deconstructing or downsizing macro-crystalline (bulk material) structures while preserving their original integrity, i.e., it moves from the general to the unique. Nanomaterials are also created from atoms and molecules in a bottom-up technique. The bottom-up approach starts with the specific and moves on to the broad.

Materials Research Forum LLC
https://doi.org/10.21741/9781644902370-2

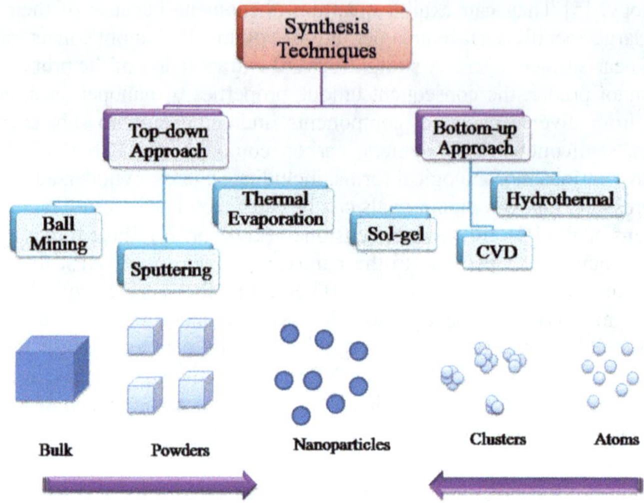

Figure 1. Top-down and bottom-up methods, and other methods for synthesis of nanomaterials.

2.1 Top-down approach

Bulk starting structures are used in top-down methods for synthesizing nanomaterials, which may be superficially manipulated in nanostructure production. In top-down methods, the starting material is generally solid. Top-down approaches are based on the grinding of the materials. Thus, the nature of this method is subtractive. Top-down approaches start with a design produced at a larger size, which is then shrunk to a nanoscale by a series of procedures. Typical examples are ball milling, etching through the mask, cutting, thermal evaporation, laser ablation, sputtering, grinding, photolithography, e-beam lithography, etc.

A) Mechanical Milling

Mechanical milling, often called ball milling, grids nanotubes into ultrafine particles. As the name implies, the ball milling technique uses a mill chamber and balls. A ball mill is a device that rotates several small irons, ceramic, flint stones, stainless steel, hardened steel, silicon carbide, or tungsten carbide balls inside a stainless-steel container. The material powder is then placed in a ball milling chamber and ground to a nanoscale. (Fig. 2) The effect of tiny microscopic hard balls in a concealed vessel causes localized high pressure during ball milling. This result in the creation of particles measuring 2 to 20 nanometers. The ball's rotational speed in the milling chamber determines the size of the particles. The present technique is low-cost for producing nanoscale materials starting from bulk

Materials Research Foundations 145 (2023) 19-53 https://doi.org/10.21741/9781644902370-2

materials. The major shortcoming of this process is the uneven shape of the particles and NPs becoming contaminated by the content of the balls due to the mechanical process. [13]

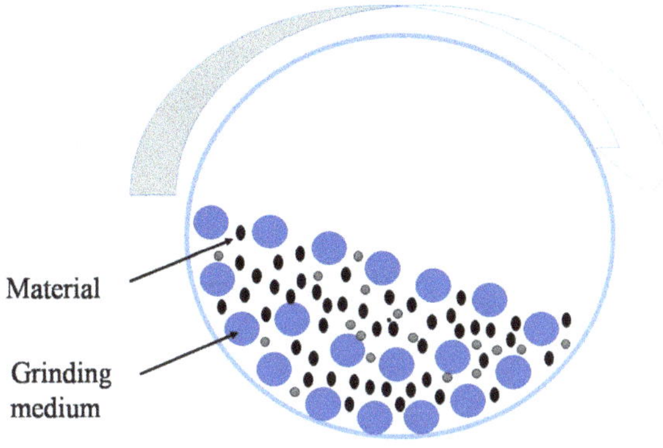

Fiure 2. Ball milling principle schematic representation

B) Electrospinning

Electrospinning is one of the most economical and effective methods for creating fibers with a micro-or nano-diameter. A spinneret, a collector plate, and a high-voltage power source are the three major parts of a standard electrospinning system. In this process, a high-voltage power source generates the electric force (generally 10 to 30 kV), and the melted or mixed polymer solution is placed in a syringe. The charged solution jet on the tip of the spinneret modifies its size to maintain the force equilibrium when subjected to a high electrostatic field. The phenomenon of the solution drop stretching into the Taylor cone results from the interaction of repulsive forces that work in the opposite direction of surface tension. [14] The solution can be fed from the spinneret at a steady rate by a syringe pump. Followed by this, a viscous fluid that has gathered on a metal collector plate experiences an electric force field, leading to nanofiber formation. Fig. 3 illustrates the basic configuration for an electrospinning device process. The electrospinning method has a couple of merits. This includes ease of operation and production of controllable micro/nanosized fibers. The main drawbacks of this method are the requirement to use hazardous solvents, a lack of cell infiltration, and an uneven distribution of cells. [15]

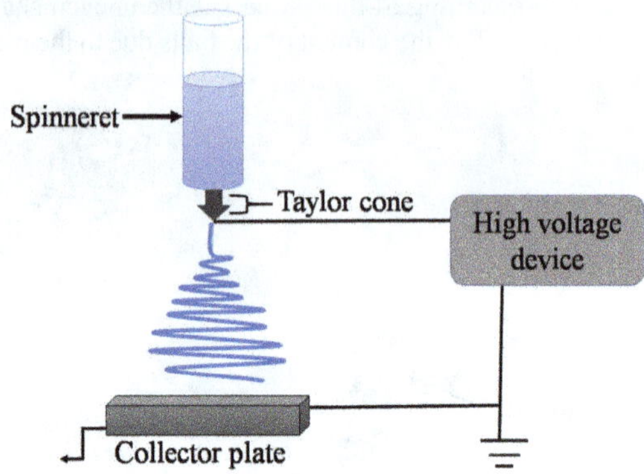

Figure 3. A schematic representation of the electrospinning process

C) Sputtering

Sputtering is a classical mechanics phenomenon that occurs when a solid substance is attacked by energetic plasma or gas particles, which causes the material to eject microscopic particles from its surface. (Fig. 4) Sputtering can be accomplished in several ways, including radio-frequency diode, magnetron, and DC diode sputtering. [16] This approach is useful because the thickness of the film can be readily controlled by establishing the parameters and modifying the deposition time. High capital costs, slow deposition rate, and high probability of impurity introduction are drawbacks of this approach.

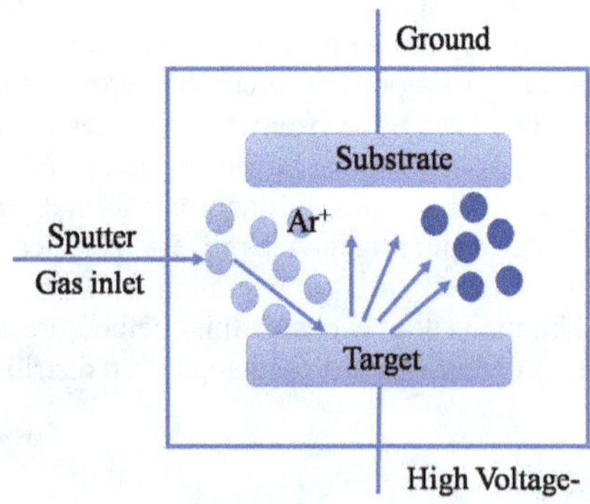

Figure 4. A schematic view of sputtering deposition technique

Materials Research Foundations 145 (2023) 19-53 https://doi.org/10.21741/9781644902370-2

D) Thermal evaporation

Physical vapor deposition (PVD) is frequently accomplished using thermal evaporation. PVD is a technique that allows for the precise enlargement of the extremely high-quality thin layer perpendicular to a substrate's plane. In this technique, the material likely deposited is sputtered from the target or evaporated from the reservoir. It is among the most basic forms of PVD, frequently using a resistive heat source to evaporate a solid substance in a vacuum atmosphere to produce a thin film. (Fig. 5) This technique is utilized for small and large-scale operations since it is simple to operate and requires little maintenance and installation. One major disadvantage of this approach for nanoparticle synthesis is the requirement for high-temperature control during deposition. The precursors and products generally have low thermal stability, which contradicts high-temperature operating conditions for the synthesis method. [17]

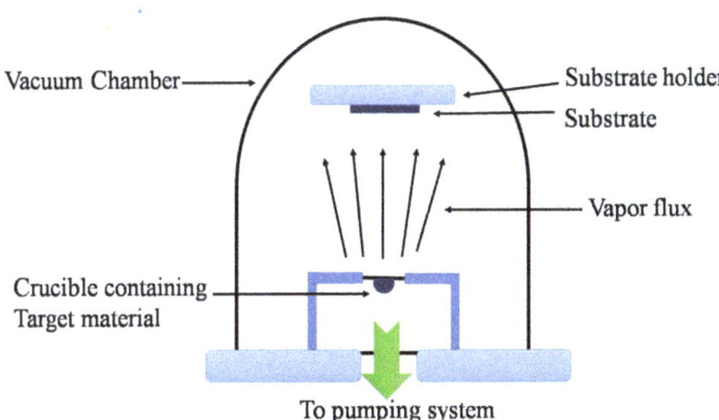

Figure 5. A schematic representation of the thermal evaporation technique

E) Lithography

A required method for the top-down production of nanomaterials is lithography. Nanolithography is a subfield of nanotechnology in which the act of imprinting, writing, or etching patterns at the minuscule level to synthesize macro/nanochips and microprocessors. A focused shaft of light or electrons is employed in lithography to create nano architecture. This method is essential because it allows for precise control over the shape and size of the materials in its synthesis. Additionally, it is a quick and economically suitable method for producing patterns throughout a whole circuit in a single step. In this process, the chip may undergo the photolithographic cycle up to 50 times to synthesize complex integrated circuits. Lithography includes photolithography, [18] nanoimprint lithography, [19] soft lithography, [20] focused ion beam lithography, [21], and scanning probe lithography. [22] X-ray and electron beam lithography has emerged as substitutes for the photolithography procedures used in top-down approaches. The electron beam method is incredibly adaptable for nanomaterial synthesis for a wide range of materials with infinite patterns. However, the synthesis of nanoparticles using this process is very expensive and slow.

F) Miscellaneous techniques

The other approaches to the top-down nanoparticle synthesis method are laser ablation and arc discharge. In the top-down approach, nanoobjects are obtained from bulk counterparts or units without atomic-level control. It is now possible to create features more minor than 100 nm, which falls under nanotechnology. This is due to several technologies that arose from traditional solid-state silicon methods for making microprocessors. A top-down design is used in giant magneto resistance (GMR)-based hard drives, which are currently commercially available. It is also possible to develop nanoelectromechanical systems (NEMS) using top-down techniques.

In the top-down method, the material is exposed to internal stress, which causes an increase in surface defects. These techniques are not appropriate for soft samples because of the stiff and hard components of the mechanical devices that are used during the synthesis of NPs. The major drawback of this method is that it requires large installations and a large amount of capital to build their setup. Therefore, these methods are pretty expensive, and the growth process is slow.

2.2 Bottom-up approach

The bottom-up technique integrates key components into larger structures using nanoscale physical and chemical processes. This method is inspired by biological systems, where natural life forces construct structures using chemical equivalents. Experts predict that we can harness nature's processes to self-integrate into superstructures. The bottom-up strategy evolved from the molecular recognition principle (i.e., self-assembly). Self-assembly refers to the process of producing more items similar to oneself. The idea behind self-assembly is to collect precursors in random places and orientations and provide energy (shaking) to allow them to sample configuration space, i.e., the shake and bake technique. "Watson-Crick base pairing and enzyme-substrate interactions are notable examples of self-assembly based on molecular recognition in biology." The vastness of this area suggests that a convergent pathway is necessary for the procedure to be finished in an acceptable amount of time. Once the precursors are in place, it might be needed to "bake" the object to strengthen the connections holding them together and permanently fix the ultimate product. Bottom-up approaches to nanomaterial production include downsizing material components to the atomic level, followed by a self-assembly process that results in the formation of nanostructures.

In bottom-up methods, the process starts with atoms or molecules solidifying to form clusters and build nanostructures by directly manipulating atoms or molecules. Bottom-up approaches entail synthesizing nanostructures of each atom, molecule, or cluster. In these

Materials Research Foundations 145 (2023) 19-53 https://doi.org/10.21741/9781644902370-2

methods, the starting material is either in a liquid state or a gaseous state. This approach includes positional assembly, self-assembly, and chemical synthesis. Self-assembled thin films can be characterized using dual polarisation interferometry. The most practical structures involve complex and improbable thermodynamic groupings of atoms. Bottom-up methods provide improved nanostructures with fewer defects or defect-free, homogeneous, long and short-range orders. The procedures listed below are examples of bottom-up approaches for NPs synthesis.

A) Chemical Vapour Deposition

A versatile and effective method for producing nanomaterials is chemical vapor deposition CVD. Even today, CVD remains one of the fascinating techniques in the microelectronics sector and is accomplished at overcoming the challenges that existing technologies entail. [23,24] With this method, Nanomaterials are produced by using relatively simple ingredients. The fundamental idea of this technique is to inject the vapor of a gaseous or liquid reactant, including chemicals and other gases required for the reaction, into the reaction chamber. (Fig. 6) The CVD process can create chemical reactions on the substrate surface. As a result, new solid materials are produced and deposited on the surface plane. The synthesis and deposition of nanoparticles are controlled by increasing the temperature, laser radiation, plasma action, or other energy sources. CVD is divided into four stages: first, the reaction gas diffuses to the material's surface, and then gas is adsorbed on the material's surface. The interaction of gases with the surface of a material causes a chemical reaction on the surface of the material, which is followed by the separation of gaseous byproducts from the material. Due to nucleation or growth occurring at the molecular level, CVD is more suited to creating dense, homogeneous coatings over irregular substrate surfaces. The deposition speed of CVD is high, and the film quality is stable. This method is frequently used in the semiconductor sector to create thin sheets. The fundamental drawback of CVD is that the substrates used in this process are generally nonvolatile, extremely toxic, and pyrophoric. [25,26]

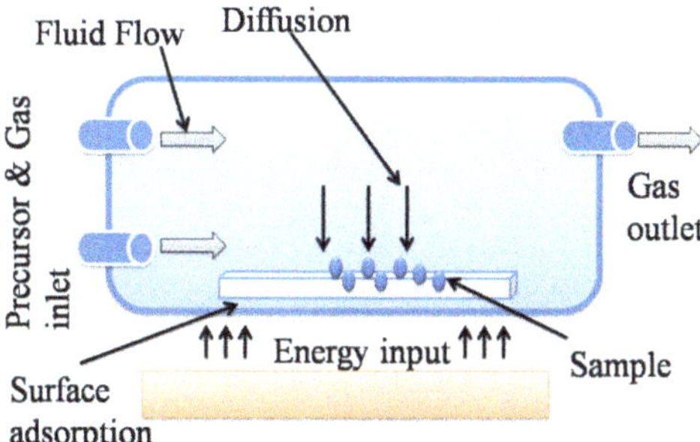

Figure 6. A schematic representation of the chemical vapour deposition process

Materials Research Forum LLC
https://doi.org/10.21741/9781644902370-2

B) Hydrothermal Synthesis

Hydrothermal synthesis involves the decomposition of a solid into tiny particles upon reacting with an aqueous solution under high pressure and temperature. This process is termed hydrothermal, as the solvent used is water. This method can be grouped as synthesis, analysis, growth of the crystal, organic waste treatment, and functional ceramic powder synthesis. This process takes place in an autoclave, a steel pressure vessel whose temperature and pressure can be maintained for a given time period. The temperature is maintained at a level above which water boils; at this temperature, the vapors get saturated. [27] Hydrothermal synthesis gained tremendous significance in the field of science and technology by low cost, uniform precipitation, benign environment, and easy setup. One disadvantage of the procedure is the inability to monitor or observe the crystal growth. Due to this, controlling the particle sizes across multiple batches is often a challenging task.

C) Sol-gel Method

Sol-gel synthesis of nanomaterials is the most popular and frequently used bottom-up method due to its ease of use. It is a fusion of the words 'gel' and 'sol.' "A sol is a colloidal solution made up of solid particles suspended in a liquid. The gel is a solid macromolecule that disperses in liquid." [28] The process consists of the steps depicted in Fig. 7. The sol-gel method produces chemical compounds in a solution. Sol transforms into a three-sided channel in the solvents due to interactions between the species and the liquid. The process is subsequently frozen, and the sol-gels are converted into deformed solids via liquid removal from the air-gel or evaporation. Sol-gel synthesis has several benefits in generating high-quality materials with homogeneity and purity at lower temperatures than conventional procedures. The major factor that limits the application of the sol-gel method of synthesis is the lengthy processing time.

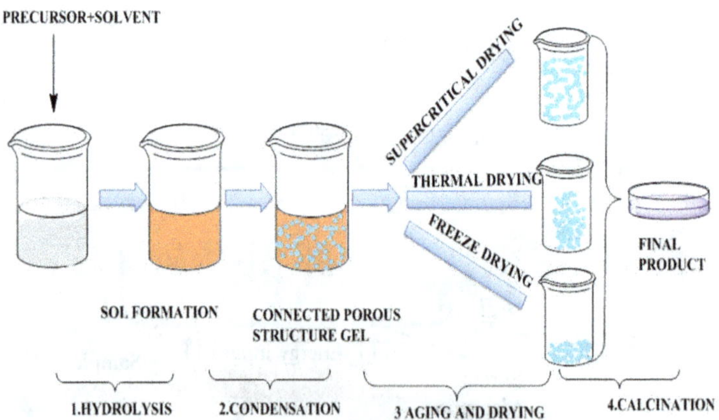

Figure 7. Sol-gel method- step-by-step process for nanoparticle synthesis

Materials Research Forum LLC
https://doi.org/10.21741/9781644902370-2

D) Other techniques of a bottom-up approach

Atomic layer deposition (ALD) is a bottom-up method that is very useful in depositing thin atomic layers on a substrate. Plasma arcing, molecular beam epitaxy, laser pyrolysis, wet synthesis, metal-organic decomposition, and self-assembly are a few additional methods employed in the bottom-up approach. Bottom-up approaches are used by high-precision actuators, which transform electrical energy into mechanical energy and vice versa to move atoms from one location to another. Micro tips such as AFM tips emboss or imprint materials. AFM tips can be used as nanoscale 'write heads' to deposit a chemical onto a surface in the desired pattern in a dip pen nanolithography process. Electron or ion beams are directed onto the surface on which the device is to be grown. To create self-assembling monolayers (i.e., single-atom-thick layers), molecules can be transferred from liquids to surfaces. Physical forces at the nanoscale combine fundamental elements to form more significant, stable structures during self-assembly. A few examples of the development of quantum dots through epitaxial growth are colloidal dispersion, CVD, PVD, etc. Bottom-up approaches can produce devices at a lower cost than top-down methods, but they become problematic by the target assembly's scale, intricacy, and complexity.

3. Unique features of nanomaterials

Compared to bulk counterparts, the characteristics of matter at the nanoscale level are noticeably different. By adjusting the nanomaterial size, the properties of nanomaterials can be customized. [29,30] The large specific surface area and smallscale of nanocrystalline materials decrease bonding energy, reducing the melting point and enthalpy of fusion. Due to dislocation and open defects, nanoparticles have a high diffusivity rate compared to other bulk lattices. Some nanoparticle solutes show enhanced solid solubility. Compared to a single palladium crystal, the solubility of hydrogen in nanocrystalline palladium (at concentrations of 10^{-3}) is enhanced by a factor of 10 to 100. [31,32]

Similar outcomes have been seen in magnesium and magnesium-based alloys widely used in hydrogen storage applications. Over the last few decades, there has been a noteworthy increase in research on magnetic nanoparticles. Their potential for use in necessary fields, including ferrofluids, spintronics, magnetic semiconductors, ultrahigh density magnetic storage equipment, nanogranular magnetic materials, Magnetic Random-Access Memory (MRAM), etc., is the primary factor for this rising attention. By passing a current through nanomaterials or by applying an electric field, nanomaterials can store significantly more energy than typical coarse-grained materials. This is due to their large surface area. [33]

NPs have received considerable attention on account of their unique optical properties. Optical properties of nanomaterials, such as absorption, emission, reflection, and transmission, are dynamic and may differ significantly from bulk crystal. Au solution appears yellow in bulk; however, at the nanoscale, it seems purple or red. The distinct colors of nano gold are induced due to the surface plasmon resonance phenomena. When light strikes the surface of a metal, it generates a surface plasmon, a collection of electrons that move back and forth in synchronization across the metallic surface. This process is referred to as the surface plasmon resonance phenomenon.[34] Some important contributing elements of NPs are the quantum confinement of electrical carriers, effective energy and charge transmission at nanoscale distances, and surface-enhanced function. Generally, when grain boundaries increase, the disordered borders' phonon scattering is enhanced. This reduces the heat conductivity of nanoparticles. Therefore, it would be reasonable to predict that nanocrystalline materials would have less thermal conductivity than conventional materials. The thermal conductivity was increased by approximately 20 % by adding copper oxide (CuO) nanoparticles (5 vol %) in water. A nanofluid containing roughly 0.3 vol % Cu nanoparticles with a mean diameter of 10 nm enhanced ethylene glycol's effective thermal conductivity by 40%. The electrical characteristics are significantly altered at the nanoscale. For instance, in its bulk form, boron is not regarded as a metal, while borophene, a two-dimensional boron network, possesses superior metallic properties. [35]

Nanomaterials' mechanical characteristics are noticeably better than their bulk counterparts due to gains in crystal perfection or decreases in crystallographic flaws. Mechanical properties characterize the material's behavior under various external loads and environmental conditions. Mechanical characteristics vary amongst different materials. The mechanical properties of classic materials typically consist of 10 components, including brittleness, strength, plasticity, rigidity, toughness, wear resistance, elasticity, flexibility, rigidity, and yield stress. Most inorganic, non-metallic nanomaterials are brittle and lack qualities like, elasticity, ductility, plasticity, toughness, and so on. Additionally, several organic compounds are flexible materials that lack rigidity and brittleness. [36] For instance, the hardness of nanocomposites with nano-Ni-Co is higher than that of Al_2O_3. [36]

4. Classification of nanomaterials

There are many ways to classify nanoparticles, some of which are mentioned here. NPs can be categorized according to their dimensions, form, surface properties, functionalization, the occurrence of chemical content, etc. The most popular method of classifying NPs based on their applications are as follows, I) special carbon-based nanomaterials; II) nanoporous materials; III) ultrathin 2D nanomaterials; IV) metal-based nanostructured materials; and V) core-shell nanoparticles. (Fig. 8)

Materials Research Forum LLC
https://doi.org/10.21741/9781644902370-2

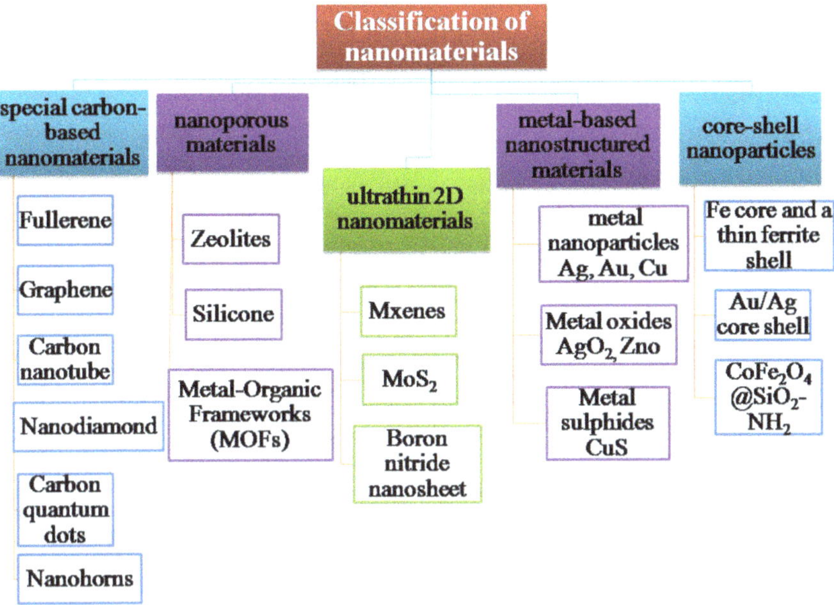

Figure 8. Classification of nanomaterials

4.1 Special carbon-based nanomaterials

The general framework of special carbon-based nanomaterials is largely made up of carbon. The unique catenation property of carbon elements accounts for many carbon-based nanomaterials. These materials take the shape of hollow spheres, sheets, ellipsoids, or tubes. Fullerenes are ellipsoidal or spherical carbon nanoparticles, whereas nanotubes are cylindrical. Carbon-based nanomaterials, key components of nanomaterials, have captured the researcher's attention. Special carbon-based nanomaterials can interact with organic molecules through non-covalent interactions, particularly van der Waals forces, hydrogen bonding, electrostatic forces, π-π-stacking, and hydrophobic interactions because of their distinctive structural features. [37] They are excellent candidates for use as adsorbents because of their non-covalent interactions and their hollow or layered nanoscale architectures. [38] There are several techniques available for creating carbon nanoparticles. The most recent interest is in using low-cost starting materials and energy-saving processes. Below are some techniques that have been used to synthesize special carbon-based nanomaterials; micromechanical cleavage [39], exfoliated graphene oxide's chemical reduction [40], epitaxial growth on SiC substrates [41], CVD [42], graphite exfoliation in the liquid phase [43], and carbon nanotube unzipping. [44] Each strategy has

Materials Research Foundations 145 (2023) 19-53
https://doi.org/10.21741/9781644902370-2

been shown to have advantages and disadvantages depending on the intended usage. Carbon-based nanoparticles have the potential to be employed for a wide array of applications. This is due to the high surface area, strong adsorption, and high binding affinity. Carbon-based nanoparticles are widely used in aerospace, defense, automobiles, energy storage applications, capacitors, better films, coatings, and electronics, such as transistors, transparent thin films, etc. In general, all of them have been made possible by the uniqueness of these materials, which is closely tied to their outstanding capabilities and unbeatable nanoscale structure. [45] Many carbon-based nanoparticles, including graphene, nanodiamonds, nanotubes, fullerenes, nanocones-disks, nanofibers, and nano horns, and their functionalized forms, have been studied in sample preparation as sorbent materials. Here is a brief discussion of some typical examples.

A) Fullerene

Fullerene, commonly known as buckminsterfullerene (C_{60}), is a carbon-based nanomaterial. Fullerene is one of the carbon allotropes with molecules of carbon atoms joined by single and double bonds. (Fig. 9) As a result, a complete or partially closed cage-like structure (a mesh made of fused rings) is created. Its remarkable antioxidant capacity makes it a promising fundamental component in many dermatological and skin care treatments. One of the valuable applications of fullerene in biomedicine is as a nanomaterial.

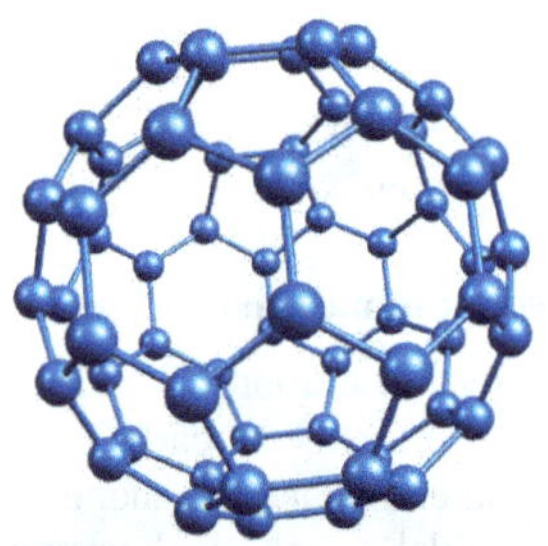

Figure 9. Structure of fullerene

B) Graphene

Graphene is the thinnest nanomaterial, consisting of a single sheet of atoms organized in a 2D honeycomb lattice nanostructure. (Fig. 10) According to its 2D structure, the typical shape of graphene nanosheets under an electron microscope is thin layers. Graphene research impacts a broad spectrum of industries, including health care, transportation, electronics, desalination, energy, and defense.

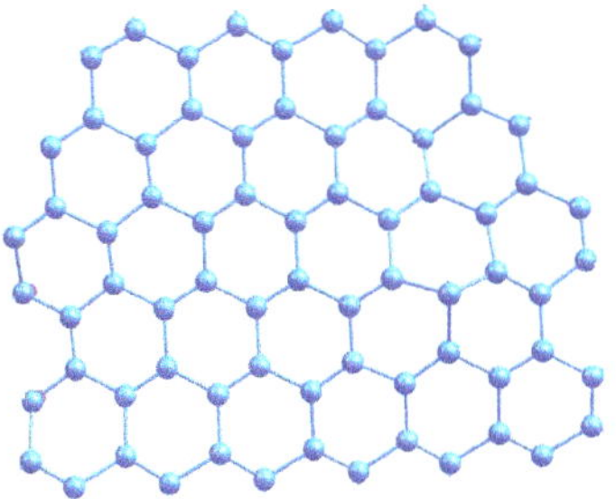

Figure 10. Structure of Graphene

C) Carbon nanotube

A carbon nanotube (CNT), also called a buckytube, is a tube formed of a carbon framework with a diameter measured in nanometers. This is one sort of carbon allotrope. Single-wall carbon nanotubes (SWCNTs) with a diameter in the nanometer range are located halfway between fullerene cages and flat graphene. (Fig. 11) Carbon nanotubes' qualities make them suitable for strengthening several structures, for example, sports equipment, body armour, cars, etc.

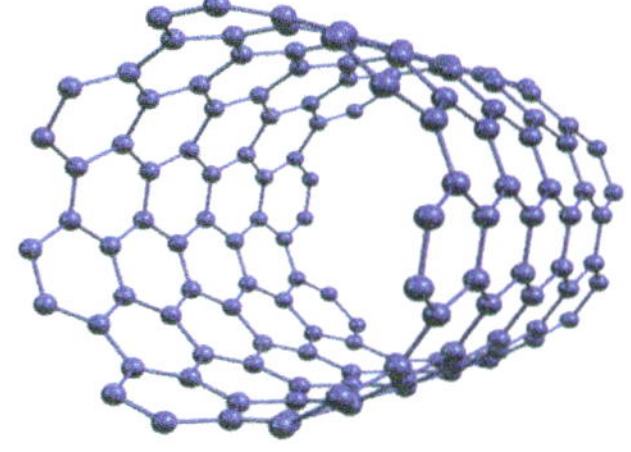

Figure 11. Structure of nanotube

D) Nanodiamond

Diamond nanoparticles, commonly referred to as nanodiamonds (NDs), are single-crystal diamonds with outstanding physical and chemical properties that are made up primarily of carbon. These are the nanoscale equivalent of sp^3 carbon, whereas sp^2 carbon is present in carbon nanotubes and fullerenes. Due to several of its characteristics, including size, shape, and biocompatibility, nanodiamonds can be utilized as antimicrobial agents. This makes them ideal for developing effective and customized nano-therapies, such as vaccinations or drug delivery systems.

Materials Research Foundations 145 (2023) 19-53 https://doi.org/10.21741/9781644902370-2

E) Carbon quantum dots

A novel class of carbon nanomaterials known as carbon quantum dots (CQDs) was discovered during the preparative electrophoresis purification of SWCNs. CQDs sometimes referred to as graphene quantum dots or carbon dots, are a possible substitute for hazardous metal-based quantum dots. CQDs are formed of discrete, quasi-spherical carbon nanoparticles with diameters below 10 nm. CQDs are employed in various applications, including photovoltaic, biosensing, chemical sensing, medical diagnosis, and biosensing.

F) Nanohorns

Conical-shaped carbon nanostructures, also known as carbon nanohorns or carbon nanocones, belong to the class of one-dimensional carbon nanostructures and have several extremely unique characteristics when compared to other carbon nanostructures. Nanohorns are graphic tubules ranging from 2 to 5 nm, larger than the 1.4 nm diameter of ordinary, and lengths ranging from 40 to 50 nm. Carbon nanohorn is a potential material for chemical and biological sensors because it makes electron transport easier.

4.2 Nanoporous materials

A subgroup of nanomaterials known as nanoporous materials includes pores that range in size of 1 to 100 nm and permit some materials to pass through while restricting others based on the size of the pores. The regular and systematic porous materials support nanoporous materials' organic and inorganic frameworks. The majority of nanoporous materials can be categorized as membranes or bulk materials. Nanoporous materials include activated carbon and zeolites, and nanoporous membranes have cell membranes. A substance with pores is known as a porous medium or material (voids). By the IUPAC classifications, the below table depicts different types of materials based on pores size. [46] (see Table 1) Nanoporous materials generally occur abundantly in nature and can also be synthesized. Due to various characteristics, including their regular structure, efficient electrical conductivity, and 3D surface morphology, nanoporous materials can be used as electrode materials. [47]

Table 1. Types of NPs based on pore size.

Sr.No.	Materials	Pore dimension range
1.	Micro porous	0.2-2 nm
2.	Meso porous	2-50 nm
3.	Macro porous	50-1000 nm

Materials Research Foundations 145 (2023) 19-53 | https://doi.org/10.21741/9781644902370-2

It is widely acknowledged that nanoporous materials offer a variety of technological benefits, including energy conversion and storage, hydrogen storage, Li-ion batteries, super capacitors, catalysis, sorption applications, drug delivery, solar cells, fuel cells, cell biology, gas purification, separation technologies, environmental remediation, water purification, desalination, separation, optical, electrical, magnetic devices, and sensors. [48] Zeolites, metal-organic frameworks (MOFs), activated carbon, covalent organic frameworks, silicates, nonsiliceous materials, ceramics, aerogels, different polymers, pillared materials, and inorganic porous hybrid materials are typical examples of natural and artificial nanoporous solids. [49] Here is a brief discussion of some nanoporous materials.

Metal-Organic Frameworks (MOFs)

MOFs are a class of crystalline substances made of metal ions and organic ligands. The most commonly employed metal ions in MOFs are Fe (III), Cd (II), Cu (II), Ag(I), Au (II), Zn (II), Hg (II), and Zr (IV). At the same time, organic linkers such as trimesic acid, terephthalic acid, or 2-methylimidazole have been excessively used. MOFs have been used as sensing systems to identify both inorganic and biological substances. Numerous studies have been published on the use of metal-organic frameworks to distinguish isotopes of elements such as hydrogen and deuterium. Micro-sized MOFs are used as sorbents in solid-phase extraction and sampling. Their usage has been effective in sample collection, light harvesting, luminescent/fluorescent sensing platforms, drug delivery, and chromatographic separation with greater precision and preconcentration.

Zeolites

Zeolites are aluminosilicate oxide nanoporous crystal forms frequently used as catalysts and adsorbents. Zeolites are porous crystalline silicates with a distinctive crystallographic structure that produces molecule-specific pores. They have a broad range of compositions, with different cations replacing some silicon atoms.

Silicone

Silicon nanoparticles (SiNPs) are inorganic nanoparticles with strong luminescence and photostability. Silicon nanoparticles (SiNPs) are metal-free quantum dots with size- and surface-tunable photoluminescence compatible with biological systems. These particular nanoparticles are in great demand for biosensor and bioimaging applications, particularly optical-based sensors.

4.3 Ultrathin 2D nanomaterials

Ultrathin two-dimensional (2D) nanomaterials refer to a new class of nanomaterials with sheet-like structures with more than 100 nm transverse dimensions and a typical thickness of less than 5 nm. [50] These nanomaterials are noticeable due to their anisotropic physical and chemical properties, unique shapes, and vast surface area. [44,45] Ultrathin (2D) nanomaterials sparked a surge in interest after Novoselov and colleagues separated graphene from graphite by the mechanical cleavage method. [53] In general, ultrathin 2D

Materials Research Foundations 145 (2023) 19-53 https://doi.org/10.21741/9781644902370-2

nanomaterials have various distinct properties compared to equivalents with other dimensionalities. The electron confinement in 2D with no interlayer Interactions in these materials, particularly single-layer nanosheets, entitles significantly more compelling electronic properties than other nanomaterials. This makes them captivating candidates for fundamental condensed matter investigations and electronic device applications. Also, their atomic thickness provides these thin materials with optical transparency and maximal mechanical and physical flexibility. Due to their transparent and flexible properties, these materials are ideal for optoelectronic applications. Finally, because of their great lateral size, ultrathin thickness, and very large surface, ultrathin 2D materials are perfect for surface-active applications. [54]

MXenes are a family of 2D inorganic chemicals in materials research. These materials comprise atomically thin layers of transition metal carbides, nitrides, or carbonitrides. A boron nitride nanosheet is a 2D crystalline form of hexagonal boron nitride with a thickness of one to a few atomic layers. The classic example is MoS_2, the most researched ultrathin 2D nanosheet. [55] When MoS_2 is thinned from bulk to monolayer, the indirect band gap of around 1.3 eV can be changed to a direct band gap of about 1.8 eV. [56] The susceptible electronic structures of ultrathin 2D nanomaterials to external stimuli, such as mechanical deformation, chemical modification, external electric fields, doping, and adsorption of other molecules or materials, are another appealing characteristic of these materials. [54] This property makes it possible to tune the electronic properties of these materials in the desired manner at a highly controllable level. [57] Developing simple, viable, and dependable methods for creating ultrathin 2D nanomaterials is critical for understanding their characteristics, functions, and applications. Great efforts have been committed to researching various synthetic techniques to make ultrathin 2D nanomaterials, driven by their interesting features and prospective applications. The following are some preparation methods for ultra-thin two-dimensional nanomaterials, mechanical cleavage [58,59], liquid exfoliation [43,60] ion-intercalation, [61,62] chemical vapor deposition growth, [63,64] wet-chemical synthesis, [65,66] modified Hummer's method, [67] and selective etching. [68]

Due to unique structural features, ultrathin 2D nanomaterials have remarkable physical, electrical, chemical, and optical capabilities that hold enormous promise for many applications. The abundance of ultrathin two-dimensional nanomaterials with diverse characteristics allows us to pick materials that are particular to the application. Ultrathin 2D nanomaterials have been investigated for use in electronics [50] [69,70], catalysis [71–73], energy storage and transformation, [74] water remediation, [75] optoelectronics, [76] sensors, gas separation, and biomedicine, [77] etc.

4.4 Metal-based nanostructured materials

Pure metals such as gold, zinc, platinum, silver, iron, titanium, cerium, and thallium and their compounds, including oxides, hydroxides, fluorides, sulfides, phosphates, and chlorides, are used to make metal nanoparticles, which are submicron-sized objects. [78] The phrase "metal nanoparticle" was created to refer to nanosized metals having length, breadth, or thickness measurements between 1 and 100 nm. The first type of nanomaterial consists of metal-based nanoparticles in their purest form, commonly called metal nanoparticles (such as silver, copper, gold, magnesium, iron, titanium, palladium, platinum, zinc, and alginate nanoparticles). Other nanomaterials include metal oxide and metal sulfide nanoparticles, such as zinc oxide, titanium dioxide, zinc sulfide, and others. Among nanomaterials, metal, doped metal, metal oxide, and metal sulfide nanoparticles are regarded as different classes. [79,80] To be conjugated with antibodies, ligands, and drugs of interest, these materials can be synthesized or modified with various chemical functional groups. As a result, a wide range of prospective applications in biotechnology, magnetic separation, target analysis, preconcentration, targeted drug administration, drug delivery, and diagnostic imaging have become viable. [81] In addition, nanomaterials made of metal sulfide and metal-organic frameworks (MOFs) have drawn much attention due to their intriguing characteristics and potential uses in various biological domains. This includes AgS, FeS, CuS nanoparticles, Cu-based, Mn-based MOF, Zn-based MOF, and other materials. These materials are frequently used in therapeutic applications such as drug discovery and antimicrobial activity. [82]

4.5 Core-shell Nanoparticles

Core-shell NPs are biphasic materials with an inner central core and an outer shell consisting of several components. These NPs have sparked the curiosity of researchers because they can display unique features due to the mixture of core and shell material. [83]. They have also been engineered such that the shell material can increase the core material's thermal stability, reactivity or oxidative stability. Strategically, these NPs can be designed to have a cheap core and a thin but expensive shell material. Such design techniques allow usage in various applications, including electrical, semiconducting materials, biomedicine, and catalysts. [77,78] Two-step solution techniques was used to synthesize most core-shell particles. This includes creating the core structure and covering it with the shell material. Gas-phase synthesis methods typically use pulsed laser deposition (PLD) or (CVD). [83] Some examples of core-shell nanomaterials are $CoFe_2O_4@SiO_2-NH_2$, Au/Ag core-shell nanoparticles, Fe core, and a thin ferrite shell nanoparticle. Core-shell nanoparticles are frequently used as chemical/bimolecular sensors, detection of drugs, and also to remove heavy metals from desired sources. These processes entail multiple phases, with the shell material often deposited onto an already-created core structure.[86,87]

5. Application of nanomaterials

Today, nanotechnology is a world-class research field. Nanomaterials have many uses, including catalysis, sensors, water treatment, nanomedicine, energy storage, military applications, sunscreens, photovoltaic cells, paints, and catalysts. (Fig.12) NPs are being utilized to manufacture products like transparent sunscreen, scratch-resistant glasses, anti-graffiti coats for walls, ceramic coatings for solar cells, anti-stain fabrics, and crack-resistant and dust-repellant paint. There are numerous applications of NPs, some of which are briefly described here.

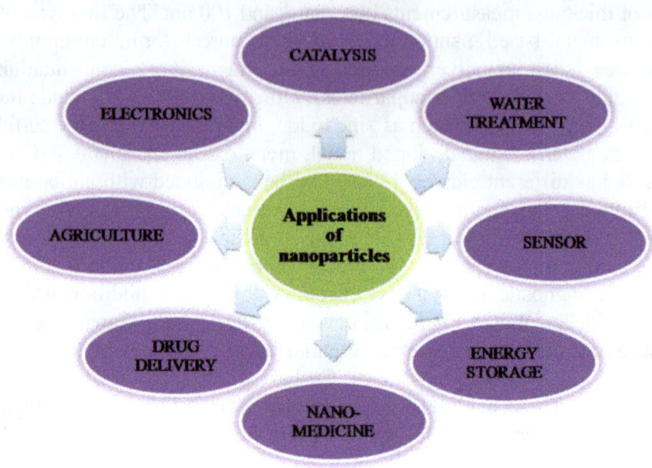

Figure 12. Applications of nanomaterials

5.1 Catalysis

Catalysis is a pioneering application of nanoparticles. For many years, several elements and materials, notably aluminum, iron, silver, rhodium, titanium dioxide, silica, and clays, have been employed as nanoscale catalysts. Because of its high activity, selectivity, and productivity, nanocatalysis has emerged as a new branch of study in recent years. Small metal nanoparticles with sizes ranging from 1 to 10 nm exhibit exceptional catalytic activity, sometimes outperforming the equivalent metal complexes. The high activity of nanocatalysts is due to many critical aspects, including the high surface-to-volume ratio, electronic effects, geometric surface effects, and quantum size impact. Some examples of nanoparticles as a catalyst are mentioned below. In 1987, Haruta discovered that gold nanoparticles smaller than 5 nm in size show their catalytic activity to speed up the oxidation of carbon monoxide (CO) to carbon dioxide (CO_2). [88] Iron oxides or other magnetic nanoparticles can operate as a stable and effective catalyst for the oxidation of alcohols, CO, vitamin precursors, etc. [89]Metal nanoparticles with catalytic capabilities can be employed to create single-walled carbon nanotubes. [90] By utilizing catalytic reagents, one may lower the temperature of a transformation, minimize reagent-based waste, and improve the selectivity of a reaction, thereby avoiding undesirable side reactions. In the absence of a nanocatalyst, the synthesis of a wide array of chemicals, including pharmaceuticals, fibers, polymers, paints, fuels, fine chemicals, lubricants, and a plethora of other value-added items necessary to humans, would not be possible. [91]

5.2 Water treatment

The use of nanoparticles in water and waste water treatment has attracted much interest. Due to their small sizes, enormous diffusivity, and specific surface areas, nanomaterials are highly reactive and have high adsorption capacities. Different nanomaterials have been successfully used to eradicate heavy metals [92], organic pollutants [93], inorganic anions [94], and microorganisms. [88,89] In recent years, the breakdown of contaminants in wastewater has been accomplished via photocatalytic degradation by TiO_2. [97–99]

5.3 Sensors

The development of high-performance electrochemical sensing tools for environmental, food safety, and medical diagnostics have been made possible by discovering a wide variety of nanomaterials. [100] Various nanomaterials have been synthesized for the electrochemical determination of some widely accepted additives and contaminants, such as hydrazine (N_2H_4), bisphenol A (BPA), malachite green (MG), caffeic acid, caffeine, sulfite (SO_3^{-2}), and nitrite, which are frequently found in food and drinks. [101] Some of the applications involving the use of NPs as the sensor are tabulated below. (Table 2) Nanotechnology significantly impacted the biosensors field due to its high sensitivity and selectivity. In this regard, fluorescent nanostructures and nanomaterials were employed for glucose monitoring. [102]

Table 2. Applications of nanoparticle-based sensors

Sr. No.	Sensor-based nanoparticles	Applications
1.	Iron oxide (Fe_3O_4) magnetic NPs	Detection of Staphylococcus aureus, Salmonella typhimurium, and *E. coli* in apple juice [103]
2.	Cadmium sulfide quantum dots (Cs-QDTs)	Detection of *E. coli* in milk samples [104]
3.	AgNPs and Cadmium (Cd) doped antimony oxide (Sb_2O_3) nanostructures	Detection of illegal additives like Melamine from Raw milk [105] [106]
4.	Hydrothermally generated zinc oxide nanorods, Polyaniline/γ-Fe_2O_3 nanocomposite, ZnO nano pencils	Liquefied Petroleum gas (LPG) sensors [107–109]
5.	Photochemically created Ag NP-embedded ZnO nanorods	Ethanol sensing [110]

Materials Research Forum LLC
https://doi.org/10.21741/9781644902370-2

5.4 Energy storage

Numerous factors contribute to nanoparticles' performance in energy storage applications. Platinum-based (Pt) nanomaterials have generated substantial interest due to their potential applications in energy-related and environmental catalysis. Pt-based nanomaterials notably increase specific surface area and expose the number of active sites. Platinum (Pt) has a rich electronic structure and good catalytic activity; hence it is preferentially used in many catalytic processes, including organic synthesis, fuel cells, petroleum refining, and hydrogen production. [111–114] Rice University researchers employed carbon nanotube coatings to prevent dendrite formation on lithium (Li) metal anodes. [115] This process might aid in developing Li-metal batteries, which could have a substantially higher capacity and charge rate than Li-ion batteries. Currently, Li-ion electrodes are found in nearly every electrical gadget, including cell phones and electric vehicles. Li-metal batteries charge faster and have around ten times more energy by volume than Li-ion batteries. [116]

5.5 Nanomedicine

The features of silver nanoparticles (Ag NPs), such as their broad-acting and solid antibacterial action, are being studied extensively. [117] Many nanosilver uses have arisen in consumer items ranging from sterilizing medical devices, disinfecting household appliances, wastewater treatments, and nanomedicine. Research interest in biocompatible and sustainable gold nanoparticles (Au NPs) for potential applications in nanomedicine has greatly expanded in last few years because of their unique size-dependent chemical, physical, electrical, and optical characteristics. Some of its important applications, such as photothermal treatment, drug administration, photodynamic gene therapy, biosensing, and biolabeling, are transforming the area of biomedicine and attracting a great deal of research interest.[118] Au NPs are non-cytotoxic in character with large surface area, allowing their surfaces to be modified with targeting molecules, making them better for other NPs for many biomedical purposes. Because it is feasible to target only the afflicted cells or parts of the body, targeted medication delivery is the most efficient therapy. This reduces the negative effects of medications. This is advantageous in cancer treatment because the medications may be administered directly to the afflicted cells with no harm to healthy cells. Fe_3O_4 [119] Quantum dots [120] and ZnO [121] are useful for targeted drug delivery.

6. Challenges and future perspectives

As discussed in earlier sections, we can conclude that nanotechnology is a technology that will revolutionize the world and medicine in the future. Although nanotechnology may appear like a future technology, many everyday items are being manufactured employing the concept of nanotechnology. For instance, the effectiveness of sunscreen has been enhanced upon employing NPs. Zinc oxide and titanium dioxide is two particular types of nanoparticles that are often included in sunscreen. [122,123] UV rays are effectively blocked by these small particles. They are commonly added to lotions and creams since they feel lighter and are quickly absorbed by the body after being applied to the skin. The long-term effects of these nanoparticles are still being researched and monitored. As described earlier, nanoparticles exhibit unique characteristics compared to their bulk counterparts. Verifying the long-term Impact of these nanoparticles about their carcinogenic effect and the effect based on their stability and adhesion qualities towards the skin becomes highly crucial and challenging. No adverse effects of sunscreen and cosmetics containing nanomaterials have yet been proven. Silica nanoparticles can be used in textiles to make water- and other liquid-resistant fabrics. Silica can be sprayed onto the fabric surface or woven into the fabric's weave to create a stain-proof or waterproof coating. Because nanotechnology is responsible for the antibacterial characteristics, nanoparticles were used in the fabric. Even though some nanoparticles in clothing can destroy bacteria, they have undergone minimal testing, poorly regulated and may be extremely dangerous to your health and the environment. [124]

Food is a necessary component of daily living and is one of the most fundamental human needs. The world's attention is primarily focused on large-scale food production through agriculture; hence, they use pesticides and fertilizers incorporated with nanoparticles to boost their food productivity. The development of next-generation pesticides and safe carriers, elimination of pollutants from soil and water bodies, restoration of soil fertility, precise water management, reclamation of salt-affected soils, prevention of the acidification of irrigated lands, and stability of erosion-prone landscapes are all made possible by nanotechnology, which is a breakthrough in agriculture.

Although there are possible concerns, nanotechnology can drastically reduce inequality and ultimately achieve the agenda for sustainable development. To adequately assess the dangers of nanomaterials, it is crucial to consider their absorption, bioavailability, and toxicity. These parameters are also controlled by particle number, stability, and size distribution. The presence of hazardous environmental pollutants in water bodies, both organic and inorganic, is a significant global issue. Nanotechnology has the potential to clean wastewater more effectively than any other technique. [125] The use of green-synthesized nanomaterials and nanocatalysts is auspicious for wastewater treatment. The following are disadvantages related to the use of nanotechnology at wastewater treatment plants.

- Nanoparticle accumulation in wastewater
- Water contamination caused by nanocomposite leaching
- Human toxicity brought on by skin contact with nanoparticles
- Issues with the disposal of nano waste after wastewater treatment
- The long-term Impact of these nanoparticles and their toxicity

Water treatment should utilise nanotechnology with extreme caution. Some nanoparticles are used because they have a great capacity for sorbing organic and inorganic pollutants. This characteristic may also cause environmental poisons to become activated that have been suppressed for a long time. The most challenging aspect of NPs is establishing methodologies for rapid, sensitive, and focused nanomaterial analysis.

Every emerging field of research has its limitations, but we can create a better future by overcoming those challenges. The most exciting future perspectives of nanotechnology are nanorobotics, sensors, self-healing structures, nano computers, medical, information technology, transportation, food safety, homeland security, energy and environmental science etc. In-depth research is currently being done in nanobiotechnology on a range of clinical uses, including disease detection, target-specific drug delivery, and molecular imaging. Such sophisticated applications of this technique to biological systems will almost certainly change the foundation of disease diagnosis, treatment, and prevention. In the coming years, there will undoubtedly be significant changes due to the additional dimension of electronic devices, as flexible devices have already started to enter the commercial market. Nanotechnologies hold out hope for developing and improving mobile and wearable technology. The fascinating topic is nanorobotics, frequently known as nanobots. Nowadays, nanobots are more than just theoretical inventions. The components consist of navigational, propulsion, and sensor systems. Currently, research is mainly concentrated on nanomotors, an essential part of propulsion. Technologies based on fuel-free and biocompatible methods must put forth much effort. [125]

References

[1] N. Baig, I. Kammakakam, W. Falath, I. Kammakakam, Nanomaterials: A review of synthesis methods, properties, recent progress, and challenges, Mater Adv. 2 (2021) 1821–1871. https://doi.org/10.1039/d0ma00807a

[2] I. Khan, K. Saeed, I. Khan, Nanoparticles: Properties, applications and toxicities, Arabian Journal of Chemistry. 12 (2019) 908–931. https://doi.org/10.1016/j.arabjc.2017.05.011

[3] R.P. Feynman, There's Plenty of Room at the Bottom, 1960

[4] N. Taniguchi, C. Arakawa, T. Kobayashi, 4. nanotechanology: on the basic concept of nanotechnology; Proceedings of the International Conference on Production Engineering; Tokyo, Japan. 26–29 August 1974., in: n.d

[5] S. Laurent, D. Forge, M. Port, A. Roch, C. Robic, L. vander Elst, R.N. Muller, Magnetic Iron Oxide Nanoparticles: Synthesis, Stabilization, Vectorization, Physicochemical Characterizations, and Biological Applications, Chem Rev. 110 (2010) 2574–2574. https://doi.org/10.1021/cr900197g

[6] J.N. Tiwari, R.N. Tiwari, K.S. Kim, Zero-dimensional, one-dimensional, two-dimensional and three-dimensional nanostructured materials for advanced electrochemical energy devices, Prog Mater Sci. 57 (2012) 724–803. https://doi.org/10.1016/j.pmatsci.2011.08.003

[7] T. Cedervall, I. Lynch, S. Lindman, T. Berggård, E. Thulin, H. Nilsson, K.A. Dawson, S. Linse, Understanding the nanoparticle-protein corona using methods to quntify exchange rates and affinities of proteins for nanoparticles, Proc Natl Acad Sci U S A. 104 (2007) 2050–2055. https://doi.org/10.1073/pnas.0608582104

[8] C.C. Fleischer, C.K. Payne, Nanoparticle-cell interactions: Molecular structure of the protein corona and cellular outcomes, Acc Chem Res. 47 (2014) 2651–2659. https://doi.org/10.1021/ar500190q

[9] M. Ramezanpour, S.S.W. Leung, K.H. Delgado-Magnero, B.Y.M. Bashe, J. Thewalt, D.P. Tieleman, Computational and experimental approaches for investigating nanoparticle-based drug delivery systems, Biochim Biophys Acta Biomembr. 1858 (2016) 1688–1709. https://doi.org/10.1016/j.bbamem.2016.02.028

[10] C.M.A. Rego, A.F. Francisco, C.N. Boeno, M. v. Paloschi, J.A. Lopes, M.D.S. Silva, H.M. Santana, S.N. Serrath, J.E. Rodrigues, C.T.L. Lemos, R.S.S. Dutra, J.N. da Cruz, C.B.R. dos Santos, S. da S. Setúbal, M.R.M. Fontes, A.M. Soares, W.L. Pires, J.P. Zuliani, Inflammasome NLRP3 activation induced by Convulxin, a C-type lectin-like isolated from Crotalus durissus terrificus snake venom, Sci Rep. 12 (2022) 4706. https://doi.org/10.1038/s41598-022-08735-7

[11] J. Cao, Y. Pan, Y. Jiang, R. Qi, B. Yuan, Z. Jia, J. Jiang, Q. Wang, Computer-aided nanotoxicology: risk assessment of metal oxide nanoparticlesvianano-QSAR, Green Chemistry. 22 (2020) 3512–3521. https://doi.org/10.1039/d0gc00933d

[12] D.N. Heo, W.K. Ko, M.S. Bae, J.B. Lee, D.W. Lee, W. Byun, C.H. Lee, E.C. Kim, B.Y. Jung, I.K. Kwon, Enhanced bone regeneration with a gold nanoparticle-hydrogel complex, J Mater Chem B. 2 (2014) 1584–1593. https://doi.org/10.1039/c3tb21246g

[13] T. Prasad Yadav, R. Manohar Yadav, D. Pratap Singh, Mechanical Milling: a Top Down Approach for the Synthesis of Nanomaterials and Nanocomposites, Nanoscience and Nanotechnology. 2 (2012) 22–48. https://doi.org/10.5923/j.nn.20120203.01

[14] Y. Long, X. Yan, X. Wang, J. Zhang, Electrospinning: The Setup and Procedure., 2019

[15] J. Hong, M. Yeo, G.H. Yang, G. Kim, Cell-electrospinning and its application for tissue engineering, Int J Mol Sci. 20 (2019). https://doi.org/10.3390/ijms20246208

[16] R.M. Pujahari, Solar cell technology, in: Energy Materials, Elsevier, 2021: pp. 27–60. https://doi.org/10.1016/B978-0-12-823710-6.00007-8

[17] A. Bashir, T.I. Awan, A. Tehseen, M.B. Tahir, M. Ijaz, Interfaces and surfaces, in: Chemistry of Nanomaterials, Elsevier, 2020: pp. 51–87. https://doi.org/10.1016/b978-0-12-818908-5.00003-2

[18] Z. Szabó, J. Volk, E. Fülöp, A. Deák, I. Bársony, Regular ZnO nanopillar arrays by nanosphere photolithography, Photonics Nanostruct. 11 (2013) 1–7. https://doi.org/10.1016/j.photonics.2012.06.009

[19] C.-W. Kuo, J.-Y. Shiu, Y.-H. Cho, P. Chen, Fabrication of Large-Area Periodic Nanopillar Arrays for Nanoimprint Lithography Using Polymer Colloid Masks, Advanced Materials. 15 (2003) 1065–1068. https://doi.org/10.1002/adma.200304824

[20] Yadong Yin, Byron Gates, Younan Xia, A Soft Lithography Approach to the Fabrication of Nanostructures of Single Crystalline Silicon with Well-Defined Dimensions and Shapes, Advance Material . 12 (2000) 1426–1430

[21] R. Matsumoto, S. Adachi, E.H.S. Sadki, S. Yamamoto, H. Tanaka, H. Takeya, Y. Takano, Maskless patterning of gallium-irradiated superconducting silicon using focused ion beam, ACS Appl Electron Mater. 2 (2020) 677–682. https://doi.org/10.1021/acsaelm.9b00781

[22] K. Xu, J. Chen, High-resolution scanning probe lithography technology: a review, Applied Nanoscience (Switzerland). 10 (2020) 1013–1022. https://doi.org/10.1007/s13204-019-01229-5

[23] I. Sayago, E. Hontañón, M. Aleixandre, Preparation of tin oxide nanostructures by chemical vapor deposition, in: Tin Oxide Materials, Elsevier, 2020: pp. 247–280. https://doi.org/10.1016/b978-0-12-815924-8.00009-8

[24] S. Bhaviripudi, E. Mile, S.A. Steiner, A.T. Zare, M.S. Dresselhaus, A.M. Belcher, J. Kong, CVD synthesis of single-walled carbon nanotubes from gold nanoparticle catalysts, J Am Chem Soc. 129 (2007) 1516–1517. https://doi.org/10.1021/ja0673332

[25] M. Mittal, S. Sardar, A. Jana, Nanofabrication techniques for semiconductor chemical sensors, Handbook of Nanomaterials for Sensing Applications. (2021) 119–137. https://doi.org/10.1016/B978-0-12-820783-3.00023-3

[26] K.V. Madhuri, Thermal protection coatings of metal oxide powders, in: Metal Oxide Powder Technologies, Elsevier, 2020: pp. 209–231. https://doi.org/10.1016/b978-0-12-817505-7.00010-5

[27] E. Suvaci, E. Özel, Hydrothermal synthesis, in: Encyclopedia of Materials: Technical Ceramics and Glasses, Elsevier, 2021: pp. 59–68. https://doi.org/10.1016/B978-0-12-803581-8.12096-X

[28] G.J. Owens, R.K. Singh, F. Foroutan, M. Alqaysi, C.M. Han, C. Mahapatra, H.W. Kim, J.C. Knowles, Sol-gel based materials for biomedical applications, Prog Mater Sci. 77 (2016) 1–79. https://doi.org/10.1016/j.pmatsci.2015.12.001

[29] R. Jose Varghese, E.H.M. Sakho, S. Parani, S. Thomas, O.S. Oluwafemi, J. Wu, Introduction to nanomaterials: Synthesis and applications, in: Nanomaterials for Solar Cell Applications, Elsevier, 2019: pp. 75–95. https://doi.org/10.1016/B978-0-12-813337-8.00003-5

Materials Research Forum LLC
https://doi.org/10.21741/9781644902370-2

[30] E. Roduner, Size matters: Why nanomaterials are different, Chem Soc Rev. 35 (2006) 583–592. https://doi.org/10.1039/b502142c

[31] B.D. Adams, A. Chen, The role of palladium in a hydrogen economy, Materials Today. 14 (2011) 282–289. https://doi.org/10.1016/S1369-7021(11)70143-2

[32] F.S. Alves, J. de A. Rodrigues Do Rego, M.L. da Costa, L.F. Lobato Da Silva, R.A. da Costa, J.N. Cruz, D.D.S.B. Brasil, Spectroscopic methods and in silico analyses using density functional theory to characterize and identify piperine alkaloid crystals isolated from pepper (Piper Nigrum L.), J Biomol Struct Dyn. 38 (2020) 2792–2799. https://doi.org/10.1080/07391102.2019.1639547

[33] B.S. Murty, P. Shankar, B. Raj, B.B. Rath, J. Murday, Unique Properties of Nanomaterials, in: Textbook of Nanoscience and Nanotechnology, Springer Berlin Heidelberg, 2013: pp. 29–65. https://doi.org/10.1007/978-3-642-28030-6_2

[34] D.M. Ledwith, A.M. Whelan, J.M. Kelly, A rapid, straightforward method for controlling the morphology of stable silver nanoparticles, J Mater Chem. 17 (2007) 2459–2464. https://doi.org/10.1039/b702141k

[35] A.J. Mannix, X.-F. Zhou, B. Kiraly, J.D. Wood, D. Alducin, B.D. Myers, X. Liu, B.L. Fisher, U. Santiago, J.R. Guest, M.J. Yacaman, A. Ponce, A.R. Oganov, M.C. Hersam, N.P. Guisinger, synthesis of borophenes: Anisotropic, two-dimensional boron polymorphs, science (1979). 350 (2015) 1513–1516. https://doi.org/10.1126/science.aad1080

[36] Q. Wu, W.S. Miao, Y. du Zhang, H.J. Gao, D. Hui, Mechanical properties of nanomaterials: A review, Nanotechnol Rev. 9 (2020) 259–273. https://doi.org/10.1515/ntrev-2020-0021

[37] B.T. Zhang, X. Zheng, H.F. Li, J.M. Lin, Application of carbon-based nanomaterials in sample preparation: A review, Anal Chim Acta. 784 (2013) 1–17. https://doi.org/10.1016/j.aca.2013.03.054

[38] K. Scida, P.W. Stege, G. Haby, G.A. Messina, C.D. García, Recent applications of carbon-based nanomaterials in analytical chemistry: Critical review, Anal Chim Acta. 691 (2011) 6–17. https://doi.org/10.1016/j.aca.2011.02.025

[39] X. Lu, M. Yu, H. Huang, R.S. Ruoff, Tailoring graphite with the goal of achieving single sheets, 1999. http://iopscience.iop.org/0957-4484/10/3/308

[40] G. Eda, G. Fanchini, M. Chhowalla, Large-area ultrathin films of reduced graphene oxide as a transparent and flexible electronic material, Nat Nanotechnol. 3 (2008) 270–274. https://doi.org/10.1038/nnano.2008.83

[41] T. Ohta, A. Bostwick, T. Seyller, K. Horn, E. Rotenberg, Controlling the electronic structure of bilayer graphene, Science (1979). 313 (2006) 951–954. https://doi.org/10.1126/science.1130681

[42] Z.Y. Juang, C.Y. Wu, A.Y. Lu, C.Y. Su, K.C. Leou, F.R. Chen, C.H. Tsai, Graphene synthesis by chemical vapor deposition and transfer by a roll-to-roll process, Carbon N Y. 48 (2010) 3169–3174. https://doi.org/10.1016/j.carbon.2010.05.001

[43] Y. Hernandez, V. Nicolosi, M. Lotya, F.M. Blighe, Z. Sun, S. De, I.T. McGovern, B. Holland, M. Byrne, Y.K. Gun'ko, J.J. Boland, P. Niraj, G. Duesberg, S. Krishnamurthy, R. Goodhue, J. Hutchison, V. Scardaci, A.C. Ferrari, J.N. Coleman, High-yield production of graphene by liquid-phase exfoliation of graphite, Nat Nanotechnol. 3 (2008) 563–568. https://doi.org/10.1038/nnano.2008.215

[44] P. Kumar, L.S. Panchakarla, C.N.R. Rao, Laser-induced unzipping of carbon nanotubes to yield graphene nanoribbons, nanoscale. 3 (2011) 2127–2129. https://doi.org/10.1039/c1nr10137d

[45] S. Nasir, M.Z. Hussein, Z. Zainal, N.A. Yusof, Carbon-based nanomaterials/allotropes: A glimpse of their synthesis, properties and some applications, Materials. 11 (2018). https://doi.org/10.3390/ma11020295

[46] G.A. Naikoo, I.U. Hassan, R.A. Dar, W. Ahmed, Development of electrode materials for high-performance supercapacitors, Emerging Nanotechnologies for Renewable Energy. (2021) 545–557. https://doi.org/10.1016/B978-0-12-821346-9.00014-6

[47] R. Mishra, J. Militky, M. Venkataraman, Nanoporous materials, in: Nanotechnology in Textiles, Elsevier, 2019: pp. 311–353. https://doi.org/10.1016/B978-0-08-102609-0.00007-9

[48] W. Ahmed, M. Booth, E. Nourafkan, Emerging Nanotechnologies for Renewable Energy, 2021

[49] S. Bhattacharyya, Y. Mastai, R. Narayan Panda, S.H. Yeon, M.Z. Hu, Advanced nanoporous materials: Synthesis, properties, and applications, J Nanomater. 2014 (2014). https://doi.org/10.1155/2014/275796

[50] C. Tan, X. Cao, X.J. Wu, Q. He, J. Yang, X. Zhang, J. Chen, W. Zhao, S. Han, G.H. Nam, M. Sindoro, H. Zhang, Recent Advances in Ultrathin Two-Dimensional Nanomaterials, Chem Rev. 117 (2017) 6225–6331. https://doi.org/10.1021/acs.chemrev.6b00558

[51] D. Chimene, D.L. Alge, A.K. Gaharwar, Two-Dimensional Nanomaterials for Biomedical Applications: Emerging Trends and Future Prospects, Advanced Materials. 27 (2015) 7261–7284. https://doi.org/10.1002/adma.201502422

[52] K.S. Novoselov, A.K. Geim, S. v. Morozov, D. Jiang, Y. Zhang, S. v. Dubonos, I. v. Grigorieva, A.A. Firsov, Electric field in atomically thin carbon films, science (1979). 306 (2004) 666–669. https://doi.org/10.1126/science.1102896

[53] A.K. Geim, K.S. Novoselov, The rise of graphene, Nat Mater. 6 (2007) 183–191. https://doi.org/10.1038/nmat1849

[54] H. Zhang, Ultrathin Two-Dimensional Nanomaterials, ACS Nano. 9 (2015) 9451–9469. https://doi.org/10.1021/acsnano.5b05040

[55] K.F. Mak, C. Lee, J. Hone, J. Shan, T.F. Heinz, Atomically thin MoS2: A new direct-gap semiconductor, Phys Rev Lett. 105 (2010). https://doi.org/10.1103/PhysRevLett.105.136805

[56] A. Splendiani, L. Sun, Y. Zhang, T. Li, J. Kim, C.Y. Chim, G. Galli, F. Wang, Emerging photoluminescence in monolayer MoS2, Nano Lett. 10 (2010) 1271–1275. https://doi.org/10.1021/nl903868w

[57] K.S. Novoselov, D. Jiang, F. Schedin, T.J. Booth, V. v. Khotkevich, S. v. Morozov, A.K. Geim, Two-dimensional atomic crystals, Proceedings of the National Academy of Sciences. 102 (2005) 10451–10453. https://doi.org/10.1073/pnas.0502848102

[58] H. Li, J. Wu, Z. Yin, H. Zhang, Preparation and Applications of Mechanically Exfoliated Single-Layer and Multilayer MoS_2 and WSe_2 Nanosheets, Acc Chem Res. 47 (2014) 1067–1075. https://doi.org/10.1021/ar4002312

[59] C.R. Dean, A.F. Young, I. Meric, C. Lee, L. Wang, S. Sorgenfrei, K. Watanabe, T. Taniguchi, P. Kim, K.L. Shepard, J. Hone, Boron nitride substrates for high-quality graphene electronics, Nat Nanotechnol. 5 (2010) 722–726. https://doi.org/10.1038/nnano.2010.172

[60] V. Nicolosi, M. Chhowalla, M.G. Kanatzidis, M.S. Strano, J.N. Coleman, Liquid Exfoliation of Layered Materials, Science (1979). 340 (2013). https://doi.org/10.1126/science.1226419

[61] M.B. Dines, Lithium intercalation via -Butyllithium of the layered transition metal dichalcogenides, Mater Res Bull. 10 (1975) 287–291. https://doi.org/10.1016/0025-5408(75)90115-4

[62] P. Joensen, R.F. Frindt, S.R. Morrison, Single-layer MoS2, Mater Res Bull. 21 (1986) 457–461. https://doi.org/10.1016/0025-5408(86)90011-5

[63] A. Reina, X. Jia, J. Ho, D. Nezich, H. Son, V. Bulovic, M.S. Dresselhaus, J. Kong, Large Area, Few-Layer Graphene Films on Arbitrary Substrates by Chemical Vapor Deposition, Nano Lett. 9 (2009) 30–35. https://doi.org/10.1021/nl801827v

[64] X. Li, W. Cai, J. An, S. Kim, J. Nah, D. Yang, R. Piner, A. Velamakanni, I. Jung, E. Tutuc, S.K. Banerjee, L. Colombo, R.S. Ruoff, Large-Area Synthesis of High-Quality and Uniform Graphene Films on Copper Foils, Science (1979). 324 (2009) 1312–1314. https://doi.org/10.1126/science.1171245

[65] J.S. Son, J.H. Yu, S.G. Kwon, J. Lee, J. Joo, T. Hyeon, Colloidal Synthesis of Ultrathin Two-Dimensional Semiconductor Nanocrystals, Advanced Materials. 23 (2011) 3214–3219. https://doi.org/10.1002/adma.201101334

[66] X.-J. Wu, X. Huang, X. Qi, H. Li, B. Li, H. Zhang, Copper-Based Ternary and Quaternary Semiconductor Nanoplates: Templated Synthesis, Characterization, and Photoelectrochemical Properties, Angewandte Chemie International Edition. 53 (2014) 8929–8933. https://doi.org/10.1002/anie.201403655

[67] W.S. Hummers, R.E. Offeman, Preparation of Graphitic Oxide, J Am Chem Soc. 80 (1958) 1339–1339. https://doi.org/10.1021/ja01539a017

[68] M. Naguib, M. Kurtoglu, V. Presser, J. Lu, J. Niu, M. Heon, L. Hultman, Y. Gogotsi, M.W. Barsoum, Two-Dimensional Nanocrystals Produced by Exfoliation of Ti3AlC2, Advanced Materials. 23 (2011) 4248–4253. https://doi.org/10.1002/adma.201102306

[69] M. Naguib, O. Mashtalir, J. Carle, V. Presser, J. Lu, L. Hultman, Y. Gogotsi, M.W. Barsoum, Two-Dimensional Transition Metal Carbides, ACS Nano. 6 (2012) 1322–1331. https://doi.org/10.1021/nn204153h

[70] G. Fiori, F. Bonaccorso, G. Iannaccone, T. Palacios, D. Neumaier, A. Seabaugh, S.K. Banerjee, L. Colombo, Electronics based on two-dimensional materials, Nat Nanotechnol. 9 (2014) 768–779. https://doi.org/10.1038/nnano.2014.207

[71] J. Xie, H. Zhang, S. Li, R. Wang, X. Sun, M. Zhou, J. Zhou, X.W.D. Lou, Y. Xie, Defect-Rich MoS_2 Ultrathin Nanosheets with Additional Active Edge Sites for Enhanced Electrocatalytic Hydrogen Evolution, Advanced Materials. 25 (2013) 5807–5813. https://doi.org/10.1002/adma.201302685

[72] D. Voiry, H. Yamaguchi, J. Li, R. Silva, D.C.B. Alves, T. Fujita, M. Chen, T. Asefa, V.B. Shenoy, G. Eda, M. Chhowalla, Enhanced catalytic activity in strained chemically exfoliated WS2 nanosheets for hydrogen evolution, Nat Mater. 12 (2013) 850–855. https://doi.org/10.1038/nmat3700

[73] X. Wang, G. Sun, P. Routh, D.-H. Kim, W. Huang, P. Chen, Heteroatom-doped graphene materials: syntheses, properties and applications, Chem. Soc. Rev. 43 (2014) 7067–7098. https://doi.org/10.1039/C4CS00141A

[74] J.-M. Tarascon, M. Armand, Issues and challenges facing rechargeable lithium batteries, nature. 414 (2001) 359–367. https://doi.org/10.1038/35104644

[75] M. Anjum, R. Miandad, M. Waqas, F. Gehany, M.A. Barakat, Remediation of wastewater using various nanomaterials, Arabian Journal of Chemistry. 12 (2019) 4897–4919. https://doi.org/10.1016/j.arabjc.2016.10.004

[76] Y.S. Zhao, H. Fu, A. Peng, Y. Ma, D. Xiao, J. Yao, Low-Dimensional Nanomaterials Based on Small Organic Molecules: Preparation and Optoelectronic Properties, Advanced Materials. 20 (2008) 2859–2876. https://doi.org/10.1002/adma.200800604

[77] M.-H. Sun, S.-Z. Huang, L.-H. Chen, Y. Li, X.-Y. Yang, Z.-Y. Yuan, B.-L. Su, Applications of hierarchically structured porous materials from energy storage and

conversion, catalysis, photocatalysis, adsorption, separation, and sensing to biomedicine, Chem Soc Rev. 45 (2016) 3479–3563. https://doi.org/10.1039/C6CS00135A

[78] S. G., U. N., N. R., K. P., V. G., C. B., A REVIEW ON PROPERTIES, APPLICATIONS AND TOXICITIES OF METAL NANOPARTICLES, International Journal of Applied Pharmaceutics. (2020) 58–63. https://doi.org/10.22159/ijap.2020v12i5.38747

[79] A.A. Dar, K. Umar, N.A. Mir, M.M. Haque, M. Muneer, C. Boxall, Photocatalysed degradation of a herbicide derivative, Dinoterb, in aqueous suspension, Research on Chemical Intermediates. 37 (2011) 567–578. https://doi.org/10.1007/s11164-011-0299-6

[80] K. Umar, M.M. Haque, N.A. Mir, M. Muneer, I.H. Farooqi, Titanium Dioxide-mediated Photocatalysed Mineralization of Two Selected Organic Pollutants in Aqueous Suspensions, Journal of Advanced Oxidation Technologies. 16 (2013). https://doi.org/10.1515/jaots-2013-0205

[81] A.A. Yaqoob, H. Ahmad, T. Parveen, A. Ahmad, M. Oves, I.M.I. Ismail, H.A. Qari, K. Umar, M.N. Mohamad Ibrahim, Recent Advances in Metal Decorated Nanomaterials and Their Various Biological Applications: A Review, Front Chem. 8 (2020). https://doi.org/10.3389/fchem.2020.00341

[82] R. Li, T. Chen, X. Pan, Metal–Organic-Framework-Based Materials for Antimicrobial Applications, ACS Nano. 15 (2021) 3808–3848. https://doi.org/10.1021/acsnano.0c09617

[83] R. Ghosh Chaudhuri, S. Paria, Core/Shell Nanoparticles: Classes, Properties, Synthesis Mechanisms, Characterization, and Applications, Chem Rev. 112 (2012) 2373–2433. https://doi.org/10.1021/cr100449n

[84] M.B. Gawande, A. Goswami, T. Asefa, H. Guo, A. v. Biradar, D.-L. Peng, R. Zboril, R.S. Varma, Core–shell nanoparticles: synthesis and applications in catalysis and electrocatalysis, Chem Soc Rev. 44 (2015) 7540–7590. https://doi.org/10.1039/C5CS00343A

[85] M. Khan, S. Mishra, D. Ratna, S. Sonawane, N.G. Shimpi, Investigation of thermal and mechanical properties of styrene–butadiene rubber nanocomposites filled with SiO_2 –polystyrene core–shell nanoparticles, J Compos Mater. 54 (2020) 1785–1795. https://doi.org/10.1177/0021998319886618

[86] A. v Nomoev, S.P. Bardakhanov, M. Schreiber, D.G. Bazarova, N.A. Romanov, B.B. Baldanov, B.R. Radnaev, V. v Syzrantsev, Structure and mechanism of the formation of core–shell nanoparticles obtained through a one-step gas-phase synthesis by electron beam evaporation, Beilstein Journal of Nanotechnology. 6 (2015) 874–880. https://doi.org/10.3762/bjnano.6.89

[87] H. Chen, L. Zhang, M. Li, G. Xie, Synthesis of Core–Shell Micro/Nanoparticles and Their Tribological Application: A Review, Materials. 13 (2020) 4590. https://doi.org/10.3390/ma13204590

[88] M. Haruta, T. Kobayashi, H. Sano, N. Yamada, Novel Gold Catalysts for the Oxidation of Carbon Monoxide at a Temperature far Below 0 °C, Chem Lett. 16 (1987) 405–408. https://doi.org/10.1246/cl.1987.405

[89] A.K. Gupta, M. Gupta, Synthesis and surface engineering of iron oxide nanoparticles for biomedical applications, Biomaterials. 26 (2005) 3995–4021. https://doi.org/10.1016/j.biomaterials.2004.10.012

[90] A. Moisala, A.G. Nasibulin, E.I. Kauppinen, The role of metal nanoparticles in the catalytic production of single-walled carbon nanotubes—a review, Journal of Physics: Condensed Matter. 15 (2003) S3011–S3035. https://doi.org/10.1088/0953-8984/15/42/003

[91] P.K. Tandon, S. Bahadur Singh, P. Kumar Tandon, Catalysis: A brief review on Nano-Catalyst, 2014. https://www.researchgate.net/publication/284727255

[92] W.-W. Tang, G.-M. Zeng, J.-L. Gong, J. Liang, P. Xu, C. Zhang, B.-B. Huang, Impact of humic/fulvic acid on the removal of heavy metals from aqueous solutions using nanomaterials: A review, Science of The Total Environment. 468–469 (2014) 1014–1027. https://doi.org/10.1016/j.scitotenv.2013.09.044

[93] J. Yan, L. Han, W. Gao, S. Xue, M. Chen, Biochar supported nanoscale zerovalent iron composite used as persulfate activator for removing trichloroethylene, Bioresour Technol. 175 (2015) 269–274. https://doi.org/10.1016/j.biortech.2014.10.103

[94] F. Liu, J. Yang, J. Zuo, D. Ma, L. Gan, B. Xie, P. Wang, B. Yang, Graphene-supported nanoscale zero-valent iron: Removal of phosphorus from aqueous solution and mechanistic study, Journal of Environmental Sciences. 26 (2014) 1751–1762. https://doi.org/10.1016/j.jes.2014.06.016

[95] R.S. Kalhapure, S.J. Sonawane, D.R. Sikwal, M. Jadhav, S. Rambharose, C. Mocktar, T. Govender, Solid lipid nanoparticles of clotrimazole silver complex: An efficient nano antibacterial against Staphylococcus aureus and MRSA, Colloids Surf B Biointerfaces. 136 (2015) 651–658. https://doi.org/10.1016/j.colsurfb.2015.10.003

[96] H. Lu, J. Wang, M. Stoller, T. Wang, Y. Bao, H. Hao, An Overview of Nanomaterials for Water and Wastewater Treatment, Advances in Materials Science and Engineering. 2016 (2016) 1–10. https://doi.org/10.1155/2016/4964828

[97] Z. Zhao, J. Sun, S. Xing, D. Liu, G. Zhang, L. Bai, B. Jiang, Enhanced Raman scattering and photocatalytic activity of TiO2 films with embedded Ag nanoparticles deposited by magnetron sputtering, J Alloys Compd. 679 (2016) 88–93. https://doi.org/10.1016/j.jallcom.2016.03.248

[98] Q. Guo, C. Zhou, Z. Ma, Z. Ren, H. Fan, X. Yang, Elementary photocatalytic chemistry on TiO_2 surfaces, Chem Soc Rev. 45 (2016) 3701–3730. https://doi.org/10.1039/C5CS00448A

[99] L. Zheng, S. Han, H. Liu, P. Yu, X. Fang, Hierarchical MoS_2 Nanosheet@TiO_2 Nanotube Array Composites with Enhanced Photocatalytic and Photocurrent Performances, Small. 12 (2016) 1527–1536. https://doi.org/10.1002/smll.201503441

[100] Kh. Brainina, N. Stozhko, M. Bukharinova, E. Vikulova, Nanomaterials: Electrochemical Properties and Application in Sensors, Physical Sciences Reviews. 3 (2018). https://doi.org/10.1515/psr-2018-8050

[101] V.S. Manikandan, B. Adhikari, A. Chen, Nanomaterial based electrochemical sensors for the safety and quality control of food and beverages, Analyst. 143 (2018) 4537–4554. https://doi.org/10.1039/C8AN00497H

[102] K.J. Cash, H.A. Clark, Nanosensors and nanomaterials for monitoring glucose in diabetes, Trends Mol Med. 16 (2010) 584–593. https://doi.org/10.1016/j.molmed.2010.08.002

[103] D. Wilson, E.M. Materón, G. Ibáñez-Redín, R.C. Faria, D.S. Correa, O.N. Oliveira, Electrical detection of pathogenic bacteria in food samples using information visualization methods with a sensor based on magnetic nanoparticles functionalized with antimicrobial peptides, Talanta. 194 (2019) 611–618. https://doi.org/10.1016/j.talanta.2018.10.089

[104] M. Zhong, L. Yang, H. Yang, C. Cheng, W. Deng, Y. Tan, Q. Xie, S. Yao, An electrochemical immunobiosensor for ultrasensitive detection of Escherichia coli O157:H7 using CdS quantum dots-encapsulated metal-organic frameworks as signal-amplifying tags, Biosens Bioelectron. 126 (2019) 493–500. https://doi.org/10.1016/j.bios.2018.11.001

[105] Jigyasa, J.K. Rajput, Bio-polyphenols promoted green synthesis of silver nanoparticles for facile and ultra-sensitive colorimetric detection of melamine in milk, Biosens Bioelectron. 120 (2018) 153–159. https://doi.org/10.1016/j.bios.2018.08.054

[106] M.M. Rahman, J. Ahmed, Cd-doped Sb2O4 nanostructures modified glassy carbon electrode for efficient detection of melamine by electrochemical approach, Biosens Bioelectron. 102 (2018) 631–636. https://doi.org/10.1016/j.bios.2017.12.007

[107] A. Sivapunniyam, N. Wiromrat, M. Myint, J. Dutta, High-performance liquefied petroleum gas sensing based on nanostructures of zinc oxide and zinc stannate. Sensors and Actuators B-Chemical. , Sensors and Actuators B-Chemical. 157 (2011) 232–239

[108] T. Sen, N.G. Shimpi, S. Mishra, R. Sharma, Polyaniline/γ-Fe2O3 nanocomposite for room temperature LPG sensing, Sens Actuators B Chem. 190 (2014) 120–126. https://doi.org/10.1016/j.snb.2013.07.091

[109] N.G. Shimpi, S. Jain, N. Karmakar, A. Shah, D.C. Kothari, S. Mishra, Synthesis of ZnO nanopencils using wet chemical method and its investigation as LPG sensor, Appl Surf Sci. 390 (2016) 17–24. https://doi.org/10.1016/j.apsusc.2016.08.050

[110] Q. Xiang, G. Meng, Y. Zhang, J. Xu, P. Xu, Q. Pan, W. Yu, Ag nanoparticle embedded-ZnO nanorods synthesized via a photochemical method and its gas-sensing properties, Sens Actuators B Chem. 143 (2010) 635–640. https://doi.org/10.1016/j.snb.2009.10.007

[111] S. Duan, Z. Du, H. Fan, R. Wang, Nanostructure Optimization of Platinum-Based Nanomaterials for Catalytic Applications, Nanomaterials. 8 (2018) 949. https://doi.org/10.3390/nano8110949

[112] O. Wolf, M. Dasog, Z. Yang, I. Balberg, J.G.C. Veinot, O. Millo, Doping and Quantum Confinement Effects in Single Si Nanocrystals Observed by Scanning Tunneling Spectroscopy, Nano Lett. 13 (2013) 2516–2521. https://doi.org/10.1021/nl400570p

[113] J.A. Sichert, Y. Tong, N. Mutz, M. Vollmer, S. Fischer, K.Z. Milowska, R. García Cortadella, B. Nickel, C. Cardenas-Daw, J.K. Stolarczyk, A.S. Urban, J. Feldmann, Quantum Size Effect in Organometal Halide Perovskite Nanoplatelets, Nano Lett. 15 (2015) 6521–6527. https://doi.org/10.1021/acs.nanolett.5b02985

[114] J. Wu, H. Yang, Platinum-Based Oxygen Reduction Electrocatalysts, Acc Chem Res. 46 (2013) 1848–1857. https://doi.org/10.1021/ar300359w

[115] R. v. Salvatierra, G.A. López-Silva, A.S. Jalilov, J. Yoon, G. Wu, A. Tsai, J.M. Tour, Suppressing Li Metal Dendrites Through a Solid Li-Ion Backup Layer, Advanced Materials. 30 (2018) 1803869. https://doi.org/10.1002/adma.201803869

[116] C. Niu, H. Pan, W. Xu, J. Xiao, J.-G. Zhang, L. Luo, C. Wang, D. Mei, J. Meng, X. Wang, Z. Liu, L. Mai, J. Liu, Self-smoothing anode for achieving high-energy lithium metal batteries under realistic conditions, Nat Nanotechnol. 14 (2019) 594–601. https://doi.org/10.1038/s41565-019-0427-9

[117] G. Franci, A. Falanga, S. Galdiero, L. Palomba, M. Rai, G. Morelli, M. Galdiero, Silver Nanoparticles as Potential Antibacterial Agents, Molecules. 20 (2015) 8856–8874. https://doi.org/10.3390/molecules20058856

[118] K. Murali, M.S. Neelakandan, S. Thomas, Biomedical Applications of Gold Nanoparticles, JSM Nanotechnol Nanomed. 6 (2018) 1064

[119] B. Chertok, B.A. Moffat, A.E. David, F. Yu, C. Bergemann, B.D. Ross, V.C. Yang, Iron oxide nanoparticles as a drug delivery vehicle for MRI monitored magnetic targeting of brain tumors, Biomaterials. 29 (2008) 487–496. https://doi.org/10.1016/j.biomaterials.2007.08.050

[120] L. Qi, X. Gao, Emerging application of quantum dots for drug delivery and therapy, Expert Opin Drug Deliv. 5 (2008) 263–267. https://doi.org/10.1517/17425247.5.3.263

[121] J.W. Rasmussen, E. Martinez, P. Louka, D.G. Wingett, Zinc oxide nanoparticles for selective destruction of tumor cells and potential for drug delivery applications, Expert Opin Drug Deliv. 7 (2010) 1063–1077. https://doi.org/10.1517/17425247.2010.502560

[122] S. Marella, A.R. Nirmal Kumar, N.V.K.V.P. Tollamadugu, Nanotechnology-based innovative technologies for high agricultural productivity: Opportunities, challenges, and future perspectives, in: Recent Developments in Applied Microbiology and Biochemistry, Elsevier, 2021: pp. 211–220. https://doi.org/10.1016/B978-0-12-821406-0.00019-9

[123] A.R.J.A. de M. Lima, A.S. Siqueira, M.L.S. Möller, R.C. de Souza, J.N. Cruz, A.R.J.A. de M. Lima, R.C. da Silva, D.C.F. Aguiar, J.L. da S.G.V. Junior, E.C. Gonçalves, In silico improvement of the cyanobacterial lectin microvirin and mannose interaction, J Biomol Struct Dyn. (2020). https://doi.org/10.1080/07391102.2020.1821782

[124] T.M. Benn, P. Westerhoff, Nanoparticle Silver Released into Water from Commercially Available Sock Fabrics, Environ Sci Technol. 42 (2008) 4133–4139. https://doi.org/10.1021/es7032718

[125] S. Campuzano, B. Esteban-Fernández de Ávila, P. Yáñez-Sedeño, J.M. Pingarrón, J. Wang, Nano/microvehicles for efficient delivery and (bio)sensing at the cellular level, Chem Sci. 8 (2017) 6750–6763. https://doi.org/10.1039/C7SC02434G

Materials Research Foundations 145 (2023) 54-91
https://doi.org/10.21741/9781644902370-3

Chapter 3

Green Chemistry Principles and Spectroscopic Methods Applied to Nanomaterials

Anindita De[1], Roopa R.A.[2], Manasa H.S.[2], Mridula Guin[1,*]

[1] Department of Chemistry and Biochemistry, Sharda University, Greater Noida 201310, India

[2] Department of Studies in Chemistry, Pooja Bhagavat Memorial Mahajana Education Centre, Mysuru, Karnataka 570016, India

mridula.guin@sharda.ac.in

Abstract

Nanomaterials have revolutionized the twenty-first century through many industrial breakthroughs. Hazardous chemical methods from non-renewable sources mostly perform the synthesis of these nanomaterials. Thus, synthesizing nanomaterials by twelve principles of green chemistry is the most demanding method that outperforms the chemical and physical synthesis methods in various aspects. The key features of the green approach are environmental friendliness, cost-effectiveness, and biocompatibility. The green principles use natural resources for nanomaterial synthesis and are currently on their way from the laboratory to commercial application. This chapter presents principles of green chemistry that are followed for nanomaterial synthesis. Recent advances in this field and overcoming the challenges to improve their commercialization have also been discussed.

Keywords

Green Chemistry, Green Synthesis, Nanomaterials, Nanotechnology, Environment Friendly, Natural Source, Toxicity

Contents

Materials Research Foundations 145 (2023) 54-91 https://doi.org/10.21741/9781644902370-3

1. Introduction

Nano-size materials and their practical uses are not uncommon in human history. The first application of nanoparticles can be traced back to the medieval ages when metallic nanoparticles could be found in strain glass windows in several European churches.

However, a detailed understanding of the properties of nanoparticles and knowledge about their controlled synthesis was developed at the end of 20th century [1]. In 1959, renowned physicist Prof. Richard Feynman predicted the possibility of having nanoscale material in his groundbreaking lecture entitled "There's Plenty of Room at the Bottom" at the California Institute of technology [2]. Later in 1974, Norio Taniguchi defined the concept 'Nanotechnology'. Since then, there has been tremendous growth in nanoscience and nano word for technology. Nowadays, the application of nanomaterial is not limited to the artistic expression of religious art only but in precision agriculture [3], the food packaging industry [4], energy storage devices [5], the biomedical field [6], catalysis [7], sensing devices [8], waste water treatment [9] to name a few. Nanomaterials unique optical, magnetic, electrical and biological properties differ from their equivalent macroscopic particle. So much is their impact in the modern research areas that nanomaterials are thought to be at the forefront of the next industrial revolution [10]. As per the data obtained from the United States Bureau of Labor Statistics and Nanotechnology Consumer Products Inventory, there are almost 1400 companies that focus on the synthesis of nanomaterial for their products. Commonly manufactured nanomaterials are zinc oxide, titanium oxide, iron oxide, CNT (carbon nanotube), gold, silver and iron [11–14]. Due to this emphasis on nanotechnology in the academic and commercial fields a lot of focus is on the synthesis of nanomaterials.

Sustainability and eco-friendliness are the guiding principles for developing next-generation nanomaterials [15,16]. Green Synthesis of nanomaterials is considered an economically favorable approach to fulfill the purpose of a secure, safe and sustainable environment. Eco-friendly synthesis of nanomaterials must fulfill certain requirements such as: evading the generation of waste, attaining atom efficiency, least utilization and generation of toxic chemicals, producing bio-degradable nano compounds with greater efficiency or at least at par with the existing compounds, least use of auxiliary substances, less demand of energy, use of renewable materials or natural resources, use of catalysts instead of stoichiometric reagents.

The green principle concept is applied in synthesizing nanomaterials as it fulfills the above requirements. In general, green synthesis of nanomaterials utilizes the chemical process or technology that takes care of the environment and improves the standard of living. The nanoparticle synthesis involves three essential conditions: a nonhazardous solvent, a reducing substance, and a nontoxic stabilizing chemical. Commercially available nanoparticles are synthesized mainly via extensive chemical methods. However, these chemical routes increase production costs and environmental risks due to using harmful and unsafe chemicals [17].

On the other hand, the biogenic method is a safe, biocompatible, eco-friendly method to synthesize nanoparticles using various natural sources such as plant parts, microbes, and waste materials [18]. The presence of phytochemicals across multiple plant parts acts as stabilizing and reducing agent during the synthesis of nanoparticles [19]. Some important phytochemicals present, especially in leaves such as aldehydes, ketones, amides, carboxylic acids, phenols, and flavones, help reduce metal salts to metal nanoparticles.

2. Green chemistry principles

Paul Anastas and John Warner first laid out the concept and the principles of green chemistry in 1998 in their book "Green Chemistry: Theory and Practice" [20]. They advised scientists, government officials, and industrialists to follow twelve principles of green chemistry that would be more sustainable and would cause the least hazard to mother nature. Over the last two decades, these principles have significantly added to this environment-friendly branch of chemistry known as green chemistry. In today's scenario, where we see the effects of climate change worldwide, the necessity of green chemistry and its principle is becoming more and more relevant. The green chemistry principles are as follows [Fig.1]:

(i) *Principle one: Prevention of waste.* According to this principle, it is better to design a chemical synthesis that will not generate waste instead of disposal/ utilization of the waste product. In green nano-synthesis, the least amount of reagent is used and chemicals are derived from natural sources, hence nonhazardous. Moreover, nanomaterials have been synthesized from agricultural waste and industrial waste.

(ii) *Principle two: atom economy.* Atom economy states that a chemical process's efficiency is related to its yield and the presence of the most reagent atoms in the product. Later in 2005, it was emphasized that atom economy should be considered in all the chemical substances and not only the primary chemical reaction [21]. This will ensure no or least waste generation.

(iii) *Principle three: Benign design.* According to this principle, any chemical synthesis should be designed to cause a minimum hazard to the environment. In green nanotechnology, many strong reagents and reducing agents are replaced by aqueous extract of plant or microorganism and thus causes the least harm to the environment.

(iv) *Principle four: safer chemicals and products.* The chemical synthesis should be designed to include nonhazardous chemicals yet be effective. At the same time, the products should be nontoxic, and their interactions with biological entities (when exposed to them) should be minimal. Information about the chemical properties and structure-activity relationships should be easily and freely available to achieve this target. So, developing a data repository digital library and non-profit, open-access scientific journals funded by government bodies is necessary. In the case of nanomaterials, the toxicity is not merely related to their composition by also to their size, surface functionality, and morphology. Thus, understanding the fate of these nanomaterials when exposed to the environment is of prime importance. Several recent review articles shed light on this topic [22–24].

(v) *Principle five: Safer Solvents and Auxiliaries.* According to this principle, chemical synthesis should try to avoid solvent use during synthesis or purification. If solvent use is unavoidable, alternative purification methods such as centrifugation and green solvents such as ionic liquids or water should be the reaction medium. In biosynthesis, the reactions are carried out in the water, and all the precursor materials are chosen to be soluble in water.

(vi) *Principle six: energy efficiency.* This principle aims to decrease the use of conventional energy sources obtained by burning fossil fuels. It also tries to minimize the use of dirty energy, such as combustion energy and any other form of energy known to contribute to the greenhouse effect. Suppose the involvement of power is necessary. Then, the design reaction in a manner that requires a minimum amount of energy. If possible, the reaction should be done at room temperature and pressure. Identification of steps that are energy consumption is also necessary. In this regard, green synthesis is better than all other synthetic procedures as the reactions occur in ambient temperature and pressure.

(vii) *Principle seven: use of renewable raw material.* The feedstock for chemical synthesis should not deplete nature. Some examples of renewable materials are waste and agro-forestry materials. Once again, biosynthesis aligns perfectly with this principle as the starting materials are derived from natural sources.

(viii) *Principle eight: reduce/avoid using derivatives.* The use of derivatives, blocking groups, protection/deprotection, and modification during a chemical synthesis unnecessarily increases reaction steps and the number of chemical substances used. Moreover, they can decrease the atom's efficiency. Hence, such type of reaction design should be avoided as much as possible. The biosynthesis of nanomaterial is a one-step, one-pot process. It does not include blocking, protective groups, etc.

(ix) *Principle nine: catalysis.* Due to their high surface area/volume ratio, nanomaterials can act as an efficient catalysts. A catalyst with a high surface area has better efficiency and thus will be preferred. Moreover, to minimize the waste generated during catalytic reaction, one should take an effective catalyst in a minimum amount that will be able to perform the chemical reaction repeatedly.

(x) *Principle ten: Final disposal.* According to this principle, the products of chemical reactions should be designed to be reused, recycled, easily removed, or biodegraded. In nanotechnology, several computer programs are now used to analyze the life cycle of nanomaterials. These programs can be utilized to design reactions involving nanomaterials for recycling or reuse [21,25].

(xi) *Principle eleven: real-time analysis.* This principle emphasizes the importance of developing analytical tools to monitor chemical reactions *in situ*. So that any harmful by-products can be eliminated immediately during the reaction. Nanomaterials have long been used as sensors due to their fast response to any recognition event [26]. Several nanomaterial-based sensors, such as plasmonic, electrochemical, surface-enhanced Raman spectroscopy, and fluorescence sensors, were developed and thoroughly studied over the last few years. Moreover, the signal generated during the recognition highly depends on its size, shape, surface functionality, etc. Thus, it is convenient to monitor chemical reactions involving nanomaterials [27].

(xii) *Principle Twelve: minimizing the possibility of accidents.* One of the ways to avoid accidents, such as fire, explosion, etc., during a chemical process is by using safer chemicals or safer forms of chemicals to avoid accidents. As nanomaterials are highly reactive, they are often embedded in a macroscopic surface for further implementation. This step can minimize the chances of releasing nanomaterial into the environment through washing, leaching, or diffusion [28].

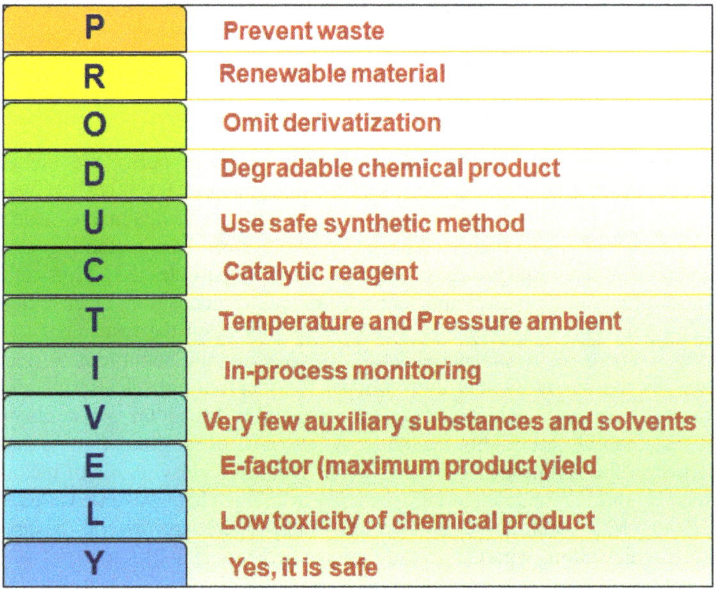

P	Prevent waste
R	Renewable material
O	Omit derivatization
D	Degradable chemical product
U	Use safe synthetic method
C	Catalytic reagent
T	Temperature and Pressure ambient
I	In-process monitoring
V	Very few auxiliary substances and solvents
E	E-factor (maximum product yield
L	Low toxicity of chemical product
Y	Yes, it is safe

Figure 1. Twelve principles of green chemistry

However, following all the principles in a single chemical synthesis is often impossible, and only some are fulfilled during the synthesis. Moreover, applications of green chemistry principles to a chemical process design are complicated because the tenets are qualitative, and there is no standard metric and minimum designation criterion to label a process "green" [22,29–32]. Consequently, the designation of "green" to a chemical reaction is often subjective and does not consider its overall impact on environmental sustainability. Only a process's environmental impact is regarded as the most important and common criterion. However, out-of-the-lab studies under non-controllable conditions are also crucial to understanding and quantifying a chemical process's impact on planetary and human health risks [29–32].

Materials Research Foundations 145 (2023) 54-91
https://doi.org/10.21741/9781644902370-3

3. Green nanosciences

There are two main approaches to nanomaterial synthesis: chemical and physical. Chemical methods include hydrothermal synthesis, sol-gel method, polyol synthesis, microemulsion technique, and chemical vapor deposition. Chemical synthesis is a bottom-up approach where nanomaterials are formed from atomic/ molecular scale entities [33]. This method requires the use of toxic chemicals. Some of the reagents are dangerous and difficult to handle. For example, chemical synthesis requires chemicals such as sodium borohydride, N, N-dimethyl formamide, and hydrazine, stabilizing agents such as cetrimonium bromide, polyamidoamine, N-methyl-D-glucamine, citric acid, poly(vinylpyrrolidone), etc. along with inorganic and organic solvents. Thus, secondary pollutant generation from any side reaction or the presence of unreacted reagent is unavoidable in chemical synthesis.

Moreover, less control over size, unknown yield, and less purity are the problems associated with chemical synthesis. Another class of nanomaterial synthesis is the physical method, which is a top-down synthesis approach. In this approach, a macro-scale material is disintegrated until its particles reach a nanoscale size. This method is an energy-consuming process that often produces polydispersed nanoparticles [33]. Physical methods involve laser ablation, physical vapor deposition, sputter deposition, electric deposition, melt mixing, etc. This environmental concern has motivated the researcher to develop novel green methods for nanomaterial synthesis, which would be more eco-friendly[22]. The above discussion shows that conventional chemical and physical methods are not environmentally friendly. Thus, innovative approaches based on green chemistry to maintain sustainability are highly desirable. A new area of nanomaterial synthesis have thus emerged in the last decade known as green synthesis/ biosynthesis of nanomaterial. In the green synthesis of nanomaterials, naturally available chemicals/microorganisms are utilized along with safer solvents such as water. There are several advantages of biosynthesis which aligns with the goals of green chemistry [Fig.2]:

(i) The use of renewable and nontoxic raw materials (principles 3,7 and 12)

(ii) They are one-pot synthesis (principle 8);

(iii) They are energy efficient. The reactions are carried out at room temperature and pressure (principle 6)

(iv) They use water or alcohol as a reaction medium (principles 3 and 12) [33]

Materials Research Foundations 145 (2023) 54-91 https://doi.org/10.21741/9781644902370-3

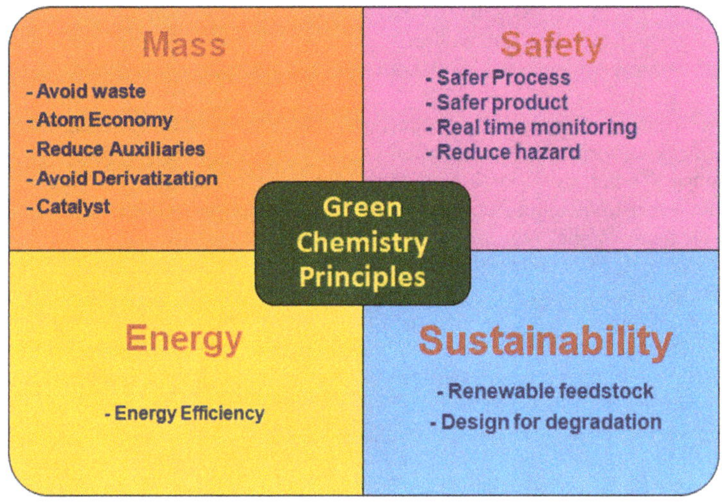

Figure 2. Broad classification of twelve principles of green chemistry

Hence, green synthesis of nanomaterials represents a low-cost, energy-efficient, simple methodology compared to traditional synthesis procedures. Additionally, biosynthesis is versatile and can be applied to various nanomaterials, such as carbon-based nanoparticles, metallic nanoparticles, and metal oxide nanoparticles. It is also possible to synthesize nanomaterials with desired morphology, surface charge, dimension, and surface functionality. Green synthesis of nanomaterials can be broadly divided into five categories:

(i) green synthesis using plant parts

(ii) green synthesis using microorganisms

(iii) green synthesis from waste material

(iv) use of green solvent

(v) use of green techniques

In the subsequent section, all these types are elaborated on with suitable examples.

3.1 Green synthesis of nanoparticles by plant parts

Plants are considered chemical factories because of their cost-efficiency and little maintenance. The nanomaterials synthesized from plants have outstanding performance in detoxifying heavy metals and controlling environmental pollution. The heavy metals are toxic even at minute concentrations [34]. Plant-derived nanoparticles are more advantageous than biological synthesis. For example, plants assisted nanoparticle synthesis route is equivalent to chemical synthesis and is very rapid than any other biosynthetic; microbial pathway approaches [35]. Various parts of plants, such as fruit, leaf, stem, and roots, produce various phytochemicals [36,37], which have been extensively used for the synthesis of nanoparticles by the green method. Synthesis of nanoparticles through natural plant extract is eco-friendly and cost-efficient because of its non-utilization of intermediate base groups. Numerous research studies of plant-based nanomaterials have shown the accumulation, detoxification, and phytoremediation of toxic metals. The eco-friendly method for removing heavy metals from aqueous plant solutions has gained considerable attention due to its great potential for eliminating pollutants and toxicity [38]. From plants, the green approach is adapted for synthesizing nanoparticles, likely, gold, silver, zinc oxide, iron, and palladium [39]. The phytocompounds, such as polyols, terpenoids, and polyphenols present in the plant extract, are responsible for the bioreduction of metallic ions [40]. Green synthesis of nanoparticles can be performed from various parts of a plant, e.g., leaves, flowers, roots, and fruits. For the synthesis of several metal nanoparticles, some important plants are used, including *Argemonemexicana* (Mexican poppy), *Acalyphaindica, Mangifera indica, Piper betle* for silver nanoparticles, *Murrayakoienigii*, Aloe vera for silver and gold nanoparticles, *Camellia sinensis* for gold nanoparticle and *Glycine max* for palladium nanoparticles [Table 1]. To the plant extract, respective reducing metal salts are added to synthesize metal nanoparticles [36,41]. The change in physical appearance, especially the change in the solution's color, unveiled the nanoparticles formation. A brief synthesis procedure of nanoparticles from plant extract is shown schematically in Fig. 3.

Table 1: Metal nanoparticles synthesized from plants

Sl No	Plants	Nanoparticle	Size of NP in nm
1	*Argemonemexicana*	Silver	30
2	*Acalyphaindica*	Silver	20-30
3	*Magniferaindica*	Silver	20
4	*Murrayakoienigii*	Silver and gold	10-25
5	*Aloevera*	Silver and gold	15.2
6	*Piper betle*	Silver	3-37
7	*Camellia sinensis*	Gold	40
8	*Glycine max*	Palladium	15

Materials Research Foundations 145 (2023) 54-91
https://doi.org/10.21741/9781644902370-3

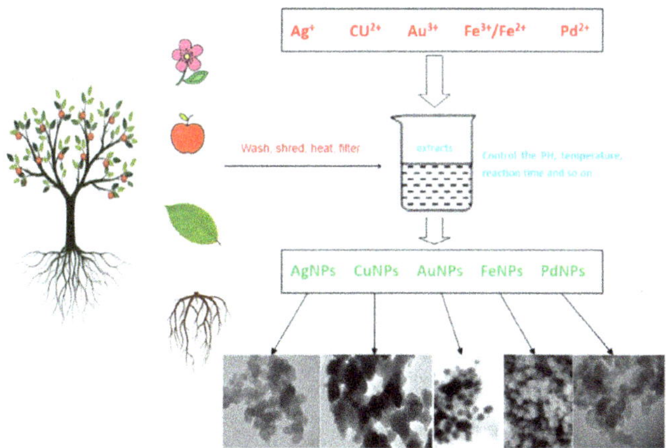

Figure 3. Synthesis of nanoparticles from plant extract [36]

3.2 Green synthesis of nanoparticles by microorganisms

Many research studies revealed that inorganic nanoparticles could be obtained from biological species. Nowadays, scientists are working on the green synthesis of nanoparticles in either intracellular or extracellular pathways. Great research articles define the production of various green-based nanoparticles via biological methods [42,43], including gold, silver, alloy, and other metal nanoparticles, oxide nanoparticles comprising magnetic and nonmagnetic oxide nanoparticles, sulfide nanoparticles, and other miscellaneous nanoparticles. Microorganisms are involved in the biological method for nanoparticle synthesis, including ion reduction of biological substances with the metal solution under the influence of enzymes as a catalyst [Fig.4]. A detailed discussion of different types of microorganisms, including bacteria, actinomycetes, fungi, yeast, virus, etc., used for synthesizing metal nanoparticles is given in the following section.

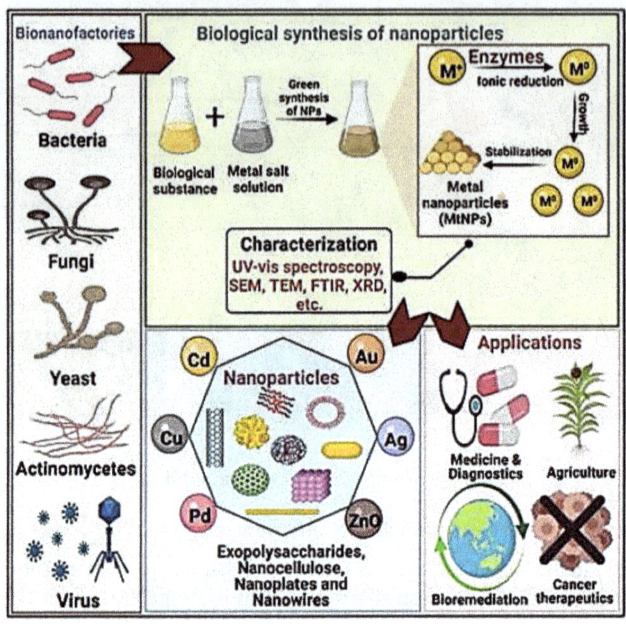

Figure 4. Schematic representation of the microbe-based synthesis of various nanoparticles, characterization, and applications [40]

3.2.1 Green synthesis of nanoparticles by bacteria

Small single-celled bacterial species have been generally utilized for viable biotechnological applications, likely bioremediation, genetic engineering, and bioleaching [44]. Prokaryotic bacteria and actinomycetes have been mostly engaged in synthesizing metal/metal oxide nanoparticles [45,46]. The metal ions are reduced from bacteria and are important candidates in biosynthesis of metallic and other novel nanoparticles.

Bacterial manipulation is used easily to synthesize green nanoparticles [45]. A few examples of bacterial strains that have been widely exploited for synthesizing metal nanoparticles, type of nanoparticle, and size are summarized in Table 2 [46].

Materials Research Foundations 145 (2023) 54-91 https://doi.org/10.21741/9781644902370-3

Table 2: List of Bacteria strains used for the synthesis of metal nanoparticles [45]

Sl. No.	Bacteria	Nanoparticle	Size in nm
1	*Bacillus subtilis*	Titanium dioxide	10-30
2	*Escherichia coli*	Silver	5-50
3	*Lactobacillus acidophilus*	Silver	45-60
4	*Mycobacterium sp.*	Gold	5-55
5	*Lactobacillus sp.*	Cadmium Sulfide	2.5-5.5
6	*Lactobacillus plantarum*	Zinc Oxide	7-19
7	*Bacillus licheniformis*	Cadmium Sulfide	20-40
8	*Lactobacillus sporogenes*	Zinc Oxide	145.7

3.2.2 Green synthesis of nanoparticles by actinomycetes

A unicellular filamentous bacteria, i.e., actinomycetes, have recently extended substantial attention for synthesizing metal nanoparticles. Nowadays, actinomycetes are recognized as superior groups among microbial species of commercial importance owing to the development of many bioactive components and extracellular enzymes through their saprophytic behavior. Different actinomycetes such as *Rhodococcus sp.*, *Gordoniaamarae*, and *Gordoniaamicalis* actinomycete are used for the biosynthesis of gold nanoparticles, and *Streptomyces sp.*, *LK3*, actinomycetes are used for silver nanoparticles synthesis [Table 3]. The synthesis of metal nanoparticles using actinomycetes involves intracellular reduction of metal ions on the surface of mycelia along with cytoplasmic membranes, leading to the formation of nanoparticles. Several researchers have documented the intracellular utilizing actinomycetes strains, which are summarized in Table 3. [46]

Table 3: List of Actinomycetes used for the synthesis of nanoparticles

Sl. No.	Actinomycetes	Nanoparticle	Size in nm
1	*Rhodococcussp. (Actinomycete)*	Gold	5-15
2	*Gordoniaamarae*	Gold	15-40
3	*Gordoniaamicalis*	Gold	5-25
4	*Streptomyces sp. LK3*	Silver	5
5	*Actinomycetes*	Silver	5-50

Materials Research Forum LLC
https://doi.org/10.21741/9781644902370-3

3.2.3 Green synthesis of nanoparticles by fungi

A eukaryotic microorganism, i.e., fungi, includes nanoparticle synthesis, which may be extracellular or intracellular. It is a biogenic route and involves the successful application of myconanotechnological methodologies. Metal salts in the mycelia, i.e., fungi, can use in the intracellular pathway and are converted into a less toxic form. The use of fungal extracts involves the extracellular pathway. The biosynthesis of various metal nanoparticles from fungi is listed in Table 4 [46]. Fungi are comparatively more practical than bacteria due to their many bioactive metabolites, high aggregation, and improved production. Different filamentous fungi have been reported to be capable in AuNP biosynthesis. It involves various methods such as fungal secreted compounds, and media components could be used to stabilize the fungal nanoparticles. The process includes metal ion reduction by nitrate-dependent reductase, and extracellular shuttle quinone was confirmed by different characterization methods, likely, UV-Visible, fluorescence, and enzymatic activity analysis.

Table 4: List of Fungi used for nanoparticle synthesis

Sl. No.	Fungi	Nanoparticle	Size in nm
1	*Penicillium sp.*	Silver	25-30
2	*Neurosporacrassa*	Silver	5-50
3	*Penicilliumoxalicum*	Silver	60-80
4	*Neurosporacrassa*	Gold	3-100
5	*Trichoderma harzianum*	Gold	32-44
6	*Aspergillus niger*	Zinc Oxide	53-69
7	*Fusarium oxysporum*	Zinc Sulfide	38
8	*Aspergillus flavus*	Tianium dioxide	62-74

3.2.4 Green synthesis of nanoparticles by yeast

Eukaryotic, single-celled microorganisms, i.e., yeast strains of different genera, are known to employ different mechanisms for nanoparticle synthesis resulting in considerable variations in size, particle position, monodispersity, and other properties. In yeast cells, the detoxification mechanism study reveals that glutathione (GSH) and two classes of metal-binding ligands-metallothioneins and phytochelatins (PC), were generated, and they play a very important role in deciding the mechanism for nanoparticle synthesis and to stabilize the resulting complexes in most of the yeast species studied. According to the resistance mechanism, yeast cells near toxic metals can change the absorbed metal ions into a nontoxic complex polymer compound. Usually, these nanoparticles synthesized in the

 Materials Research Forum LLC
https://doi.org/10.21741/9781644902370-3

yeast are referred to as "semiconductor crystals" or "quantum semiconductor crystals". Information on the yeast strains, various nanoparticles and their nanoparticle is summarized in Table 5 [46].

Table 5: Nanoparticles synthesized from yeast

Sl. No.	Yeast	Nanoparticle	Size in nm
1	*Candida albicans*	Cadmium Sulfide	50-60
2	*Baker's yeast*	Tianium dioxide	6.7
3	*Saccharomyces cerevisiae*	Tianium dioxide	12

3.2.5 Green synthesis of nanoparticles by algae

An aquatic, photosynthetic and nucleus-bearing organism, i.e., algae used for the biosynthesis of nanoparticles, is progressively becoming more and more common. *Spirulina platensis* and Sargassumcymosum has been employed in the biosynthesis of gold nanoparticles. Remarkably in this study, a blue shift in the UV absorption spectra was observed after increasing the concentration. Silver nanoparticles are synthesized from *Neochlorisoleoabundans*, *Sargassumcymosum*, and Palladium nanoparticles are obtained from *Spirulina platensis* and *Chlorella vulgaris*. Numerous algae strains, for example, *Sargassumtenerrimum*, *Turbinariaconoides*, *Laminaria japonica*, and *Acanthophoraspicifera* have been reported for synthesizing gold nanoparticles. Using *Spirulina platensis*, the synthesis of novel core (Au)-shell (Ag) nanoparticles has also been investigated. Information on the algae size of different nanoparticles is summarized below in Table 6 [46].

Table 6: Nanoparticles synthesized using algae

Sl. No.	Algae	Nanoparticle	Size in nm
1	*Neochlorisoleoabundans*	Silver	n
2	*Nostoclinckia*	Silver	5-60
3	*Spirulina platensis*	Gold	15.60–77.13
4	*Sargassumcymosum*	Gold	7 and 20
5	*Spirulina platensis*	Palladium	10-20
6	*Chlorella vulgaris*	Palladium	5-20

3.2.6 Green synthesis of nanoparticles by virus

Infectious agents, i.e., viruses, have appeared as promising candidates for the synthesis of nanoparticles for their biomedical applications due to their biocompatibility,

Materials Research Foundations 145 (2023) 54-91 https://doi.org/10.21741/9781644902370-3

biodegradability, mass production capacity, programmable scaffolds, and ease of genetic manipulation for desired characteristics. The naturally occurring nanoparticles; Viral bodies have 20–500 nanometer dimensions. Due to their robust activity, the major applications of viral nanoparticles have been in gene delivery, drug delivery, vaccines/immuno-therapeutics, and imaging. The innumerable biomedical applications of viral nanoparticles (VNPs) are observed, which are tagged with several ligands for targeting, therapeutics, or imaging agents. A similar class of materials is virus-like particles (VLPs) derived from the protein coating of the viruses [47,48]. Viral nanoparticles can be of bacteriophage, plant or animal viral origin and are dynamic, self-assembling derivatives with symmetrical and monodisperse structures. Fabrication of viral nanoparticles involves a) generation in a host body, b) further chemical conjugation and modification c) in-vitro and in-vivo analysis.

3.3 Green Synthesis of Nanoparticles from waste material

One of the major movements in the development of nanotechnology is using biodegradable wastes to synthesize nanomaterials [49,50]. The strategy of the "green" synthesis of metal nanoparticles is the use of low-cost, environmentally friendly, and renewable waste materials. This is one of the best methods to manage waste materials by reducing and reusing the waste to make our ecosystem pollution free. The redox reaction of metabolites prepares metallic nanoparticles. Plants in the agricultural and timber industry are rich in organic compounds such as aldehydes, terpene alcohols, polysaccharides, flavonoids, alkaloids, phenolic acids, and polyphenolic compounds. They may act as powerful reducing and capping agents in a single-step and one-pot reaction to produce nanoparticles (NPs) such as silver, gold, or copper [51]. Many bioactive food residues, such as fruits, vegetables, cereals, oilseeds, etc., contain secondary metabolites, mainly phenolic functional groups. In addition to phenolic acids, alkaloids, terpenoids, and flavonoids are other important secondary metabolites in biowaste. These functional groups aid in the stability and reduction process. The flavonoids in fruit residues have chelating ability and can perform reduction reactions on metal ions to produce nanoparticles. Green or biogenic or phytochemical synthesis of metallic nanoparticles is known to be a simple, ecological, cost-effective, and cheap technique. It is much safer for humans than conventional physical, chemical and electrochemical methods [52,53].

A lot of research work has been performed on various biogenic waste materials such as fruits and vegetable peels (e.g., mango peel, banana peel, pomegranate peel, sapota peel, orange peel, watermelon peel, onion peel, etc.), eggshell, groundnut shell, rice husk, coconut coir, tamarind shell, tea waste, slaughterhouse waste, marine waste, algal extract, etc. Fig.5 displays some commonly used waste materials for the synthesis of nanomaterials.

 Materials Research Forum LLC
https://doi.org/10.21741/9781644902370-3

Figure 5. Commonly used waste materials for the synthesis of nanoparticles [48]

Moreover, waste from industries, such as electronic waste, plastic waste, etc., is utilized for nanoparticle synthesis [54]. Lithium-ion battery, car battery, waste pickling liquor. Furnace slag is investigated for the preparation of various metallic nanoparticles. Waste tire rubber and plastic wastes are the primary sources for synthesizing carbon-based nanoparticles.

3.4 Use of green solvent in green synthesis of nanoparticles

Green solvent reaction conditions have been widely reported as an appropriate methodology for reducing metals using plants and microorganisms, in line with the principles of green chemistry. Many authors have considered the utility of green solvent conditions about good conversion, selectivity, and simplicity with suitable attention to physical changes occurring in the system. From crops, one can process environmentally friendly bio solvents or green solvents. For example, green solvent ethyl lactate is the ester of lactic acid is derived from processing corn. The frequent advantages of lactate esters solvents, likely being 100% biodegradable, easy to recycle, non-corrosive, non-carcinogenic, and non-ozone depleting, are commonly used in the paints and coatings industry.

The solvent used is an important parameter in nanoparticle synthesis as it provides a medium for the dissolution of starting material, transfer heat and reagent, and dispersing of the synthesized nanoparticles. Currently, available methods are mostly based on organic solvents, which harm the environment. Hence, efforts have been made to utilize more sustainable solvents for synthesis.

Water is the greenest solvent and hence the most obvious choice. However, some other green solvents are supercritical fluids such as carbon dioxide, ionic liquid, deep eutectic solvents, liquid polymers, switchable solvents, renewable solvents, etc. The dielectric constant of this solvent can be tuned near the supercritical point. Using this strategy, the solubility of the precursor can also be changed as near the supercritical end, the solubility decreases. As a result, the solution reaches a point of super-saturation where several metal nuclei can come closer and gives rise to the formation of nanoparticles. Supercritical solvent as a medium has several advantages, such as a narrow size distribution range and high dispersibility in organic solvent [55]. Ionic liquids have also been used as a green solvents. Ionic liquids have low vapor pressure and are a desirable alternative to toxic organic solvents. Moreover, they can also function as a capping agent, and thus, the use of chemicals is reduced.

3.5 Use of green techniques for green syntheis of nanoaparticles

In synthesizing nanoparticles, the green technique is employed, which refers to eco-friendly technology and is referred to as green technology. It involves the practice of technology in production processes using sustainable forms of energy, guarding the environment, and preserving natural resources. Nanomaterials manufacturing using chemical processes are known to produce dangerous pollutants and have adverse environmental effects. Although these can be diminished during the initial designing stage, it has not been seen as a priority for current industries. Conversely, industries have become more aware of environmentally conscious; these unsafe pollutants should be reduced, especially when manufacturing on a bulk scale. Using green technology in industries can minimize the environmental effects of the process [56,57].

While manufacturing nanotechnology products, green technology works with nanotechnology to develop significant and practical protocols and procedures. The risks related to nano-products are decreased by altering the production process, and green products can be used for environmental applications. This can be accomplished through direct applications, such as a) nano-enabled sensors, b) wastewater and drinking water treatment and c) remediation of hazardous waste using nanomaterials. Green technology helps environmental applications directly or indirectly by saving energy during transport, using lighter nanomaterials, or reducing waste by designing products to be smaller. Incorporating green nanotechnology and their protocols in the manufacturing process aims to make nanomaterials more eco-friendly and use nanomaterials to make current processes involving chemicals less toxic. To drive this, there are plenty of ways to do this, such as using supercritical CO_2, water, or ionic liquids to replace a volatile organic solvent. Another way to eliminate waste in the manufacturing process of nanomaterials is by employing self-assembly or templating agents [58].

Green nanotechnology involves using renewable or nontoxic replacements for non-renewable starting materials. Additionally, using other techniques like UV assisted microwaving, facile thermal and hydrothermal can aid energy conservation. Together, catalytic (engineered nanomaterials) and photocatalytic reactions can increase the manufacturing process's productivity while reducing harmful by-products. Nanosilver, which plants can absorb, is found in the soil and affects the ecosystem. According to the US Environmental Protection Agency, this is difficult because the existing tests "may not work to test the safety of nanomaterials [59] ." In addition to this, "nanomaterials have distinctive chemical properties, high reactivity, and do not dissolve in liquid."

As the world becomes more naturally aware, the use of green technology in the planning and design stage will become more predominant in the industry. Environment-friendly processes like nanotechnology and green technology can work together to create more products. However, the entire lifecycle must be considered to create truly green products.

Besides eliminating toxic chemicals/solvents, green synthesis also includes the involvement of environment-friendly methodology. Conventional synthesis involves using heat energy from energy sources such as a heating mantle, water/oil bath, and furnace. The use of microwave and sonochemical methods can decrease energy consumption by utilizing energy from microwave radiation or ultrasonic energy [55]. Thus, green synthesis of nanoparticles is a remarkably effective method of engineered nanomaterial synthesis, which is efficient and environmentally friendly.

4. Characterisation techniques of nanomaterials

Characterization techniques of nanoparticles can be divided broadly in two important categories [60].

Local probe techniques

- Scanning electron microscopy (SEM).
- Transmission electronic microscopy (TEM).
- Scanning tunneling microscopy (STM).
- Atomic-force microscopy (AFM).

Bulk-sensitive techniques

- UV–visible spectrophotometry.
- Fourier transforms infrared (FT-IR) spectroscopy.
- Electron spins resonance (ESR) spectroscopy.
- X-ray-based techniques.

4.1 Local probe techniques

4.1.1 Scanning electron microscopy (SEM)

Scanning electron microscopy is a versatile, standard, and widely used technique for visualizing and analyzing nanomaterials because of its resolution capacity of 100 Å, and some advanced instruments can achieve a resolution of 25 Å [61]. Rakesh et al. synthesized ZnO/Nb_2O_5 nanoparticles using electrochemical method and studied the surface morphology of these nanoparticles using Scanning electron microscopy. Fig.6, showed nanorods with bundle-like structures of ZnO/Nb_2O_5 nanoparticles by SEM [62]. Takahashi

 Materials Research Forum LLC
https://doi.org/10.21741/9781644902370-3

et al. synthesized biofilm containing chitosan nanoparticles made out of poly(lactide-co-glycolide) (PLGA) and characterized by using SEM [63].

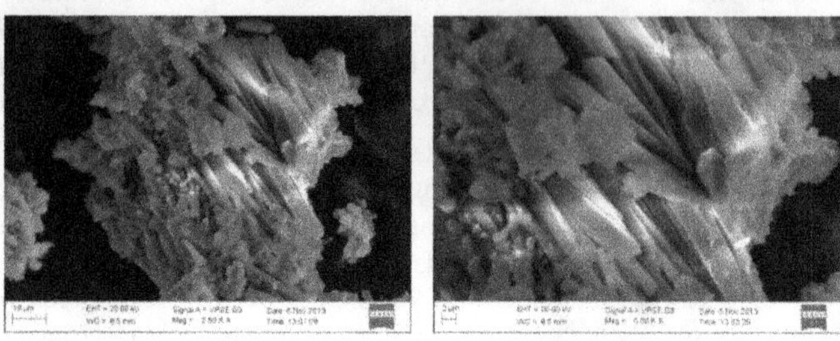

Figure 6: SEM images of ZnO/Nb_2O_5 nanoparticle [58].

4.1.2 Transmission electronic microscopy (TEM)

The transmission electronic microscopic technique is based on the interaction of high-energy electron-beam with solid particles, and the diffraction intensity provides information about crystal structure, essential information, and morphology of nanoparticles. TEM is the best technique for the study of the shape and structure of molecules and nanoparticles [64,65]; TEM method is more accurate than SEM because of greater magnification and high resolutions [66]. D. Philip synthesized Ag and Au nanoparticles using *Hibiscus sinensis* and studied the nanoparticles' morphology using TEM technology. TEM images of the colloids of the Ag and Au nanoparticle are shown in Fig. 7 [67]. Bahar Khodadadi et al. first time synthesized Pd nanoparticles (Pd NPs) supported on apricot kernel shell from the extract of the plant of *Salvia hydrangea,* which act as a catalyst to reduce many organic dyes like Rhodamine B, 4-nitrophenol, Methylene Blue, Congo red and methyl Orange at room temperature [68]. The synthesized Pd NPs/Apricot kernel shell catalyst was characterized by using TEM technique.

Materials Research Foundations 145 (2023) 54-91
https://doi.org/10.21741/9781644902370-3

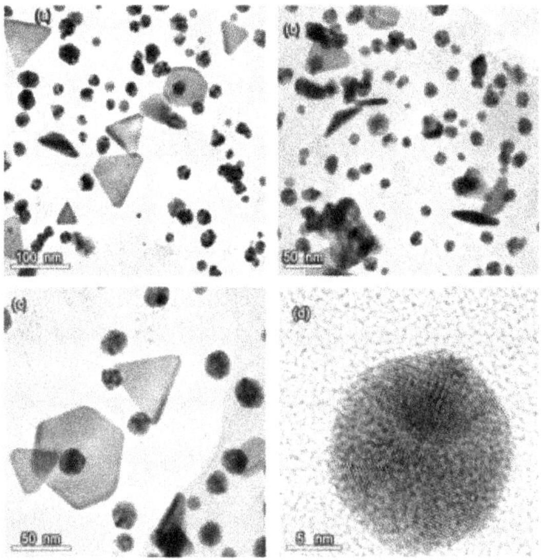

Figure 7.TEM images of Ag and Au colloids [62].

4.1.3 Scanning tunnelling microscopy (STM)

Scanning tunneling microscopy is a high-resolution microscope technique that provides information about molecules' three-dimensional structures and some spectroscopic properties. The main application of STM is the characterization of surface roughness, atomic reconstructions, cell membranes and doped semiconductors [69]. Several researchers use STM techniques for the characterization of nanoparticles. For example, Ignacio Lopez-Salido et al. grew Au nanoparticles on the surface of Highly Ordered Pyrolytic Graphite (HOPG). The film was studied using STM and X-ray Photoelectron Spectroscopy. Fig. 8 shows the STM image of the Au nanoparticle on HOPG [70]. Scanning tunneling microscopy techniques were used to study monolayer-protected gold nanoparticles by Alicia M. Jackson et al. It is the only microscopy technique capable of visualizing these molecules [71].

Materials Research Foundations 145 (2023) 54-91
https://doi.org/10.21741/9781644902370-3

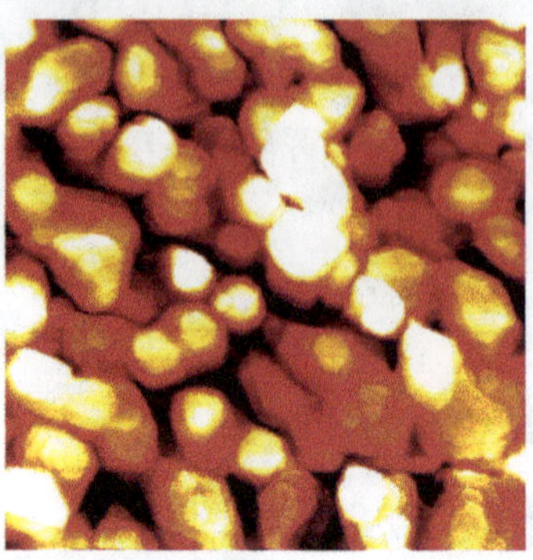

Figure 8. STM image of the Au nanoparticle on HOPG [65].

4.1.4 Atomic-force microscopy (AFM)

Atomic-force microscopy technique is closely related to Scanning tunneling microscopy, AFM technique provides high spatial resolution and three-dimensional images. Many authors prefer AFM measurement for nanoparticle analysis and characterization. AlixDubes et al. analyzed solid lipid nanoparticles derived from the amphiphilic cyclodextrin, β-CD21C6 by using Scanning electron microscopy and atomic force microscopy and compared the information obtained by these two techniques. The observed advantages of the AFM technique for the analysis of solid lipid nanoparticles are no vacuum required and simple sample preparation, while SEM technique leads to observation in a less aggregated state in this method, Fig. 9 shows AFM images of solid lipid nanoparticles [72].

Materials Research Forum LLC
https://doi.org/10.21741/9781644902370-3

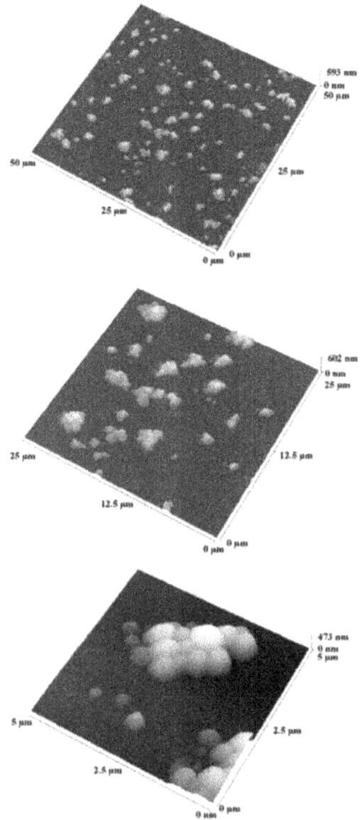

Figure 9. β-CD21C6nanoparticlesAFM images at different scan ranges [67].

4.2 Bulk sensitive techniques

4.2.1 UV–visible spectrophotometry

To investigate the optical properties, qualitative analysis, and size of nanoparticles UV–visible absorption spectroscopy plays a vital role. This instrument operates between 190-700 nm [73]. A lot of research work can be found in the literature highlighting the applications of UV–visible spectroscopy in nanoparticle characterization. Sanghiet et al. synthesized Ag nanoparticles by reduction of silver ions using white rot fungus *C. versicolor*. UV–Visible spectrum of Ag nanoparticles in fungal media and fungal mycelium media was recorded as shown in Fig. 10. It showed that absorbance of Ag nanoparticles and incubation time are directly related to each other. In fungal media, Ag nanoparticles show an absorption band at 440 nm; in the case of fungal mycelium media,

Ag nanoparticles show an absorption band at 430 nm [74]. Kumar et al. have synthesized silver nanoparticles using the green method from Morusnigra leaves extract and characterized them using UV–visible spectroscopy by monitoring the reduction of silver ions. This technique can be used to differentiate silver nanoparticles synthesized via chemical and biological routes [75].

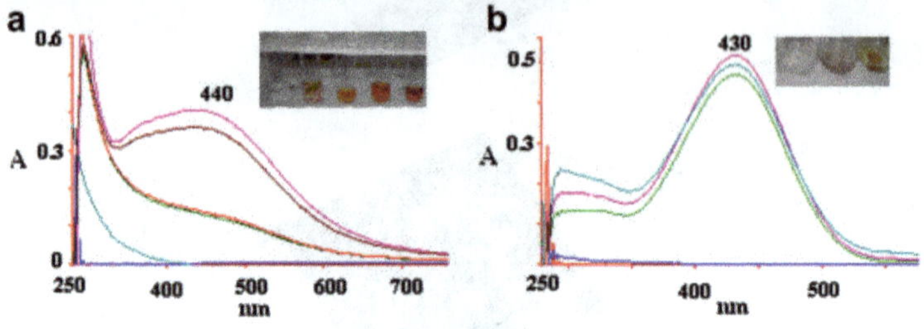

Figure 10. UV–Vis spectra of the (a) Ag nanoparticle/media (b) Ag nanoparticle /mycelium solutions

4.2.2 Fourier transforms infrared (FT-IR) spectroscopy

FT-IR spectroscopy is mainly used to analyze functional groups present in the nanoparticle sample. This spectroscopic technique is used to study the chemical composition changes during bio-synthesis and also used to study the interactions between different species. Several researchers used FT-IR technique for the characterization of nanoparticles. Caroling et al. have synthesized copper nanoparticles using green chemistry and characterized them using FT-IR [76]. This investigation used aqueous *Gooseberry* extract and biologically reducing $CuSO_4$ to synthesize copper nanoparticles. FT-IR techniques was used to analyze the presence of phenolic group and proteins in biomass before and after reduction. The analysis confirms the presence of terpenoids, OH, C=C, C-O, and many fictional groups [76].

Similarly, Fardood et al. synthesized and characterized $ZnMn_2O_4$ spinal nanoparticles for photocatalytic degradation of Congo red dye, and FTIR spectra were observed in the range of 400-4000 cm^{-1}.Mn↔O bond and Zn↔O bond in the spinel structure of the sample were confirmed by the FTIR technique that produces two bands at 621cm^{-1} and 507$cm^{-1,}$ respectively. The result is shown in Fig. 11 [77]. Other researchers focus on zinc oxide nanoparticle synthesis and characterization using *Andrographispaniculata* leaves. The FTIR spectra of terpenoid (TAP) and Zn-TAP nanoparticles exhibit a strong band at 3334.71, cm^{-1} which confirms the presence of O–H stretching vibration, 1656.36 cm^{-1} band confirms the C–O axial stretching, 1656.36 cm^{-1} band confirms the –C=C– aromatic stretching and2852.85 cm^{-1} band for the C–H stretching vibration [78].

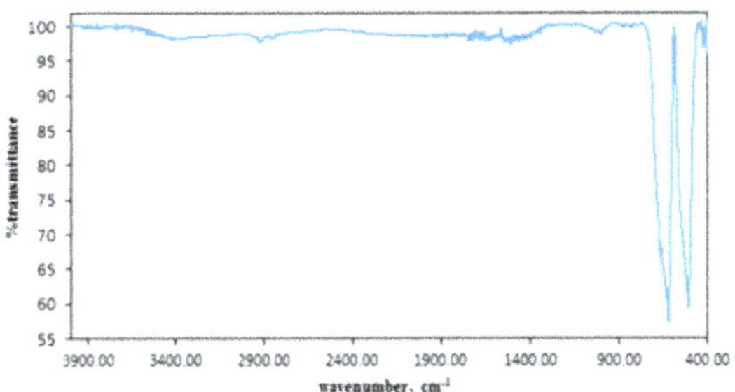

Figure 11. FTIR spectrum of $ZnMn_2O_4$ nanoparticles

4.2.3 Electron spins resonance spectroscopy (ESR)

ESR is a spectroscopic technique used to detect chemical species with unpaired electrons. Electron spins resonance spectroscopy is also called as d Electron paramagnetic resonance spectroscopy. ESR spectroscopy is highly sensitive and can identify the radicals generation in situ. Therefore, the ESR technique is extensively used in nanoscience research [78,79]. R. Elilarassi et al. successfully synthesized polycrystalline undoped and Cu-doped Zinc oxide ($Zn_{0.98}Cu_{0.02}O$) nanocrystals by solid-state reaction method, characterization of this nanocrystal was carried out by using many techniques along with the ESR technique. ESR spectra of Cu (II) in ZnO have been recorded at room temperature. This ESR spectrum gives four lines for copper, which corresponds to the parallel hyperfine splitting of the copper nucleus (I = 3/2) and also observed broad ESR signal for Cu-doped ZnO nanoparticles because of exchange interaction [80]. Manganese ferrite nanoparticles were synthesized by M. Goodarz Naseri et al. with cubic structure by using the thermal treatment method at various temperatures from 723 to 873 K, followed by calcination. Manganese ferrite nanoparticles were characterized by electron paramagnetic resonance spectroscopy, X-ray diffraction (XRD), transmission electron microscopy, etc. EPR spectra show broad and symmetrical signals at different temperatures and also calculated magnetic properties like Peak-to-peak line width (DH_{pp}), resonant magnetic field (H_r), and g factor [81]. The same analysis was done by M. GoodarzNaseri et al. for zinc ferrite nanoparticles. EPR spectra of zinc ferrite nanoparticles are shown in Fig. 12, at three different temperatures 773, 823, and 873 K. The result showed that particle sizes and saturation magnetization are inversely related, and samples exhibited super-paramagnetic behaviors [82].

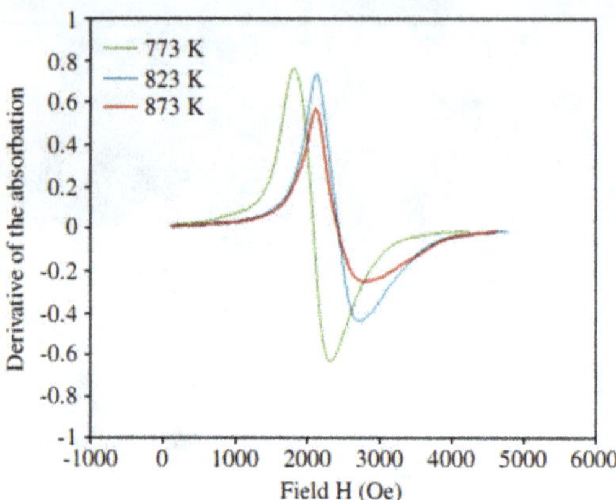

Figure 12. EPR spectrum of zinc ferrite nanoparticles calcined from 773 to 873 K [77].

4.2.4 X-ray-based techniques

X-ray Diffraction (XRD) techniques are mainly used to analyze the crystalline structure of molecules and nanocrystals. It is a well-known technique for detecting nanocrystals' size and elastic properties. Many reports utilized the XRD technique for the characterization of nanocrystals. Nath et al. synthesized cadmium selenide (CdSe) nanoparticles by using sodium cadmium chloride and hydrogen selenide precursor. X-ray diffraction (XRD) technique was used to confirm the crystalline nature of the CdSe nanoparticles with cubic Zinc blende lattice structure [Fig. 13], also calculated particle size and the elastic properties of nanocrystals [83,84]. As mentioned earlier for synthesis of manganese ferrite nanoparticles [81] and zinc ferrite nanoparticles [82] by M. Goodarz Naseri et al. with cubic structure by using thermal treatment method at various temperatures. Both methods' characterization was conducted using X-ray diffraction (XRD) techniques. The patterns shown in the XRD graph are the reflection planes (1 1 1), (2 2 0), (3 1 1), (2 2 2), (4 0 0), (4 2 2), (5 1 1), (4 4 0), and (5 3 1), which confirm the presence of single phase $MnFe_2O_4$ with a face-centered cubic structure [85]. The broad peak that occurred in the precursor does not have sharp diffraction patterns; it indicates particles are amorphous. The calcinated patterns show the reflection planes (111), (220), (311), (222), (400), (331), (422), (511), and (440); this shows that the presence of single-phase $ZnFe_2O_4$ [86].

Materials Research Forum LLC
https://doi.org/10.21741/9781644902370-3

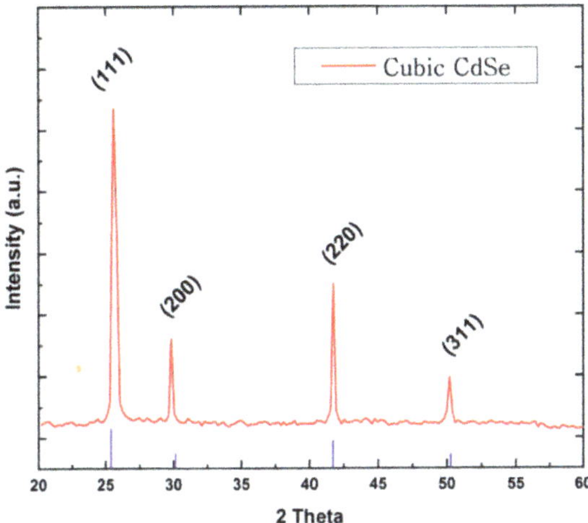

Figure 13. XRD pattern of cubic CdSe nanoparticles [78]

5. Limitations and challenges

Nanomaterials synthesized using green chemistry principles consume less energy, are cheaper, and are safe for the environment and human health. It is a promising method that can solve many issues that are faced in the case of the chemical method of synthesis. The key limitations of green synthesis are time and the plant distribution according to the geographical area, and seasonal variation leads to complexity in terms of place and time [87]. Research findings suggest that the yield and purity of green synthesized nanomaterials are very poor. The progress in this field is mainly setback due to the following five reasons: (1) Selection of material, (2) Reaction condition, (3) quality of the synthesized nanomaterials, (4) Application areas, (5) Toxicity as shown in Fig.14 [36]. The details on each of these points are discussed below.

1. Selection of material

The plant from which nanomaterials are synthesized using green principles is sometimes location-specific [88,89]. Thus production on a large scale for commercialization becomes a big challenge. Moreover, collecting raw materials from plants depends on seasonal conditions. Nanomaterials synthesized from flowers and seeds must be collected during flowering and fruiting times [90]. Arabica coffee, used for synthesizing silver nanoparticles, takes seven long years for the fully grown stage [91]. In addition, secondary materials are used for nanomaterial synthesis. In most cases, these secondary materials require costly synthetic and extraction processes adding complexity and expenses of

Materials Research Foundations 145 (2023) 54-91 https://doi.org/10.21741/9781644902370-3

synthesis cost [92]. Thus the principles of green chemistry in these situations are not obeyed completely and require attention for future improvement.

2. Reaction condition

Some reports of green synthesis of nanomaterials from plant extract require excess energy. Extremely high temperatures (about 800°C) and long reaction times do not fall under the concept of green principles [93,94]. In a few cases, it is observed that the plant extract needs preservation at very low temperatures leading to high energy consumption [95,96]. Short reaction time reduces the cost and also increases the efficiency of production. However, nanomaterials synthesized for three long days have been reported via green methods. Sometimes the pre-processing before actual synthesis, like extraction, pre-drying, etc., requires a long time. Synthesis of nanomaterials in an inert atmosphere to control oxidation of the metal nanoparticles enhances the expenditure of the synthesis process [97]. The elucidation of the mechanism of the green synthesized nanomaterials is still unknown, limiting the understanding of the synthesis process.

3. Quality of nanomaterials

Green synthesized nanoparticles have a wide variation in size and shape. Thus large-scale production of nanoparticles through green synthesis is a big challenge. Many times, the synthesized nanoparticles are of irregular shapes and clusters. Agglomeration into the heterogeneous mixture, noncrystalline form, and defects in the synthesized product are some limitations observed in their practical application [97,98]. Low conversion efficiency and low yield limit the economic benefit of nanoparticles synthesized from the green principle. Optimization and control of particle size, shape, morphology, and monodispersity of the particles need further attention.

6. Application

Nanomaterials are used for pollutant removal because of their large surface area and adsorption capacity. However, it has been observed that green synthesized nanoparticles are not very efficient in removing BOD, COD, and heavy metal pollutants. Multi-component pollutant removal is less efficient using green nanomaterials [99]. Moreover, the *in vitro* application of green nanomaterials is not studied extensively; only a few reports are available, while *in vivo* applications are missing completely [100,101].

7. Toxicity

Although green synthesized nanomaterials are highly beneficial due to its environmentally benign nature, the toxicity of the nanomaterials for long time exposure cannot be ignored. The nanomaterials can very easily penetrate deeply into living beings bodies due to their favorable size. The health of human, animal, and plant bodies is affected badly after the accumulation of nanoparticles in the internal organs. Several diseases have been reported due to the toxic effect of nanomaterials [102].

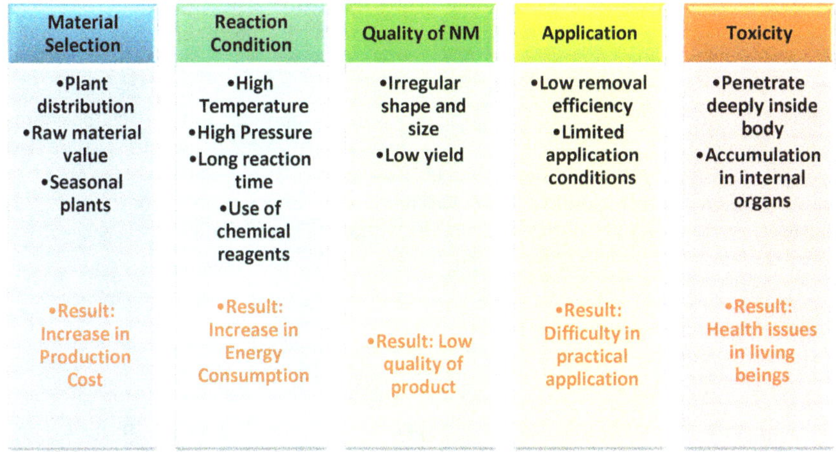

Figure 14. Challenges in green synthesis of nanomaterials[36]

8. Prospect

Synthesis of nanomaterials using green principles is an emerging field with high potential. These biogenic nanomaterials can have versatile applications with environmentally benign solutions. Thus it needs more attention to optimizing these systems for maximum benefit. More research is required to synthesize green nanomaterials from novel natural materials that are not explored much. Control of size and morphology is crucial in manipulating properties for applications in novel areas. Insufficient knowledge about the mechanism of many biogenic processes is a barrier to optimizing the nanomaterial synthesis process. Thus efforts are needed to comprehend the mechanism to develop a consistent and predictable process. To date, the green synthesis of nanomaterials is mostly confined to the laboratory scale. Upgrading it to the commercial level requires evaluating the technology for mass production. Regulatory bodies must check the safety and protection of the environment from the adverse effect of nanomaterials. The complete process cycle, from raw materials to synthesis to disposal, needs scrutiny to control toxicity risk.

Conclusions

Nanomaterials synthesized using the green principle are a growing rapidly in the research field. This method is environmentally benign, devoid of harsh chemicals, and inexpensive. The concept behind the green principle has been elaborated. This chapter comprehensively discusses the recent development in this field. Natural sources that are used for biogenic synthesis have been discussed. Various characterization techniques used by researchers are also discussed in detail. However, it has several shortcomings and challenges, including low yield, inconsistent particle size, time taking extraction process, toxicity issues etc. These issues can be addressed by further research. Finally, the future perspective for further advancement is highlighted. Scaling from the laboratory phase to commercialization requires a lot of planning and research. The development of human civilization should not compromise the safety and security of the environment and humanity. Thus, promoting green nanotechnology is the need of the hour for sustainable development of the environment and society. The progress and utilization of green nanotechnologies in various commercial sectors are possible only through cooperative support from researchers, industries, government, and policymakers. The welfare of humankind is heavily dependent on utilizing the green principle for developing green nanotechnology.

References

[1] S. Horikoshi, N. Serpone, Introduction to Nanoparticles, Microwaves Nanoparticle Synth. Fundam. Appl. (2013) 1–24. https://doi.org/10.1002/9783527648122.ch1

[2] E. Zekić, Ž. Vuković, I. Halkijević, Application of nanotechnology in wastewater treatment, Gradjevinar. 70 (2018) 315–323. https://doi.org/10.14256/JCE.2165.2017

[3] J.S. Duhan, R. Kumar, N. Kumar, P. Kaur, K. Nehra, S. Duhan, Nanotechnology: The new perspective in precision agriculture, Biotechnol. Reports. 15 (2017) 11–23. https://doi.org/10.1016/j.btre.2017.03.002

[4] Y. Huang, L. Mei, X. Chen, Q. Wang, Recent developments in food packaging based on nanomaterials, Nanomaterials. 8 (2018) 1–29. https://doi.org/10.3390/nano8100830

[5] E. Pomerantseva, F. Bonaccorso, X. Feng, Y. Cui, Y. Gogotsi, Energy storage: The future enabled by nanomaterials, Science (80-.). 366 (2019). https://doi.org/10.1126/science.aan8285

[6] R. Li, Z. Zha, Z. Miao, C.Y. Xu, Emerging 2D pnictogens for biomedical applications, Chinese Chem. Lett. 33 (2022) 2345–2353. https://doi.org/10.1016/j.cclet.2021.09.062

[7] S. Li, Z. Zhao, J. Zhao, Z. Zhang, X. Li, J. Zhang, Recent Advances of Ferro-, Piezo-, and Pyroelectric Nanomaterials for Catalytic Applications, ACS Appl. Nano Mater. 3 (2020) 1063–1079. https://doi.org/10.1021/acsanm.0c00039

[8] Y.H. Wang, L.L. He, K.J. Huang, Y.X. Chen, S.Y. Wang, Z.H. Liu, D. Li, Recent advances in nanomaterial-based electrochemical and optical sensing platforms for microRNA assays, J. Comb. Math. Comb. Comput. 144 (2019) 2849–2866. https://doi.org/10.1039/c9an00081j

[9] Anindita De, N.B. Singh, M. Guin, S. Barthwal, Water Purification by Green Synthesized Nanomaterials, Curr. Pharm. Biotechnol. 24 (2022) 101–117. https://doi.org/10.2174/1389201023666220507030548

[10] G.P. Halada, A. Orlov, Environmental degradation of engineered nanomaterials: Impact on materials design and use, Third Edit, Elsevier Inc., 2018. https://doi.org/10.1016/B978-0-323-52472-8.00011-3

[11] C.D. Jensen, N.A. Lewinski, Nanoparticle synthesis to green informatics frameworks, Curr. Opin. Green Sustain. Chem. 12 (2018) 117–126. https://doi.org/10.1016/j.cogsc.2018.08.005

[12] L.M. Skjolding, S.N. Sørensen, N.B. Hartmann, R. Hjorth, S.F. Hansen, A. Baun, Aquatic Ecotoxicity Testing of Nanoparticles—The Quest To Disclose Nanoparticle Effects, Angew. Chemie - Int. Ed. 55 (2016) 15224–15239. https://doi.org/10.1002/anie.201604964

[13] M. Fojtů, W.Z. Teo, M. Pumera, Environmental impact and potential health risks of 2D nanomaterials, Environ. Sci. Nano. 4 (2017) 1617–1633. https://doi.org/10.1039/c7en00401j

[14] M.I. Sohail, A.A. Waris, M.A. Ayub, M. Usman, M. Zia ur Rehman, M. Sabir, T. Faiz, Environmental application of nanomaterials: A promise to sustainable future, 1st ed., Elsevier B.V., 2019. https://doi.org/10.1016/bs.coac.2019.10.002

[15] R. Sanghi, Sanghi-1662.pdf, (2000) 1662–1665

[16] N.B. Singh, M.A. B.H. Susan, M. Guin, Applications of Green Synthesized Nanomaterials in Water Remediation, Curr. Pharm. Biotechnol. 22 (2020) 733–761. https://doi.org/10.2174/1389201021666201027160029

[17] D. Nath, P. Banerjee, Green nanotechnology - A new hope for medical biology, Environ. Toxicol. Pharmacol. 36 (2013) 997–1014. https://doi.org/10.1016/j.etap.2013.09.002

[18] V.A. Basiuk, E. V. Basiuk, Green processes for nanotechnology: From inorganic to bioinspired nanomaterials, in: V.A. Basiuk, E. V. Basiuk (Eds.), Green Process. Nanotechnol. From Inorg. to Bioinspired Nanomater., 2015: pp. 1–446. https://doi.org/10.1007/978-3-319-15461-9

[19] K.B. Narayanan, N. Sakthivel, Green synthesis of biogenic metal nanoparticles by terrestrial and aquatic phototrophic and heterotrophic eukaryotes and biocompatible agents, Adv. Colloid Interface Sci. 169 (2011) 59–79. https://doi.org/10.1016/j.cis.2011.08.004

[20] J.C.W. Paul T. Anastas, Green chemistry: theory and practice, Oxford University Press. New York, Naw York, 1998

[21] A.P. Dicks, A. Hent, Green Chemistry and Associated Metrics, (2015) 1–15. https://doi.org/10.1007/978-3-319-10500-0_1

[22] A. García-Quintero, M. Palencia, A critical analysis of environmental sustainability metrics applied to green synthesis of nanomaterials and the assessment

of environmental risks associated with the nanotechnology, Sci. Total Environ. 793 (2021) 148524. https://doi.org/10.1016/j.scitotenv.2021.148524

[23] P. Bardos, C. Merly, P. Kvapil, H.P. Koschitzky, Status of nanoremediation and its potential for future deployment: Risk-benefit and benchmarking appraisals, Remediation. 28 (2018) 43–56. https://doi.org/10.1002/rem.21559

[24] B. Fabiano, A.P. Reverberi, P.S. Varbanov, Safety opportunities for the synthesis of metal nanoparticles and short-cut approach to workplace risk evaluation, J. Clean. Prod. 209 (2019) 297–308. https://doi.org/10.1016/j.jclepro.2018.10.161

[25] J.A. Dahl, B.L.S. Maddux, J.E. Hutchison, Toward greener nanosynthesis, Chem. Rev. 107 (2007) 2228–2269. https://doi.org/10.1021/cr050943k

[26] A.E.D. Mahmoud, M. Fawzy, Nanosensors and Nanobiosensors for Monitoring the Environmental Pollutants, in: Top. Mining, Metall. Mater. Eng., Springer International Publishing, Chambridge, 2021: pp. 229–246. https://doi.org/10.1007/978-3-030-68031-2_9

[27] H.S. Hassan, M.F. Elkady, N.M. Serour, Intelligent nanosensors (INS) for environmental applications, in: S.K. Chaudhery Hussain (Ed.), Handb. Nanomater. Sens. Appl., Elsevier, 2021: pp. 321–344. https://doi.org/10.1016/B978-0-12-820783-3.00017-8

[28] T. Hassan, A. Salam, A. Khan, S.U. Khan, H. Khanzada, M. Wasim, M.Q. Khan, I.S. Kim, Functional nanocomposites and their potential applications: A review, J. Polym. Res. 28 (2021). https://doi.org/10.1007/s10965-021-02408-1

[29] L.M. Tufvesson, P. Tufvesson, J.M. Woodley, P. Börjesson, Life cycle assessment in green chemistry: Overview of key parameters and methodological concerns, Int. J. Life Cycle Assess. 18 (2013) 431–444. https://doi.org/10.1007/s11367-012-0500-1

[30] D. Mulvaney, Green Metrics, Wiley VCH, 2012. https://doi.org/10.4135/9781412975704.n61

[31] B.T. Reid, S.M. Reed, Improved methods for evaluating the environmental impact of nanoparticle synthesis, Green Chem. 18 (2016) 4263–4269. https://doi.org/10.1039/c6gc00383d

[32] M. Shah, D. Fawcett, S. Sharma, S.K. Tripathy, G.E.J. Poinern, Green synthesis of metallic nanoparticles via biological entities, 2015. https://doi.org/10.3390/ma8115377

[33] L.A. Kolahalam, I. V. Kasi Viswanath, B.S. Diwakar, B. Govindh, V. Reddy, Y.L.N. Murthy, Review on nanomaterials: Synthesis and applications, Mater. Today Proc. 18 (2019) 2182–2190. https://doi.org/10.1016/j.matpr.2019.07.371

[34] M. Shahid, C. Dumat, S. Khalid, E. Schreck, T. Xiong, N.K. Niazi, Foliar heavy metal uptake, toxicity and detoxification in plants: A comparison of foliar and root metal uptake, J. Hazard. Mater. 325 (2017) 36–58. https://doi.org/10.1016/j.jhazmat.2016.11.063

[35] N. Asmathunisha, K. Kathiresan, Colloids and Surfaces B : Biointerfaces A review on biosynthesis of nanoparticles by marine organisms, Colloids Surfaces B Biointerfaces. 103 (2013) 283–287. http://dx.doi.org/10.1016/j.colsurfb.2012.10.030

[36] S. Ying, Z. Guan, P.C. Ofoegbu, P. Clubb, C. Rico, F. He, J. Hong, Green synthesis of nanoparticles: Current developments and limitations, Environ. Technol. Innov. 26 (2022) 102336. https://doi.org/10.1016/j.eti.2022.102336

[37] R. de A.M.M. Neto, C.B.R.R. Santos, S.V.C.C. Henriques, L. de O. Machado, J.N. Cruz, C.H.T. d. P.T. de P. da Silva, L.B. Federico, E.H.C. d. C. de Oliveira, M.P.C.C. de Souza, P.N.B.B. da Silva, C.A. Taft, I.M. Ferreira, M.R.F.F. Gomes, Novel chalcones derivatives with potential antineoplastic activity investigated by docking and molecular dynamics simulations, J. Biomol. Struct. Dyn. (2020) 1–13. https://doi.org/10.1080/07391102.2020.1839562

[38] C.F. Carolin, P.S. Kumar, A. Saravanan, G.J. Joshiba, M. Naushad, Efficient techniques for the removal of toxic heavy metals from aquatic environment: A review, J. Environ. Chem. Eng. 5 (2017) 2782–2799. https://doi.org/10.1016/j.jece.2017.05.029

[39] J. Singh, T. Dutta, K.H. Kim, M. Rawat, P. Samddar, P. Kumar, "Green" synthesis of metals and their oxide nanoparticles: Applications for environmental remediation, J. Nanobiotechnology. 16 (2018) 1–24. https://doi.org/10.1186/s12951-018-0408-4

[40] M. Ovais, A.T. Khalil, N.U. Islam, I. Ahmad, M. Ayaz, M. Saravanan, Z.K. Shinwari, S. Mukherjee, Role of plant phytochemicals and microbial enzymes in biosynthesis of metallic nanoparticles, Appl. Microbiol. Biotechnol. 102 (2018) 6799–6814. https://doi.org/10.1007/s00253-018-9146-7

[41] V.M. Almeida, Ê.R. Dias, B.C. Souza, J.N. Cruz, C.B.R. Santos, F.H.A. Leite, R.F. Queiroz, A. Branco, Methoxylated flavonols from Vellozia dasypus Seub ethyl acetate active myeloperoxidase extract: in vitro and in silico assays, J. Biomol. Struct. Dyn. 40 (2022) 7574–7583. https://doi.org/10.1080/07391102.2021.1900916

[42] B. Koul, A.K. Poonia, D. Yadav, J.O. Jin, Microbe-mediated biosynthesis of nanoparticles: Applications and future prospects, Biomolecules. 11 (2021). https://doi.org/10.3390/biom11060886

[43] S. Ghosh, R. Ahmad, M. Zeyaullah, S.K. Khare, Microbial Nano-Factories: Synthesis and Biomedical Applications, Front. Chem. 9 (2021). https://doi.org/10.3389/fchem.2021.626834

[44] M. Gericke, A. Pinches, Microbial production of gold nanoparticles, Gold Bull. 39 (2006) 22–28. https://doi.org/10.1007/BF03215529

[45] K.N. Thakkar, S.S. Mhatre, R.Y. Parikh, Biological synthesis of metallic nanoparticles, Nanomedicine Nanotechnology, Biol. Med. 6 (2010) 257–262. https://doi.org/10.1016/j.nano.2009.07.002

[46] N. Shreyash, S. Bajpai, M.A. Khan, Y. Vijay, S.K. Tiwary, M. Sonker, Green Synthesis of Nanoparticles and Their Biomedical Applications: A Review, ACS Appl. Nano Mater. 4 (2021) 11428–11457. https://doi.org/10.1021/acsanm.1c02946

[47] Y.H. Chung, H. Cai, N.F. Steinmetz, Viral nanoparticles for drug delivery, imaging, immunotherapy, and theranostic applications, Adv. Drug Deliv. Rev. 156 (2020) 214–235. https://doi.org/10.1016/j.addr.2020.06.024

[48] C.M.A. Rego, A.F. Francisco, C.N. Boeno, M. V Paloschi, J.A. Lopes, M.D.S. Silva, H.M. Santana, S.N. Serrath, J.E. Rodrigues, C.T.L. Lemos, R.S.S. Dutra, J.N. da Cruz, C.B.R. dos Santos, S. da S. Setúbal, M.R.M. Fontes, A.M. Soares, W.L. Pires, J.P. Zuliani, Inflammasome NLRP3 activation induced by Convulxin, a C-type lectin-like isolated from Crotalus durissus terrificus snake venom, Sci. Rep. 12 (2022) 1–17. https://doi.org/10.1038/s41598-022-08735-7

[49] V.P. Aswathi, S. Meera, C.G.A. Maria, M. Nidhin, Green synthesis of nanoparticles from biodegradable waste extracts and their applications: a critical review, Nanotechnol. Environ. Eng. (2022). https://doi.org/10.1007/s41204-022-00276-8

[50] S.M. Abdelbasir, K.M. McCourt, C.M. Lee, D.C. Vanegas, Waste-Derived Nanoparticles: Synthesis Approaches, Environmental Applications, and Sustainability Considerations, Front. Chem. 8 (2020) 1–18. https://doi.org/10.3389/fchem.2020.00782

[51] P.R. Reddy, S.D. Ganesh, N. Saha, O. Zandraa, P. Sáha, Ecofriendly Synthesis of Silver Nanoparticles from Garden Rhubarb (Rheum rhabarbarum), J. Nanotechnol. 2016 (2016). https://doi.org/10.1155/2016/4964752

[52] R.S.R. Isaac, G. Sakthivel, C. Murthy, Green synthesis of gold and silver nanoparticles using averrhoa bilimbi fruit extract, J. Nanotechnol. 2013 (2013). https://doi.org/10.1155/2013/906592

[53] L.D. Do Nascimento, A.A.B. de Moraes, K.S. da Costa, J.M.P. Galúcio, P.S. Taube, C.M.L. Costa, J.N. Cruz, E.H. de A. Andrade, L.J.G. de Faria, Bioactive natural compounds and antioxidant activity of essential oils from spice plants: New findings and potential applications, Biomolecules. 10 (2020) 1–37. https://doi.org/10.3390/biom10070988

[54] M. Canle, M.I. Fernández, J.A. Santaballa, Applications of Nanomaterials in Environmental Remediation, 2021. https://doi.org/10.1201/9781003129042-1

[55] H. Duan, D. Wang, Y. Li, Green chemistry for nanoparticle synthesis, Chem. Soc. Rev. 44 (2015) 5778–5792. https://doi.org/10.1039/c4cs00363b

[56] E. Grossman, Tiny materials, big questions: How green is nanotechnology?, (2016)

Materials Research Forum LLC
https://doi.org/10.21741/9781644902370-3

[57] A.R.J.A. de M. Lima, A.S. Siqueira, M.L.S. Möller, R.C. de Souza, J.N. Cruz, A.R.J.A. de M. Lima, R.C. da Silva, D.C.F. Aguiar, J.L. da S.G.V. Junior, E.C. Gonçalves, In silico improvement of the cyanobacterial lectin microvirin and mannose interaction, J. Biomol. Struct. Dyn. (2020). https://doi.org/10.1080/07391102.2020.1821782

[58] S. Wong, B. Karn, Ensuring sustainability with green nanotechnology, Nanotechnology. 23 (2012) 11–13. https://doi.org/10.1088/0957-4484/23/29/290201

[59] U.S. Environmental Protection Agency, Research on Nanomaterials, Us Epa. (2016). https://www.epa.gov/chemical-research/research-nanomaterials

[60] N.S. J. Herrera, Microscopic and Spectroscopic Characterization of Nanoparticles, in: Drug Deliv. Nanoparticles Formul. Charact., 2020: pp. 259–271. https://doi.org/10.3109/9781420078053-18

[61] H.Y. Joseph I. Goldstein, Practical Scanning Electron Microscopy, Springer US, 1975. https://doi.org/10.1007/978-1-4613-4422-3

[62] R. , S. Ananda, N.M.M. Gowda, K.R. Raksha, Synthesis of Niobium Doped ZnO Nanoparticles by Electrochemical Method: Characterization, Photodegradation of Indigo Carmine Dye and Antibacterial Study, Adv. Nanoparticles. 03 (2014) 133–147. https://doi.org/10.4236/anp.2014.34018

[63] C. Takahashi, N. Ogawa, Y. Kawashima, H. Yamamoto, Observation of antibacterial effect of biodegradable polymeric nanoparticles on Staphylococcus epidermidis biofilm using FE-SEM with an ionic liquid, Microscopy. 64 (2015) 169–180. https://doi.org/10.1093/jmicro/dfv010

[64] M. Joshi, A. Bhattacharyya, S.W. Ali, Characterization techniques for nanotechnology applications in textiles, Indian J. Fibre Text. Res. 33 (2008) 304–317

[65] F.S. Alves, J. de A. Rodrigues Do Rego, M.L. Da Costa, L.F. Lobato Da Silva, R.A. Da Costa, J.N. Cruz, D.D.S.B. Brasil, Spectroscopic methods and in silico analyses using density functional theory to characterize and identify piperine alkaloid crystals isolated from pepper (Piper Nigrum L.), J. Biomol. Struct. Dyn. 38 (2020) 2792–2799. https://doi.org/10.1080/07391102.2019.1639547

[66] B. Schaffer, U. Hohenester, A. Trügler, F. Hofer, High-resolution surface plasmon imaging of gold nanoparticles by energy-filtered transmission electron microscopy, Phys. Rev. B - Condens. Matter Mater. Phys. 79 (2009) 1–4. https://doi.org/10.1103/PhysRevB.79.041401

[67] D. Philip, Green synthesis of gold and silver nanoparticles using Hibiscus rosa sinensis, Phys. E Low-Dimensional Syst. Nanostructures. 42 (2010) 1417–1424. https://doi.org/10.1016/j.physe.2009.11.081

[68] B. Khodadadi, M. Bordbar, M. Nasrollahzadeh, Green synthesis of Pd nanoparticles at Apricot kernel shell substrate using Salvia hydrangea extract:

Catalytic activity for reduction of organic dyes, J. Colloid Interface Sci. 490 (2017) 1–10. https://doi.org/10.1016/j.jcis.2016.11.032

[69] L.E.C. Van De Leemput, H. Van Kempen, Scanning tunnelling microscopy, Reports Prog. Phys. 55 (1992) 1165–1240. https://doi.org/10.1088/0034-4885/55/8/002

[70] I. Lopez-Salido, D.C. Lim, R. Dietsche, N. Bertram, Y.D. Kim, Electronic and geometric properties of Au nanoparticles on Highly Ordered Pyrolytic Graphite (HOPG) studied using X-ray Photoelectron Spectroscopy (XPS) and Scanning Tunneling Microscopy (STM), J. Phys. Chem. B. 110 (2006) 1128–1136. https://doi.org/10.1021/jp054790g

[71] A.M. Jackson, Y. Hu, P.J. Silva, F. Stellacci, From homoligand- to mixed-ligand-monolayer-protected metal nanoparticles: A scanning tunneling microscopy investigation, J. Am. Chem. Soc. 128 (2006) 11135–11149. https://doi.org/10.1021/ja061545h

[72] A. Dubes, H. Parrot-Lopez, W. Abdelwahed, G. Degobert, H. Fessi, P. Shahgaldian, A.W. Coleman, Scanning electron microscopy and atomic force microscopy imaging of solid lipid nanoparticles derived from amphiphilic cyclodextrins, Eur. J. Pharm. Biopharm. 55 (2003) 279–282. https://doi.org/10.1016/S0939-6411(03)00020-1

[73] S.L. Upstone, Ultraviolet/Visible Light Absorption Spectrophotometry in Clinical Chemistry, Encycl. Anal. Chem. (2006) 1699–1714. https://doi.org/10.1002/9780470027318.a0547

[74] R. Sanghi, P. Verma, Biomimetic synthesis and characterisation of protein capped silver nanoparticles, Bioresour. Technol. 100 (2009) 501–504. https://doi.org/10.1016/j.biortech.2008.05.048

[75] S. Parvathy, B.R. Venkatraman, Synthesis, characterization and antibacterial potential of silver nanoparticles using Solanum erianthum (D. Don) leaf extract, 1 (2015) 16–24. https://doi.org/10.5176/2301-3761_ccecp15.43

[76] G. Mishra, V. Yadav, D.A. Saxena, Biosynthesis of Copper Nanoparticles Using Aqueous Ficus Racemosa Extract- Characterization and Study of Antimicrobial Effects, Am. J. Pharm. Heal. Res. 7 (2019) 63–76. https://doi.org/10.46624/ajphr.2019.v7.i11.004

[77] S.T. Fardood, F. Moradnia, A. Ramazani, Green synthesis and characterisation of ZnMn2O4 nanoparticles for photocatalytic degradation of Congo red dye and kinetic study, Micro Nano Lett. 14 (2019) 986–991. https://doi.org/10.1049/mnl.2019.0071

[78] S. Kavitha, M. Dhamodaran, R. Prasad, M. Ganesan, Synthesis and characterisation of zinc oxide nanoparticles using terpenoid fractions of Andrographis paniculata leaves, Int. Nano Lett. 7 (2017) 141–147. https://doi.org/10.1007/s40089-017-0207-1

[79] J.J. Yin, P.P. Fu, Application of Electron Spin Resonance to Study Food Anitoxidative and Prooxidative Activities, in: G.A.W. María Guðjónsdóttir, Peter S Belton (Ed.), Magn. Reson. Food Sci. Challenges a Chang. World, Royal Society of Chemistry, 2009: pp. 213–221. https://doi.org/10.1039/9781847559494-00213

[80] R. Elilarassi, G. Chandrasekaran, Structural, optical and magnetic characterization of Cu-doped ZnO nanoparticles synthesized using solid state reaction method, J. Mater. Sci. Mater. Electron. 21 (2010) 1168–1173. https://doi.org/10.1007/s10854-009-0041-y

[81] M. Goodarz Naseri, E. Bin Saion, H.A. Ahangar, M. Hashim, A.H. Shaari, Synthesis and characterization of manganese ferrite nanoparticles by thermal treatment method, J. Magn. Magn. Mater. 323 (2011) 1745–1749. https://doi.org/10.1016/j.jmmm.2011.01.016

[82] M.G. Naseri, E.B. Saion, M. Hashim, A.H. Shaari, H.A. Ahangar, Synthesis and characterization of zinc ferrite nanoparticles by a thermal treatment method, Solid State Commun. 151 (2011) 1031–1035. https://doi.org/10.1016/j.ssc.2011.04.018

[83] D. Nath, F. Singh, R. Das, X-ray diffraction analysis by Williamson-Hall, Halder-Wagner and size-strain plot methods of CdSe nanoparticles- a comparative study, Mater. Chem. Phys. 239 (2020) 122021. https://doi.org/10.1016/j.matchemphys.2019.122021

[84] R.A. Costa, J.N. Cruz, F.C.A. Nascimento, S.O.S.G. Silva, S.O.S.G. Silva, M.C. Martelli, S.M.L. Carvalho, C.B.R. Santos, A.M.J.C. Neto, D.S.B. Brasil, Studies of NMR, molecular docking, and molecular dynamics simulation of new promising inhibitors of cruzaine from the parasite Trypanosoma cruzi, Med. Chem. Res. 28 (2019) 246–259. https://doi.org/10.1007/s00044-018-2280-z

[85] J.P. Singh, R.C. Srivastava, H.M. Agrawal, R.P.S. Kushwaha, 57Fe Mössbauer spectroscopic study of nanostructured zinc ferrite, Hyperfine Interact. 183 (2008) 221–228. https://doi.org/10.1007/s10751-008-9756-z

[86] C.N. Chinnasamy, A. Narayanasamy, N. Ponpandian, K. Chattopadhyay, H. Guérault, J.M. Greneche, Magnetic properties of nanostructured ferrimagnetic zinc ferrite, J. Phys. Condens. Matter. 12 (2000) 7795–7805. https://doi.org/10.1088/0953-8984/12/35/314

[87] S. Iravani, Green synthesis of metal nanoparticles using plants, Green Chem. 13 (2011) 2638–2650. https://doi.org/10.1039/c1gc15386b

[88] E. Turunc, R. Binzet, I. Gumus, G. Binzet, H. Arslan, Green synthesis of silver and palladium nanoparticles using Lithodora hispidula (Sm.) Griseb. (Boraginaceae) and application to the electrocatalytic reduction of hydrogen peroxide, Mater. Chem. Phys. 202 (2017) 310–319. https://doi.org/10.1016/j.matchemphys.2017.09.032

[89] K. Tahir, S. Nazir, B. Li, A. Ahmad, T. Nasir, A.U. Khan, S.A.A. Shah, Z.U.H. Khan, G. Yasin, M.U. Hameed, Sapium sebiferum leaf extract mediated synthesis of

palladium nanoparticles and in vitro investigation of their bacterial and photocatalytic activities, J. Photochem. Photobiol. B Biol. 164 (2016) 164–173. https://doi.org/10.1016/j.jphotobiol.2016.09.030

[90] S.S. Sana, L.K. Dogiparthi, Green synthesis of silver nanoparticles using Givotia moluccana leaf extract and evaluation of their antimicrobial activity, Mater. Lett. 226 (2018) 47–51. https://doi.org/10.1016/j.matlet.2018.05.009

[91] V. Dhand, L. Soumya, S. Bharadwaj, S. Chakra, D. Bhatt, B. Sreedhar, Green synthesis of silver nanoparticles using Coffea arabica seed extract and its antibacterial activity, Mater. Sci. Eng. C. 58 (2016) 36–43. https://doi.org/10.1016/j.msec.2015.08.018

[92] G. Li, Y. Li, Z. Wang, H. Liu, Green synthesis of palladium nanoparticles with carboxymethyl cellulose for degradation of azo-dyes, Mater. Chem. Phys. 187 (2017) 133–140. https://doi.org/10.1016/j.matchemphys.2016.11.057

[93] J. Nasiri, M. Rahimi, Z. Hamezadeh, E. Motamedi, M.R. Naghavi, Fulfillment of green chemistry for synthesis of silver nanoparticles using root and leaf extracts of Ferula persica: Solid-state route vs. solution-phase method, J. Clean. Prod. 192 (2018) 514–530. https://doi.org/10.1016/j.jclepro.2018.04.218

[94] A. Muthuvel, M. Jothibas, C. Manoharan, Synthesis of copper oxide nanoparticles by chemical and biogenic methods: photocatalytic degradation and in vitro antioxidant activity, Nanotechnol. Environ. Eng. 5 (2020). https://doi.org/10.1007/s41204-020-00078-w

[95] N. González-Ballesteros, S. Prado-López, J.B. Rodríguez-González, M. Lastra, M.C. Rodríguez-Argüelles, Green synthesis of gold nanoparticles using brown algae Cystoseira baccata: Its activity in colon cancer cells, Colloids Surfaces B Biointerfaces. 153 (2017) 190–198. https://doi.org/10.1016/j.colsurfb.2017.02.020

[96] M. Khatami, I. Sharifi, M.A.L. Nobre, N. Zafarnia, M.R. Aflatoonian, Waste-grass-mediated green synthesis of silver nanoparticles and evaluation of their anticancer, antifungal and antibacterial activity, Green Chem. Lett. Rev. 11 (2018) 125–134. https://doi.org/10.1080/17518253.2018.1444797

[97] M. Leili, M. Fazlzadeh, A. Bhatnagar, Green synthesis of nano-zero-valent iron from Nettle and Thyme leaf extracts and their application for the removal of cephalexin antibiotic from aqueous solutions, Environ. Technol. (United Kingdom). 39 (2018) 1158–1172. https://doi.org/10.1080/09593330.2017.1323956

[98] L. Xu, Y.Y. Wang, J. Huang, C.Y. Chen, Z.X. Wang, H. Xie, Silver nanoparticles: Synthesis, medical applications and biosafety, Theranostics. 10 (2020) 8996–9031. https://doi.org/10.7150/thno.45413

[99] X. Weng, X. Jin, J. Lin, R. Naidu, Z. Chen, Removal of mixed contaminants Cr(VI) and Cu(II) by green synthesized iron based nanoparticles, Ecol. Eng. 97 (2016) 32–39. https://doi.org/10.1016/j.ecoleng.2016.08.003

[100] P.C. Nagajyothi, M. Pandurangan, D.H. Kim, T.V.M. Sreekanth, J. Shim, Green Synthesis of Iron Oxide Nanoparticles and Their Catalytic and In Vitro Anticancer Activities, J. Clust. Sci. 28 (2017) 245–257. https://doi.org/10.1007/s10876-016-1082-z

[101] P.C. Nagajyothi, P. Muthuraman, T.V.M. Sreekanth, D.H. Kim, J. Shim, Green synthesis: In-vitro anticancer activity of copper oxide nanoparticles against human cervical carcinoma cells, Arab. J. Chem. 10 (2017) 215–225. https://doi.org/10.1016/j.arabjc.2016.01.011

[102] R.G. Saratale, G.D. Saratale, H.S. Shin, J.M. Jacob, A. Pugazhendhi, M. Bhaisare, G. Kumar, New insights on the green synthesis of metallic nanoparticles using plant and waste biomaterials: current knowledge, their agricultural and environmental applications, Environ. Sci. Pollut. Res. 25 (2018) 10164–10183. https://doi.org/10.1007/s11356-017-9912-6

Materials Research Forum LLC
https://doi.org/10.21741/9781644902370-4

Chapter 4

Top-down and Bottom-up Approaches for Synthesis of Nanoparticles

Supriya Tripathy[1], Jolina Rodrigues[1] and Navinchandra Gopal Shimpi[1,*]

[1] Laboratory of Materials Science, Department of Chemistry, University of Mumbai, Santacruz (East), Mumbai-400098, India

navin_shimpi@rediffmail.com

Abstract

There are numerous potential uses as nanoparticles produced from green methods in the medicinal and environmental sciences with a reduction of harmful chemicals and solvents. To synthesize nanoparticles use of safe organic resources like plants, and microorganisms producing nanoparticles for numerous applications such as catalysis, sensing, electronics, photonics and medicine. Aiming to minimize or prevent waste generated from reactions while maintaining efficacy. In this chapter, as well as the production of plant-mediated nanoparticles and some current uses of these materials, such as gold, silver, copper, palladium, platinum, zinc oxide, and titanium dioxide, the fundamentals of green chemistry have been discussed.

Keywords

Nanoparticles, Green Synthesis, Carbon-Based Nanomaterials, Metal-Based Nanomaterials

Contents

Materials Research Foundations 145 (2023) 92-130 https://doi.org/10.21741/9781644902370-4

Materials Research Foundations 145 (2023) 92-130 https://doi.org/10.21741/9781644902370-4

1. Introduction

Nanotechnology is a branch of material science that deals with the synthesis and study of its physical and chemical properties with an average particle size of 1-100nm. These materials have wide applications and are used in various fields of science and technology due to their properties. The concern for the environment and the surrounding safety increased rapidly during the late 90s. As a result, certain environmentally conscious terms were introduced in the field of chemistry, such as clean, environmental, green, sustainable, and benign chemistry. These newly introduced terms became the topic of interest for the researchers and chemists to define the meaning and complete understanding of the concept involved. Considering the importance amongst all aspects, the present chapter deals with green chemistry, which covers the meaning of green chemistry, the twelve principles of green chemistry, green synthesis of nanomaterials, the application of green synthesis with challenges, and future perspectives [1].

Cathcart (1990) was probably the first chemist to use the term green chemistry, wherein he discussed the merits and demerits of the significant growth of the Irish chemical industry. The turning point came in 1996 when *Anastas* and *Williamson* (1996 a,b) published a matter entitled Green Chemistry: An overview in green chemistry: Designing Chemistry for the environment [2]. European Association for Chemical and Molecular Sciences members held a meeting on 28th February 2008 in Tundo about Green and Sustainable Chemistry where they commented: *" no consensus in the scientific community,"* which was further modified by *Winterton* in 2001 that *"chemistry that calls itself green is necessarily going to lead to pollution prevention"* [3]. Sustainable development aims to reduce the effect caused due to the substances used and produced. It also aims at judicial use of resources monitoring energy usage and production. As it takes years and years to form non-renewable resources like fossil fuels and petroleum products and due to excessive use of these, the rate of ozone depletion has increased and to avoid depletion rate, the transition or shift from non-renewable to renewable sources should be accelerated by considering the demand to fulfill the human requirements, limited access and rising cost [4]. Chemistry plays a vital role in ensuring the use of chemicals, materials, and energy used by future generations to be more sustainable than the current due to the increasing worldwide demand for environmentally friendly processes for developing novel and cost-effective approaches to prevent pollution. To achieve this sustainability, green chemistry was introduced, which involved utilizing principles to either reduce or eliminate the production or use of toxic materials by designing, manufacturing, or applying chemical products. Although certain principles are similar and seem to be common sense, it forms a designed framework to modify the process concerning the need. There is rapid growth in the field of green chemistry by focusing on methods to design environmentally friendly products and processes, which will be economical in the long term [4]. Sustainable development stresses providing goods and services for the growing population without sacrificing the needs of future generations and environmental quality. It is expected that by 2050, the world's population will reach 10.7 billion, which is nearly double, so the demand for goods will also be set double, thereby promoting the growth of chemical industries

Materials Research Foundations 145 (2023) 92-130
https://doi.org/10.21741/9781644902370-4

coincident with the rising population. Due to population growth, there is an impact on the global environment tied to chemical processes or products, such as

- Loss of biological species from their habitat
- Water pollution
- Ozone layer depletion
- Climate change
- The rise in sea level
- Pollutants in ecosystem

Green chemistry promotes the elimination of hazardous organic solvents wherever possible or the replacement of hazardous solvents with the potential use of safer solvents (water, fluorous and ionic liquids, supercritical media). Catalysis also plays a vital role in green chemistry, which is energy efficient, selective, and economical by providing solutions to important problems, mainly in industries. The emerging area of green chemistry is not only demonstrates the growth in positive aspects of chemistry but also the success in terms of quantitative benefit to human health and the environment is paramount as well, merely a drop in the ocean as compared to the potential. To reach full potential, by creating awareness and by adapting to the methodologies and techniques, the development of green chemistry is necessary and is promoted. [5] Nanomaterials and their synthesis have great attention due to their tiny size, while the excellent ratio of surface to volume, which shows peculiar differences in their physical and chemical properties (catalytic activity, mechanical properties, biological properties, thermal properties, electrical conductivity, optical absorption, melting point) concerning bulk of same chemical composition.[6,7] Hence, the design and production of novel applications can be accomplished by monitoring the shape and size of the nanometer scale. A wide range of applications of nanomaterials concerning their size and shape-dependent nature ranges from bio-sensing and catalysts to optics, antimicrobial activity, computer transistors, chemical sensors, electrometers, and wireless electronic logic and memory schemes. These nanoparticles have applications in medical imaging, nanocomposites, filters, drug delivery, and hyperthermia of tumors [8–12]. For instance, geranium quantum dots of size less than 10nm can be produced in fixed proportion for a novel optoelectronic device for single electron transistors (SETs) and light emitters [13].

Metal-based nanoparticles also have wide applications in the field of medicine and pharmacy. For instance, gold and silver-based nanoparticles are commonly used in biomedical and nano-biotechnology (one of the most promising interdisciplinary fields[14]. Recent development in the formation of nanoparticles shows a peculiar role of microorganisms and biological systems in the production of metal-based nanomaterials due to their growing success, ease of formation, and environmentally friendly method (green chemistry) without involving harsh, toxic, and expensive chemicals [15–18].

Materials Research Foundations 145 (2023) 92-130 https://doi.org/10.21741/9781644902370-4

The characteristics of production of highly stable and monodispersed nanoparticles are as follows

Selection of suitable organisms

To determine the potential of organisms before selecting their intrinsic properties like enzyme activities and biochemical pathways are studied. For instance, plants with heavy metal accumulation and detoxification serve as the best candidate for nanoparticle synthesis.

Requisite conditions for enzyme activity and cell growth

The growth is associated with factors like nutrients, light, temperature, pH, and buffer strength, which should be optimized to promote growth. The presence of these substrates in sub-toxic levels from the start of the development would eventually accelerate enzymatic activities.

Reaction conditions

For the industrial-scale synthesis of metal-based nanomaterials, the production rate and yield play a vital role in maintaining and monitoring the reaction conditions. Bio-reduction conditions like substrate concentration, the bio-catalyst concentration, electron donor and its concentration, pH, temperature, exposure time, buffer strength, mixing speed, and light is monitored and optimized [19]. However, some scientists have used complementary aspects like visible light, microwave irradiation, and boiling to affect the morphology, size, and rate of reaction.

2. Principles of sustainable and green chemistry

Green or sustainable chemistry deals with reducing or eliminating the production and use of toxic substances. The growth and development in science and technology have advanced and are rapidly increasing daily as per the living standard of humanity. Such accelerated progress leads to drastic changes like environmental degradation, depletion of the ozone layer, climate change, and global warming affecting life on earth. So to overcome such shortcomings, green chemistry came into the picture to balance the judicial use of natural resources considering economic growth and environmental conservation [20]. There is a slight difference between Green Chemistry and Environmental Chemistry as the latter deals with the identification of the source, elucidation of mechanism, and quantifying the problems associated with the environment. In contrast, green chemistry deals with designing an alternative pathway or technology safe enough to overcome the shortcomings of Environmental chemistry [21].

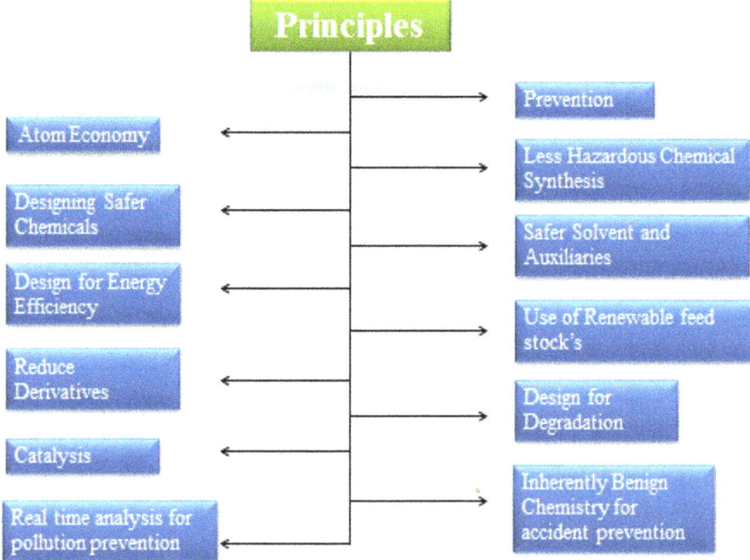

Figure 1. Green chemistry principles

Green chemistry aims to protect the environment using methods like catalysis, bio-catalysis, alternative renewable raw materials, reaction media, and reaction conditions. The basis of the green chemistry framework stands on twelve principles used to create a safe passage to design molecules, materials, reactions, and processes.[22,23] Paul Anastas and John Warner, in their book named *"Green Chemistry Theory and Practice book, 1998"* explain every term related to the development of the green chemistry principle. The principle highlights the removal of dangerous and hazardous substances (Fig. 1) as follows

Prevention:

The first principle suggests that it is relatively more efficient and wise to prevent the generation of waste and toxins than to treat or clean afterward. The prevention process is more viable for humans and the environment as it is considered cheaper and less tedious than eliminating waste after generation. For instance, the over-exploitation of natural resources like coal and petroleum products and extensive usage results in the release of sulfur and nitrogen oxides to the atmosphere by contributing as a main source of acid rain [24].

Materials Research Foundations 145 (2023) 92-130 https://doi.org/10.21741/9781644902370-4

Atom economy

The principle of Atom Economy is initially linked to waste management as it involves maximum utilization of raw materials and production of desired product, generating the least amount of byproducts. When we look for the proper meaning of atom economy, we say it is the ratio of the relative molecular masses of the product we desire to the relative molecular masses of all the reactants participating in the reaction expressed in percentages. This means we need to design the synthesis process so that the generation of waste material is reduced and the amount of product we desire is maximized; for instance, the ancient method (boots process) of synthesis of ibuprofen had low economic cost as the input of raw material was confined only up to 40% while in the newly developed method (Hoechst synthesis) involving three steps utilizes almost all (up to 99%) the raw materials used thereby eliminating the generation of waste material. This new method is cost-effective with a high atom economy, contributes to preventing pollution, and, subsequently, is environmentally friendly [25].

Less hazardous chemical synthesis

This principle involves the design of chemical processes and methods to be done so that the product we obtain has less toxicity or is dangerous to the environment and human health. Most of the synthesis we carry out, or design involves multiple steps wherein we use a lot of toxic chemicals and reagents. Despite the product being less toxic, the risk of contamination becomes high due to multiple-step synthesis. It is a task in green chemistry to redesign such environmentally friendly methods. Biological enzymes can replace toxic chemicals, making the processes cleaner and cheaper. For instance, *Asahi Kasei*'s polycarbonate synthesis is simpler when toxic carbonyl dichloride ($COCl_2$) is replaced with carbon dioxide (CO_2) resulting in the removal of dichloromethane (CH_2Cl_2) as solvent promoting green synthesis as it is more toxic compared to carbon dioxide (CO_2) [26].

Designing safer chemicals

Clearly to achieve this goal of maintaining the particular function and efficiency for designing safer products is a task that can be accomplished by having knowledge not only about chemistry but also the principles of toxicology and environment. As the name suggests, the principle deals with designing chemicals with high selective nature while carrying out the specific desired function, thereby minimizing the toxicity aim is to minimize the usage of chemicals which are carcinogenic, mutagenic, and neurotoxic, ensuring the risk and toxicity caused by such chemicals is reduced. The application of this principle in various fields is not hidden from all. It is widely used to develop insecticides and pesticides specific to certain organisms (target specific). Another example is the production of polymers of polyphenylene sulfone (PPSU), a new-age engineering plastic used in underground trains and indoor airplanes, as it provides environment safety without mechanical and flame-resistant properties [27].

Materials Research Foundations 145 (2023) 92-130 https://doi.org/10.21741/9781644902370-4

Safer solvents and auxiliaries

Using auxiliaries, solvents, and separating agents should be avoided wherever possible. If it is supposed to be used, then the substances having nature as nontoxic, non-hazardous, non-carcinogenic, non-explosive materials should be used. For any given chemical reaction, the use of auxiliaries and solvents is predominant. So by reducing the number of such chemicals in use, we are generating less waste and keeping the environment clean and safe. The conventional solvents used are primarily toxic, corrosive, and flammable, and their recovery involves a large amount of energy loss; therefore, the development of environment-friendly solvents is necessary concerning current demand. The substitution of conventional solvents with environment-friendly solvents is done based on worker safety, environment safety, process safety, and sustainability. The most important characteristics these solvents should possess are that they must be easy to use and recycle, physically and chemically stable, and have low volatility. For instance, certain chemicals like pyridine, dichloromethane, chloroform, and dimethoxyethane are some of the chemicals that are harmful for both health and the environment. Instead of these solvents, green solvents like methyl ethyl ketone, propan-2-ol, butan-1-ol, and acetone can be used as they are less toxic [28].

Design for energy efficiency

Energy efficiency is one of a chemical reaction's basic or fundamental roots. As the name suggests, this principle deals with energy efficiency and the requirements of a chemical reaction. The energy efficiency and need for a chemical reaction should be designed so that by consuming less energy, the process can proceed to make it environment-friendly, efficient and viable. It is expected that these kinds of processes should be carried out at an ambient temperature and pressure as the major aim of this principle is to save energy or, in other words, to exploit every bit of power to the fullest supplied to complete the same chemical process minimizes the use by avoiding wastage, the chemical industries and institutions have adopted various methods. Also, catalysts can be used to carry out certain chemical reactions to lower the activation energy of the reaction and thereby minimize the energy requirement [29].

Use of renewable feedstock

Renewable resources came into the picture when the concept of sustainable development was enlightened due to rapid growth, worldwide demand, and development. With the increase in population, the requirement of growth and production increased to overcome that, and the principle of use of renewable feedstocks was promoted. We depended on non-renewable sources like coal, petroleum, and natural gases for ages. Still, due to their over-usage and exploitation, the levels of non-renewable resources are coming down day by day, and the formation of this coal and petroleum requires years and years. Instead of these non-renewable energy sources, we can use renewable resources as a predominately available alternative.[21]

Materials Research Foundations 145 (2023) 92-130 https://doi.org/10.21741/9781644902370-4

Reduce derivatives

The central concept of green chemistry revolves around synthesizing the target molecule in an environment-friendly method. The term derivatization means modifying the chemical and physical state of chemical species through eiting a blocking group or through protection or deprotection. This principle of green chemistry highlights the concept of minimizing or avoiding the excessive formation of derivatives as the formation and removal of the exact needs extra reagents as the steps increases and can contribute to generating waste to avoid the formation of derivatives; the physical-chemical processes can be replaced by biological processes for synthesis as the waste produced in such cases can easily be degraded with the help of biological processes of waste management [30].

Catalysis

A catalyst is a chemical substance that enhances the reaction rate by lowering the activation energy of a reaction. It is used in small quantities and recovered by the reaction's end[31]. If you consider stoichiometric reagents, they are used in excess amounts and are difficult or not recovered at the end of the reaction [32]. The principle behind this is to use less energy, reduce the generation of waste, and minimize the use of water. All catalysts are enzymatic, homogeneous, or heterogeneous; the major role is to accelerate the reaction without affecting relative energy between the products, reactants, or the participating reaction. Catalyst has high chemical selectivity, specificity, and stereochemistry, but the only drawback is they lack heat sensitivity and poor stability.

Design for Degradation

This principle deals with designing the reaction so that the desired product and waste material obtained are not harmful in nature and must be easily degraded by natural means without accumulating in the environment. This principle's goal is to avoid forming harmful toxic substances and recycle the waste as much as possible to return to production [33].

Real-time analysis for pollution prevention

This principle focuses on analyzing the methods of chemical synthesis and their analysis to monitor the production of the desired product. It also aims to prevent the formation of toxic, dangerous substances for human health and the environment. This also includes the development of analytical methods considering the in-process monitoring and control before the formation of toxins at any stage. Previous methods used were tedious and used analysis of the sample in bulk quantity, along with the use of solvents and energy in large amounts. But with the advances in technology, the sample size to analyze has been reduced, and the amount of solvent employed is also minimized.

Inherently benign chemistry for accident prevention

The agenda behind this principle is that the substances used in the process and the ones produced should not be prone to cause any accidents. Selecting safe substances minimizes the potential for chemical accidents, explosions, and accidental fires. For instance, the widely used supercritical CO_2 replaced the organic solvent.

Materials Research Foundations 145 (2023) 92-130
https://doi.org/10.21741/9781644902370-4

3. Synthesis of nanoparticles via green route

Nanoparticles can be synthesized in various ways, but their synthesis is categorized into two main type's namely traditional methods and green methods. The traditional methods have its own benefits such as extensive scalability [34], extensive control over morphology of nanoparticles [35–37], energy storage or conservation [38–40], and target specific disease therapy [41,42]. Despite having a lot of advantages traditional method of synthesis also employs a lot of demerits like these methods involves organic solvents for the synthesis which not only contributes to neurobehavioral and reproductive health related stress but also due to the use of drastic conditions (high pressure and heat) for the synthesis process also compromises the health and the working conditions [43–48]. While working with organic solvents the concern for volatile vapors also arises due to their low boiling point along with the concern of excessive production of carbon dioxide which is responsible for environmental changes like global warming and greenhouse effect [49–51]. As a result of these irreversible risks involved in synthesis using traditional form for both the environment and the scientist these methods have fallen out of favor irrespective of the benefits they provide. Due to these short comings the concept of green synthesis came into picture considering the climatic condition, fast growing and developing world (Fig. 2). Moving towards green chemistry and green techniques for nanomaterial production have been recently brought into picture to avoid toxic, hazardous, and non-eco-friendly substances which promoted scientist and researchers to develop keen interest towards biological approaches to synthesize nanomaterials [52].

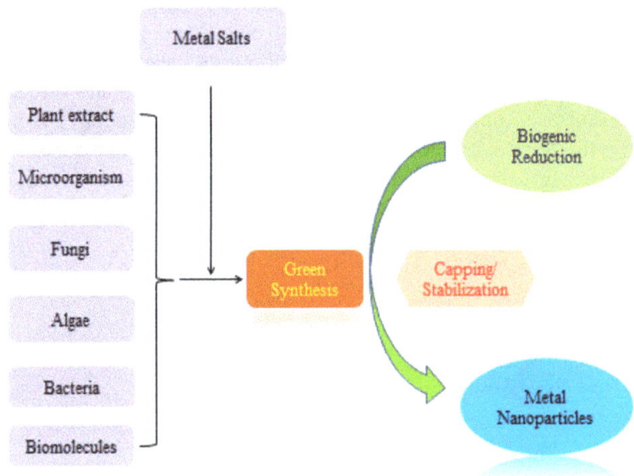

Figure 2. Flow chart of synthesis of nanoparticles via green route

Green synthesis brings a clean, safe, cost effective, sustainable and environment friendly method to synthesize nanoparticles [53]. Green method of synthesis includes replacing hazardous toxic chemicals while developing new strategies to minimize or eradicate harmful chemicals and byproducts hazardous to health and environment [52]. Green methods include microorganisms like bacteria, algae, fungi, yeast species as well as plants. The final size and the morphology of the nanoparticles are determined by different active molecules and precursors like metal salts that are employed while synthesis. Green methods provide a lot of benefits from microbial action to natural reducing and stabilizing properties [54]. The synthesis of nanomaterials can be done using physical, chemical and biological approaches. Physical method includes cracking a superstructure and restricting their nanosize margin and chemical method includes reaction of chemical compounds. Physical methods are tedious compared to chemical methods whereas chemical methods use toxins and hazardous chemicals. Some of these methods are condensation, sol-gel method, chemical vapor deposition and other biochemical approaches [55]. Biological methods for synthesis includes synthesis of nanoparticles using plant extracts, microorganisms, (bacteria, fungi, yeast, algae) and bio-molecules (enzymes, sugar and proteins) and also whole cells [52].

The most widely used methods for synthesis are top down approach and bottom up approach. The top down approach to synthesize nanoparticles involves larger bulk materials (superstructure) broken into very small nano-sized particles (atomic or molecular level), while the bottom up approach involves individual atoms brought together to form larger nanomaterials [54]. Top down method includes few techniques like lithographic, etching, sputtering and grinding in a ball mill, while bottom up method includes growing of nanoparticles from simpler molecules. Size and shape of the nanoparticle can be monitored by using target precursors concentrations, reaction media or conditions (temperature, pH), and surface functionalization along with templates utilization, a bottom up approach. Surface functionalization has a crucial aspect when synthesizing nanomaterials with respect to certain applications like high chemical activity and large surface promotes aggregation. Hence, aggregation is avoided by functionalization of nanoparticles using certain chemicals to increase their stability during storage, transportation and application (Fig. 3) [56].

Materials Research Forum LLC
https://doi.org/10.21741/9781644902370-4

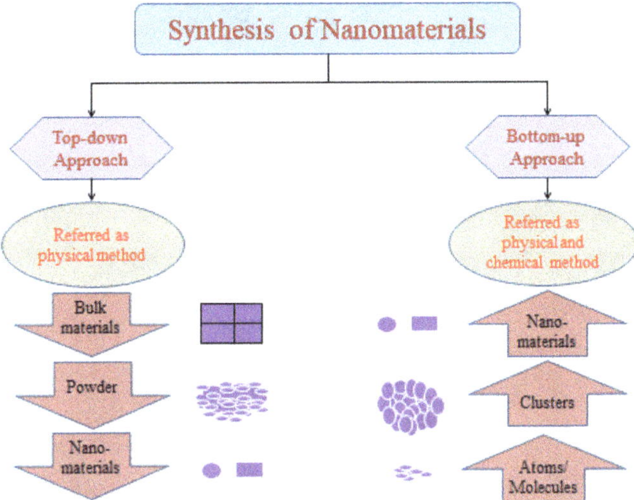

Figure 3. Approaches and steps of nanomaterials synthesis

3.1 Top down approach

The top down method includes the breakdown of huge bulk matter into tiny nano sized particle. It is a subtractive method includes breaking of superstructure bulk matter and confining it to smaller structures by chemical or mechanical means. Some techniques that involves top down approaches are mechanical grinding, attrition, milling [52]. Top down approaches are simple yet they produced irregular shaped extremely small particles and since there is a difficulty face to maintain a proper size and shape of the particles is the major downfall of this approach [57]. The following are few methods, which are employed under top down approach.

3.1.1 Mechanical milling or ball milling process

Ball milling is a fundamental and effective top-down method that produces nanomaterials by transferring kinetic energy from the grinding media to the material that is being reduced. This method is useful to produce various nanoparticles and metal alloys using different materials. The interaction of balls, and between balls and vessel walls, produce drastic phase transformation at high temperature and pressure.

 Materials Research Forum LLC
https://doi.org/10.21741/9781644902370-4

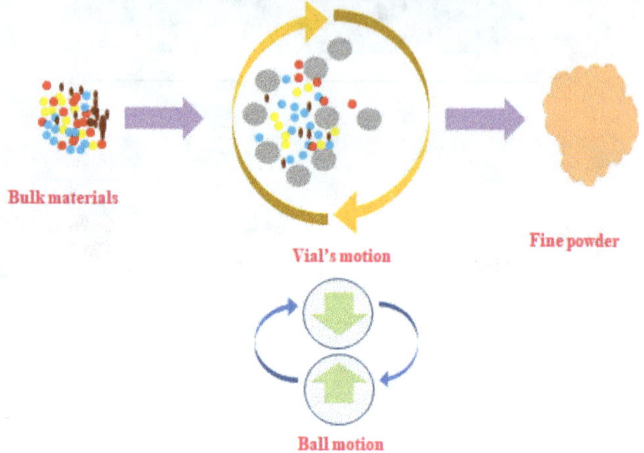

Figure 4. Schematics of ball milling process

The ball milling process have repercussions on various factors such as crystal size, particle size and surface, mechanical dislocation and modification of meta-stable phases. Milling process can be used for growth and compression of the size of nanoparticles as well as change in structure, agglomeration of nanoparticle, mixing of two or more phases [58–63]. There are certain merits of these techniques which includes high capacity, safety, simplicity, reliability, long term maintenance and universality [64–66]. With merits come certain limitations like energy wastage, destructive/destroying/disordering crystal structure, noise and contamination while working, sturdy weight (Fig. 4).

3.1.2 Thermal evaporation

In this method chemical link in a molecule breaks down due to heat as the process is endothermic [67]. Thermal evaporation methods give rise to monodispersed suspensions promoting production of inorganic nanoparticles [68]. In the center of heating zone an alumina crucible having powder is placed where a resistive heating method is employed with a 20 cm consistent gap between source and substrate is maintained. Substrates are cleaned with acetone and ethanol before placing it in the evaporator. PID (proportional integral derivative) controller and a digital display used to measure and control the temperature at source and substrate respectively. Deposition time is controlled by a shutter fixed at the entrance of crucible while glass slides before being used for deposition is cleaned in soap solution followed by acetone and de-ionized water ultrasonically [69–72].

Materials Research Foundations 145 (2023) 92-130
https://doi.org/10.21741/9781644902370-4

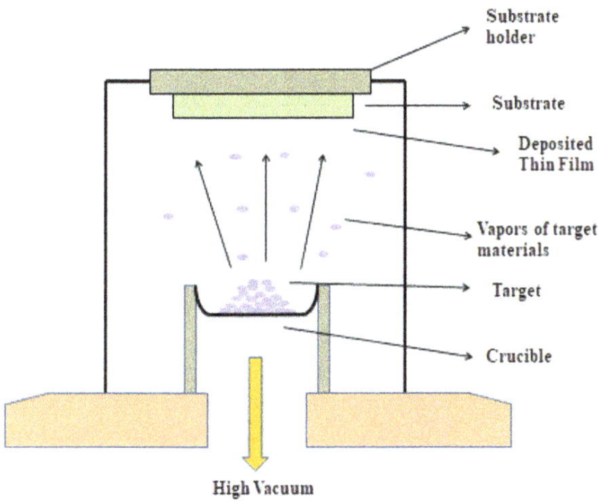

Figure 5. Schematics thermal evaporation method

Thermal evaporation method has a lot of advantages for synthesizing and fabrication of nanomaterials, for instance, fabrication can be carried out in solvent free mode. Similar to advantages we also have certain short comings to thermal evaporation method like low quality source material, difficulty in depositing alloy compositions and combinations (Fig. 5) [73–75].

3.1.3 Laser ablation

A powerful electromagnetic beam that has been increased by encouraging radiation emission is referred as laser and the hypothesis was first introduced by *Einstein* [76]. Laser ablation technique for the synthesis of nanoparticles from diverse solvents is quite a simple process [67]. In 1960, *Mainman* was the first to build practical laser, now used in multiple fields like information transmission, industry, medical treatment, and military services [77]. Removal of molecules from surface of substrate to create micro/nano structure is done with the help of pulsed laser which indeed has wide applications in metals, ceramics, polymers and glasses. A substance is eliminated from substrate by focusing a laser beam, via which the energy is absorbed in order to melt or evaporate or vanish, ablation occurs. In the entire laser machining applications laser ablation is constant which basically means the process which deals with both vaporization as well as melts ejection [76]. Traditional micro-processing methods are replaced by laser due to processing speed which is widely used in various techniques by irradiating solid target dipped in liquid by laser beam to produce nanoparticles altering its particle size, morphology, composition by adjusting laser settings and liquid medium [78–80]. This method has a lot of merits like it promotes the production of ligand-free noble particles and energy loss is low, with merits there comes short coming of the method which involves high energy requirement due to high efficiency in ablation, due to dispersion of laser source in industrial scale the capability to form nanoparticles reduces [81].

3.1.4 Lithographic methods

This method of synthesis of nanoparticle require expensive equipment and instrument, and are also dreadful with respect to energy requirement but are capable enough to make micron sized particles. Since ages lithographic techniques were used for producing computers and printed circuits but now a days we use nanoimprint lithography method which is different from the conventional lithographic, initial step involves formation of template material followed by stamping of soft polymeric method to form patterns. The formation of template matric requires latex sphere through nanospheres lithography method. Various types of lithographic techniques are used now a day out of which some are photo-lithography (proximity printing and projection printing), electron beam lithography, soft lithography, nanoimprint lithography, focused ion lithography, and dip pin lithography [82].

3.2 Bottom up approach

The bottom up approach is an additive method where smaller units (atoms or molecules) serves as building blocks, combine to form nanostructures. Some techniques that involves bottom up approaches are sol-gel method, chemical vapor deposition, plasma spraying, micro-emulsion method and laser ablation. Chemical method involves bottom up approach to synthesize nanomaterials [52]. Bottom up approach is also described as constructive technique as it involves formation of an assembly using smaller units. Quite the opposite of top down approach, the output have well defined shape, size and chemical composition as, a self-assembly of atoms and molecules as building units are formed [57].

3.2.1 Chemical vapor deposition

Chemical vapor deposition is a successful and a distinct method for the synthesis of nanoparticles, in the era of microelectronics, chemical vapor deposition is still one of the most promising synthesis methods since decades overcoming all the limitations employed by modern technology. In this method simple materials are used, and in reaction chamber, a thin coating of gaseous reactant is placed on the surface of substrate, and when the gas comes in contact with the heated substrate surface, chemical reaction occurs. In order to increase the deposition process and to maintain the stability there are various parameters that need to be controlled for instance, gas phase reactants delivery, and enclosed reaction chamber to be employed, proper discharge of gases, monitoring reaction pressure, energy source employed for chemical reactions, cleaning of exhaust gases to accomplish safety and nontoxic goals and automation process [83–85]. The chemical vapor deposition method have various advantages for nanostructure growth and fabrications such as finest degree of control, dense and pure material production, repeatable synthesis process,

 Materials Research Forum LLC
https://doi.org/10.21741/9781644902370-4

regulation in growth rate, production of hard, robust, homogeneous and pure nanoparticles, large scale fabrication with purity, fine crystal quality, and few substrate flaws, and parameter control can result in regulating crystal structure, morphology, and orientation. With advantages come disadvantages of using chemical vapor deposition method like poisonous, flammable or explosive precursors, expensive precursors, raise fabrication costs, poisonous gaseous byproducts, and limited substrate use due to high deposition temperature. [83,86,87]

3.2.2 Hydrothermal

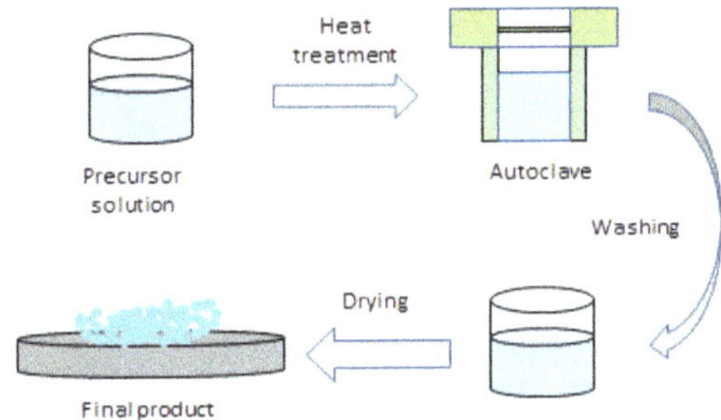

Figure 6. Schematics of hydrothermal technique

A high-temperature, high-pressure reaction between a solid substance and an aqueous solution in a reaction vessel followed by deposition of small particles is termed as hydrothermal synthesis. This method employs solution reaction-based approach where water is used as a solvent; it is carried out in an autoclave (steel pressure vessel) by adjusting temperature and pressure. To attain vapor saturation the temperature of the vessel is increase beyond the boiling point of water. Hydrothermal process works as a great achievement in the field of science and technology as it employs homogeneous precipitation, cost effective, environment friendly, easy scale up and pure final product. The morphology of crystal produced under hydrothermal conditions depends on the growth conditions of the crystal. To monitor the morphology of crystal produced, the pressure of the system is controlled which indeed is dependent on the vapor pressure of the main reaction composition. By this hydrothermal process we can prepare solids such as luminescence phosphorus, super-ionic conductors, and microporous crystals as well as it can also help to synthesize distinct condensed material such as thin films, nm particles and gels [88–94]. High temperature and pressure, reaction time and solvent type plays a key role in hydrothermal process of synthesis [95]. Hydrothermal processes of synthesis also have certain merits for instance, the size of the nanoparticles can be controlled precisely, and nanocrystals produced have high crystalline nature [88,96]. Hydrothermal method of synthesis also have limitations such as the instrument that is the autoclave used for this process is expensive, and while synthesizing the nanoparticles the growth of the crystal cannot be monitored directly (Fig. 6) [88,91,96].

3.2.3 Co-precipitation

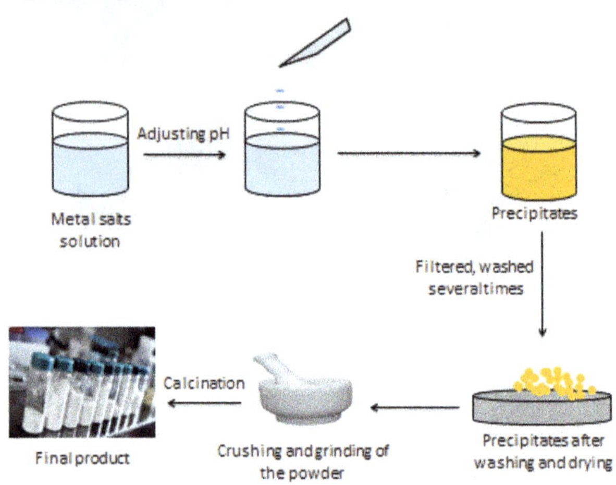

Figure 7. Schematics of co-precipitation technique

One of the ancient wet chemical processes to synthesize nanomaterials is co-precipitation method which is simplest, basic and widely used method. The precipitate obtained at the end of synthesis consists of both product and impurities which are eventually separated using filtration and washing. This method is employed to obtain composition of two or more cations in a homogeneous solution having an advantage to form homogeneous nanomaterials with small sizes and size distribution. The method involves drop by drop addition of one solution into another solution (containing dissolved precipitating agents) directly. Widely used precipitants in this approach are hydroxides, chlorides, carbonates and oxalates. The advantages of this method include energy efficiency, easy and quick method, and homogeneity of particle is maintained throughout. Similarly the disadvantages includes difference in precipitation rates, impurities produced along with product numerous chemicals are employed which is neither environment friendly but is also hazardous to human health (Fig. 7) [89,97–102].

Materials Research Forum LLC
https://doi.org/10.21741/9781644902370-4

3.2.4 Sol-gel method

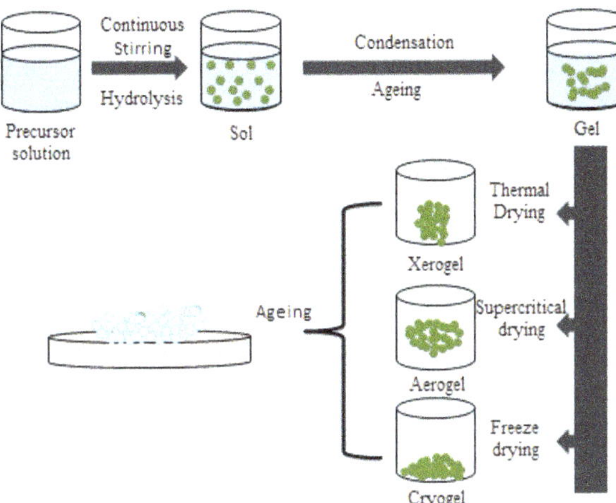

Figure 8. Schematics of sol-gel technique

The word sol-gel consists of two words one is sol and the other is gel. It is one of the easiest methods to synthesis nanoparticles as it is simple. A colloidal solution made with the help of solid particles suspended in it is sol while, a solid macromolecule dissolved in liquid is gel. This sol gel technique involves few steps namely hydrolysis followed by polycondensation followed by aging followed by drying and at the end calcination. In the first step the precursor (metal alkoxides) undergoes hydrolysis in presences of water or alcohol forming hydroxide solution. The reaction medium determines the approach for instance if the medium is water then the approach is called as aqueous sol-gel approach but if the medium is an organic solvent then the approach is called as non-aqueous sol-gel approach. The second step involves the removal of water and alcohol forming metal oxide bonds through condensation of neighboring molecules. Increased solvent viscosity caused by polycondensation creates porous structures that retain a gel-like liquid phase. In the third step the gel network re-precipitates due to continuous polycondensation within the localized solution resulting in reduction of porosity of colloidal particles and increasing thickness between them, hence it is called aging. The gel is then dried which is very difficult as the separation of water and organic components disrupts the structure and so drying is carried out. Final step involves calcination to get rid of water and residues from the sample. There are multiple advantages for this method like high product purity, cost effective, homogeneous character of the produced material, and simple technique to synthesize complicated nanostructures. Similarly disadvantages includes tedious process,

Materials Research Foundations 145 (2023) 92-130
https://doi.org/10.21741/9781644902370-4

health hazardous due to use of organic chemicals and post treatment to purify sample is mandatory (Fig. 8). [67,89,97–99,103,104]

3.2.5 Pyrolysis

The most popular method for synthesis of nanoparticles in industries is pyrolysis where the precursor (liquid or solid state) used to synthesize nanomaterial is burned using flame. In order to recover the nanoparticles formed the precursor is transferred into furnace at high pressure. High temperature employed in this process of pyrolysis supports easy evaporation of the precursor and to achieve that we use laser or plasma sometimes. The advantages of pyrolysis includes that this process is simple, cost effective, continuous, efficient with high yield of the nanoparticles [105].

3.2.6 Sputtering

Sputtering is widely used method among bottom up approaches as it involves non-thermal vaporization method which can be achieved at <0.67 Pa, maintained by vacuum pump. The components of sputtering method involve (Fig. 9):

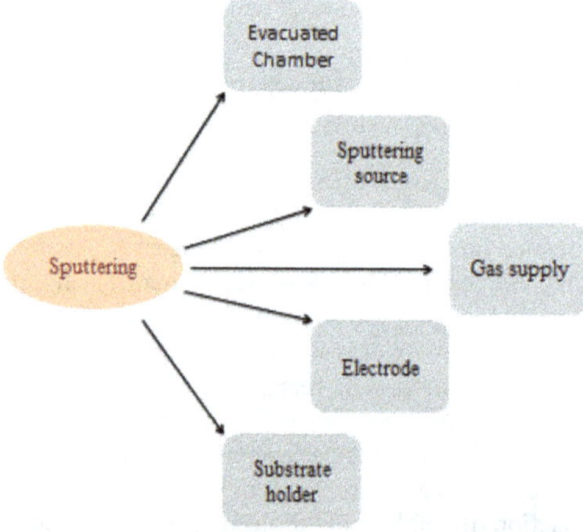

Figure 9. Components of sputtering method

The method involves deposition of nanomaterials on the surface of substrate by particle ejection using high energy ions. This method employs surface coating, thin layer deposition and surface etching applications [106–109]. The simplest source for sputtering is plasma ions produced by applying electric potential between the electrodes in gas phase, and when

Materials Research Foundations 145 (2023) 92-130 https://doi.org/10.21741/9781644902370-4

the glow discharge sources [110]. The advantages of sputtering method is it works at low temperature by coating large surface area uniformly on elements, alloys, and compounds [111,112]. Just like the advantages there are certain short comings to this method for instance, low purity, energy conversion into heat which should be removed and sputtering rates are low with respect to thermal evaporation [112,113].

4. Role of plants in synthesis of green nanoparticles

Synthesis of green nanoparticles is the simplest, cost-effective, reliable, reproducible resulting in production of more stable products. Microorganisms can be employed for the synthesis of nanoparticles via green route but the process is quite slow and tedious with only limited number of size and shapes possible compared to the synthesis route based on plant based materials. Using green method for synthesis the fewer parameters like high pressure, energy, temperature and toxic chemicals can be avoided or completely eliminated making the process environment friendly. Now a days the researchers and scientists are hooked on to engaging them to avoid using synthetic method for synthesis as plant based materials not only produce highly stable nanoparticles but also involve in straight forward reaction to scale up along with low risk of contamination by impurities (Fig. 10).

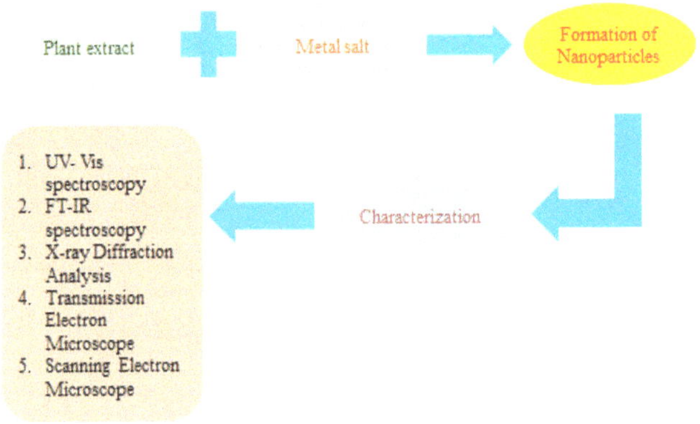

Figure 10. Synthesis and characterization of green nanoparticles

Hence, plants and their parts are used to synthesize nanoparticles as an alternative route for synthesis of nanoparticles instead of synthetic route due to the tremendous advantages these methods bring, out of which the most important advantage or impact is nature friendly method they employ. These methods have a lot of merits and so are the new topic of interest for researchers and their extensive work using plants and their parts, some benefits are larger production due to availability of the precursor in bulk quantity most of the time, scaling up can also be easily carried out, apart from that this method is budget friendly as the precursor or source material is available in nature, lastly it is an environment friendly method. Nanoparticles synthesized using extract of *green tea* leaves [114–116], *Terminalia chebula* fruit extract [117],extract of *Oolong* and *black tea* leaves [118,119], extract of banana peel and *Colocasia esculenta* leaves [120], extract of *Sorghum bran* [121], and extract of *Eucalyptus* leaves [122] have already been used (Fig. 11).

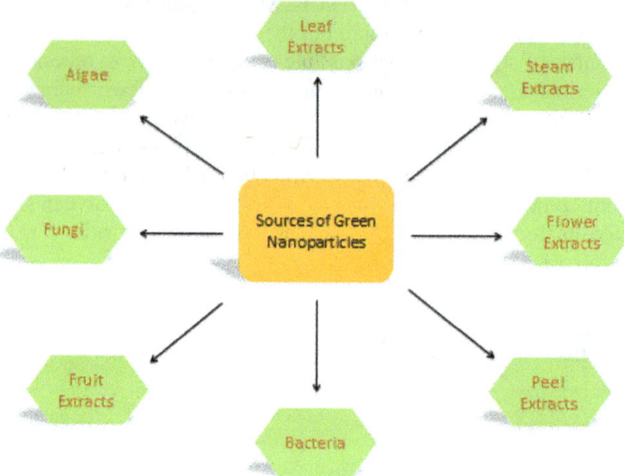

Figure 11. Plant and microorganisms for the source of nanoparticles

5. Role of microorganisms in green synthesis of nanoparticles

5.1. Microbes

Microbes have excellent defense mechanism as the bacterial cells have resistance for reactive ions in the surrounding which indeed is responsible for the synthesis of nanoparticles as high concentration of ions are toxic for bacterial cells in general. In order to prevent their death due to hostile condition, their cellular machinery promotes the conversion of highly reactive ion into stable atoms. This peculiar property exhibited by bacterial cell is exploited by researchers and scientist for bio-synthesis of nanoparticles. Cell destruction may occur if the nanoparticles are produced in high concentration. Microbes live in an ambient condition irrespective of variation in parameters like pH, temperature, and pressure but it has certain limitations which arise due to potential contamination caused by the reaction period between metal nanoparticles and organisms, which could cause cell structure to be destroyed [123,124].

5.2 Algae

The synthesis of nanoparticles synthesized by algae based precursor is rapidly growing due to its variety of applications and its distinct characteristics. Synthesis of gold stable nanoparticles using *Sargassum wightii* was demonstrated [125] . The only limitation this process employs is limited size and number of nanoparticles will be cooperative to carry out the applications.

5.3 Fungi

Using fungi based precursor is one of the most prominent and also most popular means of synthesizing nanoparticles due to the ease to handling biomass, their economically affordability as well as an efficient source of extracellular enzymes leading to large scale production of enzymes. But due to contamination it sometimes invites the creation of genetic mutation or genetic manipulation of organisms [126]. The synthesis rate is slow to handle the entire biomass produced, research are going to overcome these short comings.

6. Types of green nanoparticles

Variety of nanomaterials are found in nature and for our convenience they have been categorized in various class such as metals, metal oxides, carbon based, polymers, nanocomposites, but majorly it is classified as zero dimensional which involves clusters, one dimensional which involves nanotubes, fibers, rods, similarly two dimensional which includes films and coats and lastly it is three dimensional which includes poly-crystals.

The framework or the backbone or the basic structure of these types of nanomaterials is carbon atoms and hence they are called as carbon based nanoparticles. These nanoparticles are special as they exhibit distinctive properties like different hybridization as well as allotropic forms. These nanoparticles are very sensitive for commotion in synthesis parameters and factors allowing tailored manipulation on a large extent are unmatched till date by inorganic based nanostructures [127,128]. Most of the substances found on the earth surface consist of carbon and due to which the synthesis and applications of these distinct carbon based nanoparticles have numerous applications and stands as one of the most promising areas of science and technology drawing the attention of young scientist and researchers [129]. The synthesis of carbon based nanoparticles by synthetic route have been practiced since decades but the switch towards greener approach is seen as it supports eco-friendly nature and is also cost effective [130]. In order to synthesize these carbon based nanoparticles researchers and scientist are using either a green precursor or green condition for the synthesis [131]. The structural conformation and its hybridization determine the physicochemical and electronic properties of carbon based nanomaterials [127]. At high temperature and high pressure carbon thermodynamically favors tetrahedral sp^3 conformation of a diamond but at lower heat of formation it favors planar sp^2

conformation forming graphite sheets. Graphite sheets are more stable thermodynamically compared to two dimensional forms[127]. During the formation of nanoparticles through top-down approach, the planar graphite sheet curvature produces strain energy compensating the reduction of thermodynamically unfavorable dangling bonds of graphite layers [132]. Based on shape of nanoparticles it sub-categorized for instance, Carbon Nanotubes (CNT) are tube like shaped, nano-horns are horn shaped particles, fiber like shape give rise to nanofibers and ellipsoids or sphere like structure belongs to the group named as fullerenes [133].

6.1 Carbon nanotubes

Based on their structural characteristics, carbon nanotubes are divided into two groups: single-walled carbon nanotubes (SWCNTs) and multi-walled carbon nanotubes (MWCNTs). The diameter of these synthetic nanometers ranges from a few micrometres to millimetres, with lengths as long as 550 mm [134,135].

6.2 Fullerenes

The molecular form of carbon or the C_n clusters ($n>20$) of carbon arranged in spherical shape is popularly known as fullerene family. It consist of a cage like structure having 12 five membered rings and an unspecified number of six membered rings[136].

6.3 Graphenes

A two dimensional hexagonal crystal lattice having mono-layer framework of carbon atom consisting of sp^2 hybridized atoms are known as graphenes [136].

6.4 Carbon nanofibers

Graphene nanofoils are used to create the thread-like nanostructures known as carbon nanofibers, which are twisted into a cup or cone shape as opposed to the more conventional cylindrical tube shape [136].

7. Nanoparticles applications

Nanotechnology is a rapidly developing field of study with multiple applications in areas like biomedicine, physical science, chemical science, and environmental research. The following consist of general discussion on wide range of applications of green nanoparticles (Fig. 12) [137] :

7.1 Antimicrobial agents

Numerous studies have been conducted and carried out to determine the antimicrobial activities of green nanoparticles. In recent studies, it was reported that the positively charged silver nanoparticles (AgNPs) have more affinity to negatively charged bacterial cells. The mechanism of this process can both be extra-cellular as well as intracellular. The extra-cellular mechanism by green nanoparticles includes aggregation of silver nanoparticles on the surface of bacterial cell inhibiting cellular transport and cell signaling while the intracellular mechanism deals with altering the bacterial activity by entering the cell cytoplasm of bacteria through cell membrane releasing silver ions. Similar to bacterial cell synthesis the silver nanoparticles can be synthesized by *Melia azedarach* leaf extract showing antifungal activities against *Verticillium dahlia* in *Solamum melongena L* (728 eggplants) both *in vitro* and *in vivo* medium. The results of the action of silver nanoparticles on *Verticillium dahlia* in *in vitro* conditions with concentration 60 mg/L reduced the growth effectively while *in vivo* condition only 20 mg/L showed growth reduction by 87% causing vascular discoloration by 97% compared to control. In another study of antibacterial activity of biogenic iron oxide nanoparticles synthesized by the extract of *Skimmia laureola* on pathogen *Ralstonia solanacearun* was carried out it was found that the growth is inhibited by cell degradation at nanoparticle concentration 6mg/L [137–139].

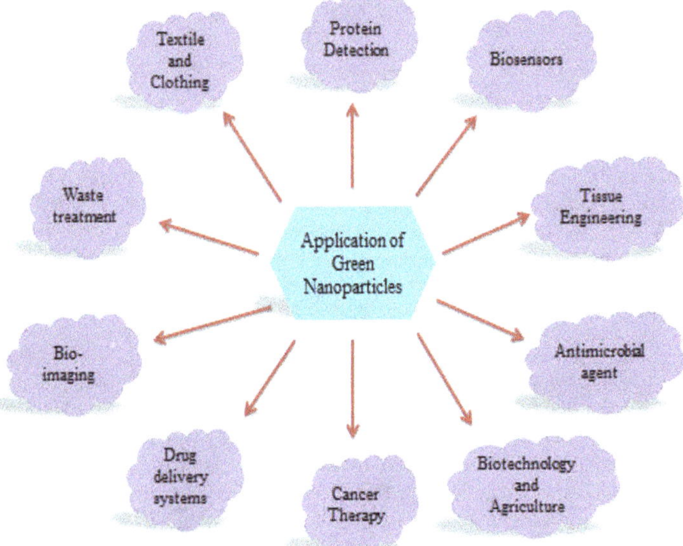

Figure 12. Applications of green nanoparticles

7.2 Cellular imaging

In vivo self-bio-imaging of cancer cells using fluorescent gold nanocluster synthesized intracellularly by reducing chloroauric acid inside the cytoplasm of cancerous cell into gold nanoparticle (AuNP) which helps in diagnosing cancerous cells. Similarly carbon nanoparticles synthesized using advance green method yields blue coloured photoluminescence nanoparticles having low cytotoxicity, bio-compatible, high water

Materials Research Foundations 145 (2023) 92-130 https://doi.org/10.21741/9781644902370-4

solubility and quantum yield serving as a good candidate for cellular imaging showing cellular uptake by Human HeLa cells [137].

7.3 Catalyst

Gold nanoparticles produced from the egg shell extract of *Anas platyrhynchos* was used as a catalyst to remove toxic dye (*Eosin Y*) based on photodegradation method resulted removal of 96.1% of dye. Spherical shaped gold nanoparticles was synthesized with the help of *Eucommia ulmoides* bark extract used for decolorization of two model compounds (reactive yellow 179 and Congo red) in $NaBH_4$ presence acts as an efficient catalyst by reducing both the model compounds within 20 minutes [137].

7.4 Remediation of environmental pollutants

Green nanoparticles have shown a promising potential to adsorb/co-precipitate/reduce/oxidize the contamination arises from soil and water because of their high surface area to mass ratio. Various iron precipitates like magnetite (Fe_3O_4), siderite ($FeCO_3$) and vivianite ($Fe_3(PO_4).8H_2O$) are extracted from microbial consortium serve as a useful tool to act as an adsorbent for As, Cr, Cu, and Zn generated from, waste water. Swine water remediation with high concentration of N and P, is removed in presence of iron nanoparticles produced using extract of Eucalyptus leaves and to prevent eutrophication of water. Removal of halogenated herbicide (quinclorac), by synthesizing green gold particles using yeast (Saccharomyces cerevisiae) is employed, which basically converts quinclorac to 8-quinoline-carboxylic acid hence confirms dechlorination [137].

7.5 Sensor

The field of nanoscience have tremendous popularity as this field is not only associated with other subjects but also have wide applications on large scale such as high sensitive sensor used for the determination of various environmental parameters. *Murraya koenigi* extract helped to synthesize silver nanoparticles are highly sensitive towards calorimetric detection of mercury (II). The detection of mercury (II) is based on a linear relation between intensity of band of surface plasmon resonance and it various concentration (50 nM-500μM) of mercury (II). Additionally mercury (II) in silver nanoparticle solution, disappearing the colour and SPR band reduction shifting towards blue region [137].

7.6 Oily sewage re-mediating agent

Oil sewage in marine ecosystem makes the life of aquatic organisms hostile which in directly is affects human life and mother nature, in this process Melamine sponge came into picture with the qualities such as it is highly stable, low costing, highly porous, with low surface energy and strong adhesion. The merits of using Melamine sponge is that it helps in complete removal of oily sewage associated with selective absorption of layered oil/water mixture having really high separation efficiency (99.9%) and fair permeation flux (1300 Lm-2h-1). the determination and removal of total petroleum hydrocarbons (TPHs) from water can be achieved by the iron nanoparticles formed from Vaccinium floribundum, as the iron nanoparticles provide the condition to reduce or to remove total petroleum hydrocarbons by 88.34% [137].

7.7 Antimicrofouling agent

In recent times studies shows the use of biogenic nanomaterials as anti-fouling agents in health care facilities, food production and processing industries, shipping industries and for membrane based separation techniques to forbid the formation of bio-film. The antimicrofouling capacity of gold and silver nanoparticles derived from *Turbinaria conoides* was compared and the results shows that silver nanoparticles are more skilled in controlling biofilm formed by *E. coli*, *Salmonella sp.*, *Serratia liquefaciens* and *Aeromonas hydrophilia*. The silver nanoparticles derived from *Turbinaria ornate* exhibited maximum inhibition for *E. coli* (71.9%) out of 15 bio-film under study. Synthesis of silver nanoparticles produced by *Bacillus vallismortis* and its action against micro-algal strains and marine bio-film forming bacteria was studied which indicated reduction in the formation of bio-film at 0.5-1 nM concentration [137].

8. Challenges and future perspectives

Nanomaterials and nanoscience is the topic of interest amongst the researchers and scientist, due to growing demand worldwide, due to this the synthesis of nanoparticles and advance applications in various field of science have drawn the attention of the researchers. As per the survey literature surveys various bio-molecules plays key role as a capping agent for the synthesis of nanomaterials, which includes several aspects to look into which includes temperature, particle size, concentration of extract, pH and the effect of synthesis. Size variation of nanoparticles may arise due to various factors like the concentration of capping agent, pH alternations, the size and shape of nanoparticles can be monitored controlled through investigating various aspects and parameters via experimental methods. For instance, pH of the reaction mixture can be measured at the initial phase of the reaction as well as throughout the reaction process followed by end. When the pH of the mixture is acidic (low pH) the agglomeration by nucleation of nanoparticles occurs, while at high pH (alkaline) the nanoparticles produced are stable. Hence, acidic medium leads to formation of nanoparticles with poor stability and will form clusters that agglomerate, while alkaline medium leads to formation of pearl like stable nanoparticles consisting large diameter and the process is a fast process [118,140,141]. At high temperature the growth dynamics would be fast in the same span of time but the major drawback here would be the formation of defects thereby affecting the quality of crystal. Nucleation process controls the size of the nanoparticles, the smaller the time for nucleation the better size controlled nanoparticles will be formed. Optimization of pH, temperature, time of reaction, ratio of solvent is necessary for production of nanoparticles. Surface charge is also an important aspect to study and characterize nanoparticles as the nature and intensity of surface charge determines the interaction of nanoparticles with its environment. Due to the growing applications of nanoparticles in various field have also increased their demand and production and with repeated production of nanoparticles there comes the risk of

Materials Research Forum LLC
https://doi.org/10.21741/9781644902370-4

contamination of environment. In order to protect the environment the greener methods of synthesis is practiced and promoted by the scientist and the researchers which is a simple process, being cost effective and environment friendly minimizing the harmful toxins production which in a way required less efforts and is less tedious as it doesn't involve elimination of toxins. The principle of green chemistry covers all aspects related to environment and safe practices but the swing towards green chemistry also required the recycling and reusing of energy sources and chemicals so as to make it cost effective improving the quality of life, human health and environment. Recent studies focus on the synthesis, characterization and behavior of nanomaterials along with their wide applications in various fields of science and technology including pharmacology and medicinal sciences, agriculture, water treatment, air treatment, electronics and telecommunication. The technique used to synthesize these materials is now considered to be available in the local vicinity and must be techno-economically feasible. Also the biochemical modification in the synthesis process has now begun a sustainable evolution of green methods for synthesis. It is expected that the shortcomings or the challenges faced by this generation will be soon endured in near future due to advancement of instruments, methods and techniques along with modernization, deep understanding and fast development.

References

[1] J.A. Linthorst, An overview: origins and development of green chemistry, Found Chem. 12 (2010) 55–68. https://doi.org/10.1007/s10698-009-9079-4

[2] cathcart, green chemistry in the emerald, Isle.Chem.Ind. 5 (1990) 684–687

[3] W. Leitner, Toward Benign Ends, Science (1979). 284 (1999) 1780–1781. https://doi.org/10.1126/science.284.5421.1780b

[4] István T. Horváth6, P.T. Anastas, Introduction: Green Chemistry, Chem Rev. 107 (2007) 2167–2168. https://doi.org/10.1021/cr0783784

[5] R.E. Benedick, Human Population and Environmental Stresses in the Twenty-first Century INTRODUCTION: PEOPLE AND THEIR ENVIRONMENT, n.d

[6] M.-C. Daniel, D. Astruc, Gold Nanoparticles: Assembly, Supramolecular Chemistry, Quantum-Size-Related Properties, and Applications toward Biology, Catalysis, and Nanotechnology, (2004). https://doi.org/10.1021/cr030698

[7] K. Bogunia-Kubik, M. Sugisaka, From molecular biology to nanotechnology and nanomedicine, Biosystems. 65 (2002) 123–138. https://doi.org/10.1016/S0303-2647(02)00010-2

[8] V.P. Zharov, J.-W. Kim, D.T. Curiel, M. Everts, Self-assembling nanoclusters in living systems: application for integrated photothermal nanodiagnostics and nanotherapy, Nanomedicine. 1 (2005) 326–345. https://doi.org/10.1016/j.nano.2005.10.006

[9] M. Tan, G. Wang, Z. Ye, J. Yuan, Synthesis and characterization of titania-based monodisperse fluorescent europium nanoparticles for biolabeling, J Lumin. 117 (2006) 20–28. https://doi.org/10.1016/j.jlumin.2005.04.004

[10] H.-Y. Lee, Z. Li, K. Chen, A.R. Hsu, C. Xu, J. Xie, S. Sun, X. Chen, PET/MRI Dual-Modality Tumor Imaging Using Arginine-Glycine-Aspartic (RGD)–Conjugated Radiolabeled Iron Oxide Nanoparticles, Journal of Nuclear Medicine. 49 (2008) 1371–1379. https://doi.org/10.2967/jnumed.108.051243

[11] D. Pissuwan, S.M. Valenzuela, M.B. Cortie, Therapeutic possibilities of plasmonically heated gold nanoparticles, Trends Biotechnol. 24 (2006) 62–67. https://doi.org/10.1016/j.tibtech.2005.12.004

[12] S. Panigrahi, S. Kundu, S. Ghosh, S. Nath, T. Pal, General method of synthesis for metal nanoparticles, Journal of Nanoparticle Research. 6 (2004) 411–414. https://doi.org/10.1007/s11051-004-6575-2

[13] W.M. Liao, W.T. Lai, P.W. Li, M.T. Kuo, P.S. Chen, M.J. Tsai, 5th IEEE Conference on nanotechnology, in: 2005: pp. 549–552

[14] S. Iravani, Green synthesis of metal nanoparticles using plants, Green Chemistry. 13 (2011) 2638. https://doi.org/10.1039/c1gc15386b

[15] A. Ahmad, P. Mukherjee, S. Senapati, D. Mandal, M.I. Khan, R. Kumar, M. Sastry, Extracellular biosynthesis of silver nanoparticles using the fungus Fusarium oxysporum, Colloids Surf B Biointerfaces. 28 (2003) 313–318. https://doi.org/10.1016/S0927-7765(02)00174-1

[16] S.S. Shankar, A. Rai, B. Ankamwar, A. Singh, A. Ahmad, M. Sastry, Biological synthesis of triangular gold nanoprisms, Nat Mater. 3 (2004) 482–488. https://doi.org/10.1038/nmat1152

[17] U. Kumar Sur, B. Ankamwar, S. Karmakar, A. Halder, P. Das, Green synthesis of Silver nanoparticles using the plant extract of Shikakai and Reetha, Mater Today Proc. 5 (2018) 2321–2329. https://doi.org/10.1016/j.matpr.2017.09.236

[18] J. Huang, Q. Li, D. Sun, Y. Lu, Y. Su, X. Yang, H. Wang, Y. Wang, W. Shao, N. He, J. Hong, C. Chen, Biosynthesis of silver and gold nanoparticles by novel sundried *Cinnamomum camphora* leaf, Nanotechnology. 18 (2007) 105104. https://doi.org/10.1088/0957-4484/18/10/105104

[19] H. Korbekandi, R.M. Jouneghani, S. Mohseni, M. Pourhossein, S. Iravani, Synthesis of silver nanoparticles using biotransformations by *Saccharomyces boulardii*, Green Processing and Synthesis. 3 (2014) 271–277. https://doi.org/10.1515/gps-2014-0035

[20] A. Ivanković, Review of 12 Principles of Green Chemistry in Practice, International Journal of Sustainable and Green Energy. 6 (2017) 39. https://doi.org/10.11648/j.ijrse.20170603.12

[21] Mohd Wahid, Faizan Ahmad, Nafees Ahmad, Green Chemistry: Principle and its Application, in: 2017: pp. 395–399

[22] M. Jukic, S. Djakovic, Z. Filipovic-Kovacevic, v. Kovac, J. Vorkapic-Furac, Dominant trends of green chemistry, . . Kem Ind. 54 (2005) 255–272

[23] D. Margetic, Mechanic-chemical organic reactions without the use of solvents, Kem Ind. 54 (2005) 351–358

[24] G. S. Sodhi, Fundamental Concepts of Environmental Chemistry, Alpha Science International, 2005

[25] D. Mijin, M. Stankovic, S. Petrovic, Ibuprofen: Synthesis, production and properties, Hem Ind. 57 (2003) 199–214. https://doi.org/10.2298/HEMIND0305199M

[26] T. Welton, Solvents and sustainable chemistry, Proceedings of the Royal Society A: Mathematical, Physical and Engineering Sciences. 471 (2015) 20150502. https://doi.org/10.1098/rspa.2015.0502

[27] H.W. Hill, D.G. Brady, Properties, environmental stability, and molding characteristics of polyphenylene sulfide, Polym Eng Sci. 16 (1976) 831–835. https://doi.org/10.1002/pen.760161211

[28] J. Kärkkäinen, M. Siponen, H. Mantila, J. Kostamovaara, V. Senior, A. Timo, L. Communications Officer, E. Stjerna, S. Eriksson, O. Vuolteenaho, K. Nurkkala, A scientiae rerum naturalium a 480 preparation and characterization of some ionic liquids and their use in the dimerization reaction of 2-methylpropene scientiae rerum naturalium humaniora technica medica scientiae rerum socialium scripta academica oeconomica editor in chief editorial secretary, in: 2007

[29] M.I. Hoffert, K. Caldeira, G. Benford, D.R. Criswell, C. Green, H. Herzog, A.K. Jain, H.S. Kheshgi, K.S. Lackner, J.S. Lewis, H.D. Lightfoot, W. Manheimer, J.C. Mankins, M.E. Mauel, L.J. Perkins, M.E. Schlesinger, T. Volk, T.M.L. Wigley, Advanced Technology Paths to Global Climate Stability: Energy for a Greenhouse Planet, Science (1979). 298 (2002) 981–987. https://doi.org/10.1126/science.1072357

[30] Z. Findrik Blazevic, Bioreactivity Technique I, 2013

[31] Paul T. Anastas, Handbook of green chemistry , Green Catalysis. 1 (2013)

[32] Roger Arthur Sheldon, Isabel Arends, Ulf Hanefeld, Introduction: Green Chemistry and catalysis, 2007

[33] R. T. Williams, Human health pharmaceuticals in the environment: an introduction, Allen Press/ACG Publishing, 2005

[34] K. Wegner, B. Schimmöller, B. Thiebaut, C. Fernandez, T.N. Rao, Pilot Plants for Industrial Nanoparticle Production by Flame Spray Pyrolysis, KONA Powder and Particle Journal. 29 (2011) 251–265. https://doi.org/10.14356/kona.2011025

[35] J.C. Ion, Laser Processing of Engineering Materials: Principal Procedure and Industrial Application, Elsevier: Oxford, 2006

[36] H. Zeng, X.-W. Du, S.C. Singh, S.A. Kulinich, S. Yang, J. He, W. Cai, Nanomaterials via Laser Ablation/Irradiation in Liquid: A Review, Adv Funct Mater. 22 (2012) 1333–1353. https://doi.org/10.1002/adfm.201102295

[37] V. Amendola, M. Meneghetti, What controls the composition and the structure of nanomaterials generated by laser ablation in liquid solution?, Phys. Chem. Chem. Phys. 15 (2013) 3027–3046. https://doi.org/10.1039/C2CP42895D

[38] J. Leng, Z. Wang, J. Wang, H.-H. Wu, G. Yan, X. Li, H. Guo, Y. Liu, Q. Zhang, Z. Guo, Advances in nanostructures fabricated *via* spray pyrolysis and their applications in energy storage and conversion, Chem Soc Rev. 48 (2019) 3015–3072. https://doi.org/10.1039/C8CS00904J

[39] A. Aboulouard, B. Gultekin, M. Can, M. Erol, A. Jouaiti, B. Elhadadi, C. Zafer, S. Demic, Dye sensitized solar cells based on titanium dioxide nanoparticles synthesized by flame spray pyrolysis and hydrothermal sol-gel methods: a comparative study on photovoltaic performances, Journal of Materials Research and Technology. 9 (2020) 1569–1577. https://doi.org/10.1016/j.jmrt.2019.11.083

[40] P. Pawinrat, O. Mekasuwandumrong, J. Panpranot, Synthesis of Au–ZnO and Pt–ZnO nanocomposites by one-step flame spray pyrolysis and its application for photocatalytic degradation of dyes, Catal Commun. 10 (2009) 1380–1385. https://doi.org/10.1016/j.catcom.2009.03.002

[41] J.G. Walter, S. Petersen, F. Stahl, T. Scheper, S. Barcikowski, Laser ablation-based one-step generation and bio-functionalization of gold nanoparticles conjugated with aptamers, J Nanobiotechnology. 8 (2010) 21. https://doi.org/10.1186/1477-3155-8-21

[42] S. Salmaso, P. Caliceti, V. Amendola, M. Meneghetti, J.P. Magnusson, G. Pasparakis, C. Alexander, Cell up-take control of gold nanoparticles functionalized with a thermoresponsive polymer, J Mater Chem. 19 (2009) 1608. https://doi.org/10.1039/b816603j

[43] D.R. Joshi, N. Adhikari, An Overview on Common Organic Solvents and Their Toxicity, J Pharm Res Int. (2019) 1–18. https://doi.org/10.9734/jpri/2019/v28i330203

[44] M. Tobiszewski, J. Namieśnik, F. Pena-Pereira, Environmental risk-based ranking of solvents using the combination of a multimedia model and multi-criteria decision analysis, Green Chemistry. 19 (2017) 1034–1042. https://doi.org/10.1039/C6GC03424A

[45] P.A. Akinyemi, C.A. Adegbenro, T.O. Ojo, O. Elugbaju, Neurobehavioral Effects of Organic Solvents Exposure Among Wood Furniture Makers in Ile-Ife, Osun State, Southwestern Nigeria, J Health Pollut. 9 (2019). https://doi.org/10.5696/2156-9614-9.22.190604

[46] R. Mueller, R. Jossen, S.E. Pratsinis, M. Watson, M.K. Akhtar, Zirconia Nanoparticles Made in Spray Flames at High Production Rates, Journal of the American Ceramic Society. 87 (2004) 197–202. https://doi.org/10.1111/j.1551-2916.2004.00197.x

[47] R. Strobel, A. Baiker, S.E. Pratsinis, Aerosol flame synthesis of catalysts, Advanced Powder Technology. 17 (2006) 457–480. https://doi.org/10.1163/156855206778440525

[48] W.Y. Teoh, R. Amal, L. Mädler, Flame spray pyrolysis: An enabling technology for nanoparticles design and fabrication, Nanoscale. 2 (2010) 1324. https://doi.org/10.1039/c0nr00017e

[49] K.-M. Choi, T.-H. Kim, K.-S. Kim, S.-G. Kim, Case Study, J Occup Environ Hyg. 10 (2013) D1–D5. https://doi.org/10.1080/15459624.2012.734274

[50] P. Caramazana, P. Dunne, M. Gimeno-Fabra, J. McKechnie, E. Lester, A review of the environmental impact of nanomaterial synthesis using continuous flow hydrothermal synthesis, Curr Opin Green Sustain Chem. 12 (2018) 57–62. https://doi.org/10.1016/j.cogsc.2018.06.016

[51] L. Pourzahedi, M.J. Eckelman, Comparative life cycle assessment of silver nanoparticle synthesis routes, Environ Sci Nano. 2 (2015) 361–369. https://doi.org/10.1039/C5EN00075K

[52] P. Aarthye, M. Sureshkumar, Green synthesis of nanomaterials: An overview, Mater Today Proc. 47 (2021) 907–913. https://doi.org/10.1016/j.matpr.2021.04.564

[53] M. Huston, M. DeBella, M. DiBella, A. Gupta, Green Synthesis of Nanomaterials, Nanomaterials. 11 (2021) 2130. https://doi.org/10.3390/nano11082130

[54] A. Sivaraj, V. Kumar, R. Sunder, K. Parthasarathy, G. Kasivelu, Commercial Yeast Extracts Mediated Green Synthesis of Silver Chloride Nanoparticles and their Anti-mycobacterial Activity, J Clust Sci. 31 (2020) 287–291. https://doi.org/10.1007/s10876-019-01626-4

[55] M. Tulinski, M. Jurczyk, Nanomaterials Synthesis Methods, in: Metrology and Standardization of Nanotechnology, Wiley-VCH Verlag GmbH & Co. KGaA, Weinheim, Germany, 2017: pp. 75–98. https://doi.org/10.1002/9783527800308.ch4

[56] N. Shamim, K. Virender, Sustainable Nanotechnology and the Environment: Advances and Achievements, Green Synthesis of Nanomaterials: Environmental Aspects, 2013

[57] N. Abid, A.M. Khan, S. Shujait, K. Chaudhary, M. Ikram, M. Imran, J. Haider, M. Khan, Q. Khan, M. Maqbool, Synthesis of nanomaterials using various top-down and bottom-up approaches, influencing factors, advantages, and disadvantages: A review, Adv Colloid Interface Sci. 300 (2022) 102597. https://doi.org/10.1016/j.cis.2021.102597

[58] D.B. Ravnsbaek, L.H. Sørensen, Y. Filinchuk, F. Besenbacher, T.R. Jensen, Screening of Metal Borohydrides by Mechanochemistry and Diffraction, Angewandte Chemie. 124 (2012) 3642–3646. https://doi.org/10.1002/ange.201106661

[59] R. Černý, P. Schouwink, Y. Sadikin, K. Stare, L. Smrčok, B. Richter, T.R. Jensen, Trimetallic Borohydride $Li_3MZn_5(BH_4)_{15}$ (M = Mg, Mn) Containing Two Weakly Interconnected Frameworks, Inorg Chem. 52 (2013) 9941–9947. https://doi.org/10.1021/ic401139k

[60] D.B. Ravnsbæk, E.A. Nickels, R. Černý, C.H. Olesen, W.I.F. David, P.P. Edwards, Y. Filinchuk, T.R. Jensen, Novel Alkali Earth Borohydride $Sr(BH_4)_2$ and Borohydride-Chloride $Sr(BH_4)Cl$, Inorg Chem. 52 (2013) 10877–10885. https://doi.org/10.1021/ic400862s

[61] M.B. Ley, D.B. Ravnsbæk, Y. Filinchuk, Y.-S. Lee, R. Janot, Y.W. Cho, J. Skibsted, T.R. Jensen, $LiCe(BH_4)_3Cl$, a New Lithium-Ion Conductor and Hydrogen Storage Material with Isolated Tetranuclear Anionic Clusters, Chemistry of Materials. 24 (2012) 1654–1663. https://doi.org/10.1021/cm300792t

[62] J. Huot, D.B. Ravnsbæk, J. Zhang, F. Cuevas, M. Latroche, T.R. Jensen, Mechanochemical synthesis of hydrogen storage materials, Prog Mater Sci. 58 (2013) 30–75. https://doi.org/10.1016/j.pmatsci.2012.07.001

[63] M.M. Verdian, K. Raeissi, M. Salehi, Electrochemical impedance spectroscopy of HVOF-sprayed NiTi intermetallic coatings deposited on AISI 1045 steel, J Alloys Compd. 507 (2010) 42–46. https://doi.org/10.1016/j.jallcom.2010.07.132

[64] O.D. Neikov, Mechanical Crushing and Grinding, in: Handbook of Non-Ferrous Metal Powders, Elsevier, 2009: pp. 47–62. https://doi.org/10.1016/B978-1-85617-422-0.00002-1

[65] L.H. Li, Y. Chen, G. Behan, H. Zhang, M. Petravic, A.M. Glushenkov, Large-scale mechanical peeling of boron nitride nanosheets by low-energy ball milling, J Mater Chem. 21 (2011) 11862. https://doi.org/10.1039/c1jm11192b

[66] D.G. R. Janot, Ball-milling: the behavior of graphite as a function of the dispersal media, Carbon N Y, 40 (2002) 2887–2896

[67] I. Ijaz, E. Gilani, A. Nazir, A. Bukhari, Detail review on chemical, physical and green synthesis, classification, characterizations and applications of nanoparticles, Green Chem Lett Rev. 13 (2020) 223–245. https://doi.org/10.1080/17518253.2020.1802517

[68] M. Salavati-Niasari, F. Davar, N. Mir, Synthesis and characterization of metallic copper nanoparticles via thermal decomposition, Polyhedron. 27 (2008) 3514–3518. https://doi.org/10.1016/j.poly.2008.08.020

[69] Y. Guo, X. Fu, Z. Peng, Controllable synthesis of MoS2 nanostructures from monolayer flakes, few-layer pyramids to multilayer blocks by catalyst-assisted thermal

evaporation, J Mater Sci. 53 (2018) 8098–8107. https://doi.org/10.1007/s10853-018-2103-0

[70] A. Basak, A. Hati, A. Mondal, U.P. Singh, S.K. Taheruddin, Effect of substrate on the structural, optical and electrical properties of SnS thin films grown by thermal evaporation method, Thin Solid Films. 645 (2018) 97–101. https://doi.org/10.1016/j.tsf.2017.10.039

[71] Q. Shi, Q. Wang, D. Zhang, Q. Wang, S. Li, W. Wang, Q. Fan, J. Zhang, Structural, optical and photoluminescence properties of Ga2O3 thin films deposited by vacuum thermal evaporation, J Lumin. 206 (2019) 53–58. https://doi.org/10.1016/j.jlumin.2018.10.005

[72] R. Keshav, M.G. Mahesha, Investigation on performance of CdTe solar cells with CdS and bilayer ZnS/CdS windows grown by thermal evaporation technique, Int J Energy Res. 45 (2021) 7421–7435. https://doi.org/10.1002/er.6325

[73] D.M. Mattox, Physical vapor deposition (PVD) processes, Metal Finishing. 100 (2002) 394–408. https://doi.org/10.1016/S0026-0576(02)82043-8

[74] A. Bashir, T.I. Awan, A. Tehseen, M.B. Tahir, M. Ijaz, Interfaces and surfaces, in: Chemistry of Nanomaterials, Elsevier, 2020: pp. 51–87. https://doi.org/10.1016/B978-0-12-818908-5.00003-2

[75] S. Wang, X. Li, J. Wu, W. Wen, Y. Qi, Fabrication of efficient metal halide perovskite solar cells by vacuum thermal evaporation: A progress review, Curr Opin Electrochem. 11 (2018) 130–140. https://doi.org/10.1016/j.coelec.2018.10.006

[76] S. Ravi-Kumar, B. Lies, X. Zhang, H. Lyu, H. Qin, Laser ablation of polymers: a review, Polym Int. 68 (2019) 1391–1401. https://doi.org/10.1002/pi.5834

[77] H. Huang, J. Lai, J. Lu, Z. Li, Pulsed laser ablation of bulk target and particle products in liquid for nanomaterial fabrication, AIP Adv. 9 (2019) 015307. https://doi.org/10.1063/1.5082695

[78] R. Zhou, S. Lin, Y. Ding, H. Yang, K. Ong Yong Keng, M. Hong, Enhancement of laser ablation via interacting spatial double-pulse effect, Opto-Electronic Advances. 1 (2018) 18001401–18001406. https://doi.org/10.29026/oea.2018.180014

[79] N. Mintcheva, S. Yamaguchi, S.A. Kulinich, Hybrid TiO2-ZnO Nanomaterials Prepared Using Laser Ablation in Liquid, Materials. 13 (2020) 719. https://doi.org/10.3390/ma13030719

[80] N. Mintcheva, A.A. Aljulaih, S. Bito, M. Honda, T. Kondo, S. Iwamori, S.A. Kulinich, Nanomaterials produced by laser beam ablating Sn-Zn alloy in water, J Alloys Compd. 747 (2018) 166–175. https://doi.org/10.1016/j.jallcom.2018.02.350

[81] M. Sportelli, M. Izzi, A. Volpe, M. Clemente, R. Picca, A. Ancona, P. Lugarà, G. Palazzo, N. Cioffi, The Pros and Cons of the Use of Laser Ablation Synthesis for the

Production of Silver Nano-Antimicrobials, Antibiotics. 7 (2018) 67. https://doi.org/10.3390/antibiotics7030067

[82] I. Ijaz, E. Gilani, A. Nazir, A. Bukhari, Detail review on chemical, physical and green synthesis, classification, characterizations and applications of nanoparticles, Green Chem Lett Rev. 13 (2020) 223–245. https://doi.org/10.1080/17518253.2020.1802517

[83] I. Sayago, E. Hontañón, M. Aleixandre, Preparation of tin oxide nanostructures by chemical vapor deposition, in: Tin Oxide Materials, Elsevier, 2020: pp. 247–280. https://doi.org/10.1016/B978-0-12-815924-8.00009-8

[84] S. Bhaviripudi, E. Mile, S.A. Steiner, A.T. Zare, M.S. Dresselhaus, A.M. Belcher, J. Kong, CVD Synthesis of Single-Walled Carbon Nanotubes from Gold Nanoparticle Catalysts, J Am Chem Soc. 129 (2007) 1516–1517. https://doi.org/10.1021/ja0673332

[85] L. Sun, G. Yuan, L. Gao, J. Yang, M. Chhowalla, M.H. Gharahcheshmeh, K.K. Gleason, Y.S. Choi, B.H. Hong, Z. Liu, Chemical vapour deposition, Nature Reviews Methods Primers. 1 (2021) 5. https://doi.org/10.1038/s43586-020-00005-y

[86] M. Adachi, S. Tsukui, K. Okuyama, Nanoparticle Synthesis by Ionizing Source Gas in Chemical Vapor Deposition, Jpn J Appl Phys. 42 (2003) L77–L79. https://doi.org/10.1143/JJAP.42.L77

[87] Y.B. Pottathara, Y. Grohens, V. Kokol, N. Kalarikkal, S. Thomas, Synthesis and Processing of Emerging Two-Dimensional Nanomaterials, in: Nanomaterials Synthesis, Elsevier, 2019: pp. 1–25. https://doi.org/10.1016/B978-0-12-815751-0.00001-8

[88] P.G. Jamkhande, N.W. Ghule, A.H. Bamer, M.G. Kalaskar, Metal nanoparticles synthesis: An overview on methods of preparation, advantages and disadvantages, and applications, J Drug Deliv Sci Technol. 53 (2019) 101174. https://doi.org/10.1016/j.jddst.2019.101174

[89] A.V. Rane, K. Kanny, V.K. Abitha, S. Thomas, Methods for Synthesis of Nanoparticles and Fabrication of Nanocomposites, in: Synthesis of Inorganic Nanomaterials, Elsevier, 2018: pp. 121–139. https://doi.org/10.1016/B978-0-08-101975-7.00005-1

[90] S. Ghasaban, M. Atai, M. Imani, Simple mass production of zinc oxide nanostructures via low-temperature hydrothermal synthesis, Mater Res Express. 4 (2017) 035010. https://doi.org/10.1088/2053-1591/aa5dcc

[91] G. Yang, S.-J. Park, Conventional and Microwave Hydrothermal Synthesis and Application of Functional Materials: A Review, Materials. 12 (2019) 1177. https://doi.org/10.3390/ma12071177

[92] B.P. Kafle, Introduction to nanomaterials and application of UV–Visible spectroscopy for their characterization, in: Chemical Analysis and Material

Characterization by Spectrophotometry, Elsevier, 2020: pp. 147–198. https://doi.org/10.1016/B978-0-12-814866-2.00006-3

[93] R. Dorey, Routes to thick films, in: Ceramic Thick Films for MEMS and Microdevices, Elsevier, 2012: pp. 35–61. https://doi.org/10.1016/B978-1-4377-7817-5.00002-X

[94] Y.X. Gan, A.H. Jayatissa, Z. Yu, X. Chen, M. Li, Hydrothermal Synthesis of Nanomaterials, J Nanomater. 2020 (2020) 1–3. https://doi.org/10.1155/2020/8917013

[95] E.B. Denkbaş, E. Çelik, E. Erdal, D. Kavaz, Ö. Akbal, G. Kara, C. Bayram, Magnetically based nanocarriers in drug delivery, in: Nanobiomaterials in Drug Delivery, Elsevier, 2016: pp. 285–331. https://doi.org/10.1016/B978-0-323-42866-8.00009-5

[96] N. Asim, S. Ahmadi, M.A. Alghoul, F.Y. Hammadi, K. Saeedfar, K. Sopian, Research and Development Aspects on Chemical Preparation Techniques of Photoanodes for Dye Sensitized Solar Cells, International Journal of Photoenergy. 2014 (2014) 1–21. https://doi.org/10.1155/2014/518156

[97] Z. Vaseghi, A. Nematollahzadeh, Nanomaterials, in: Green Synthesis of Nanomaterials for Bioenergy Applications, Wiley, 2020: pp. 23–82. https://doi.org/10.1002/9781119576785.ch2

[98] L.A. Kolahalam, I.V. Kasi Viswanath, B.S. Diwakar, B. Govindh, V. Reddy, Y.L.N. Murthy, Review on nanomaterials: Synthesis and applications, Mater Today Proc. 18 (2019) 2182–2190. https://doi.org/10.1016/j.matpr.2019.07.371

[99] N. Wang, J.Y.H. Fuh, S.T. Dheen, A. Senthil Kumar, Synthesis methods of functionalized nanoparticles: a review, Biodes Manuf. 4 (2021) 379–404. https://doi.org/10.1007/s42242-020-00106-3

[100] U.P.M. Ashik, S. Kudo, J. Hayashi, An Overview of Metal Oxide Nanostructures, in: Synthesis of Inorganic Nanomaterials, Elsevier, 2018: pp. 19–57. https://doi.org/10.1016/B978-0-08-101975-7.00002-6

[101] K. Ravichandran, P.K. Praseetha, T. Arun, S. Gobalakrishnan, Synthesis of Nanocomposites, in: Synthesis of Inorganic Nanomaterials, Elsevier, 2018: pp. 141–168. https://doi.org/10.1016/B978-0-08-101975-7.00006-3

[102] H. Li, B.-S. Yang, Model evaluation of particle breakage facilitated process intensification for Mixed-Suspension-Mixed-Product-Removal (MSMPR) crystallization, Chem Eng Sci. 207 (2019) 1175–1186. https://doi.org/10.1016/j.ces.2019.07.030

[103] M. Parashar, V.K. Shukla, R. Singh, Metal oxides nanoparticles via sol–gel method: a review on synthesis, characterization and applications, Journal of Materials Science: Materials in Electronics. 31 (2020) 3729–3749. https://doi.org/10.1007/s10854-020-02994-8

[104] N. Baig, I. Kammakakam, W. Falath, Nanomaterials: a review of synthesis methods, properties, recent progress, and challenges, Mater Adv. 2 (2021) 1821–1871. https://doi.org/10.1039/D0MA00807A

[105] R. D'Amato, M. Falconieri, S. Gagliardi, E. Popovici, E. Serra, G. Terranova, E. Borsella, Synthesis of ceramic nanoparticles by laser pyrolysis: From research to applications, J Anal Appl Pyrolysis. 104 (2013) 461–469. https://doi.org/10.1016/j.jaap.2013.05.026

[106] S. Anu Mary Ealia, M.P. Saravanakumar, A review on the classification, characterisation, synthesis of nanoparticles and their application, IOP Conf Ser Mater Sci Eng. 263 (2017) 032019. https://doi.org/10.1088/1757-899X/263/3/032019

[107] O. Oluwatosin Abegunde, E. Titilayo Akinlabi, O. Philip Oladijo, S. Akinlabi, A. Uchenna Ude, Overview of thin film deposition techniques, AIMS Mater Sci. 6 (2019) 174–199. https://doi.org/10.3934/matersci.2019.2.174

[108] M. Maqbool, I. Ahmad, H.H. Richardson, M.E. Kordesch, Direct ultraviolet excitation of an amorphous AlN:praseodymium phosphor by codoped Gd3+ cathodoluminescence, Appl Phys Lett. 91 (2007) 193511. https://doi.org/10.1063/1.2809607

[109] M.T. Nguyen, T. Yonezawa, Sputtering onto a liquid: interesting physical preparation method for multi-metallic nanoparticles, Sci Technol Adv Mater. 19 (2018) 883–898. https://doi.org/10.1080/14686996.2018.1542926

[110] A. Bengtson, V. Hoffmann, M. Kasik, K. Marshall, Analytical Glow Discharges: Fundamentals, Applications, and New Developments, in: Encyclopedia of Analytical Chemistry, John Wiley & Sons, Ltd, Chichester, UK, 2017: pp. 1–40. https://doi.org/10.1002/9780470027318.a9426

[111] P. Savale, Comparative study of various chemical deposition methods for synthesis of thin films: a review, Asian J. Res. Chem. 11 (2018) 195

[112] S. Shahidi, B. Moazzenchi, M. Ghoranneviss, A review-application of physical vapor deposition (PVD) and related methods in the textile industry, The European Physical Journal Applied Physics. 71 (2015) 31302. https://doi.org/10.1051/epjap/2015140439

[113] W. Zeng, N. Chen, W. Xie, Research progress on the preparation methods for VO_2 nanoparticles and their application in smart windows, CrystEngComm. 22 (2020) 851–869. https://doi.org/10.1039/C9CE01655D

[114] G.E. Hoag, J.B. Collins, J.L. Holcomb, J.R. Hoag, M.N. Nadagouda, R.S. Varma, Degradation of bromothymol blue by 'greener' nano-scale zero-valent iron synthesized using tea polyphenols, J Mater Chem. 19 (2009) 8671. https://doi.org/10.1039/b909148c

[115] T. Shahwan, S. Abu Sirriah, M. Nairat, E. Boyacı, A.E. Eroğlu, T.B. Scott, K.R. Hallam, Green synthesis of iron nanoparticles and their application as a Fenton-like catalyst for the degradation of aqueous cationic and anionic dyes, Chemical Engineering Journal. 172 (2011) 258–266. https://doi.org/10.1016/j.cej.2011.05.103

[116] N.C. da R. Galucio, D. de A. Moysés, J.R.S. Pina, P.S.B. Marinho, P.C. Gomes Júnior, J.N. Cruz, V.V. Vale, A.S. Khayat, A.M. do R. Marinho, Antiproliferative, genotoxic activities and quantification of extracts and cucurbitacin B obtained from Luffa operculata (L.) Cogn, Arabian Journal of Chemistry. 15 (2022) 103589. https://doi.org/10.1016/j.arabjc.2021.103589

[117] K. Mohan Kumar, B.K. Mandal, K. Siva Kumar, P. Sreedhara Reddy, B. Sreedhar, Biobased green method to synthesise palladium and iron nanoparticles using Terminalia chebula aqueous extract, Spectrochim Acta A Mol Biomol Spectrosc. 102 (2013) 128–133. https://doi.org/10.1016/j.saa.2012.10.015

[118] L. Huang, X. Weng, Z. Chen, M. Megharaj, R. Naidu, Green synthesis of iron nanoparticles by various tea extracts: Comparative study of the reactivity, Spectrochim Acta A Mol Biomol Spectrosc. 130 (2014) 295–301. https://doi.org/10.1016/j.saa.2014.04.037

[119] Y. Kuang, Q. Wang, Z. Chen, M. Megharaj, R. Naidu, Heterogeneous Fenton-like oxidation of monochlorobenzene using green synthesis of iron nanoparticles, J Colloid Interface Sci. 410 (2013) 67–73. https://doi.org/10.1016/j.jcis.2013.08.020

[120] S. Thakur, N. Karak, One-step approach to prepare magnetic iron oxide/reduced graphene oxide nanohybrid for efficient organic and inorganic pollutants removal, Mater Chem Phys. 144 (2014) 425–432. https://doi.org/10.1016/j.matchemphys.2014.01.015

[121] E.C. Njagi, H. Huang, L. Stafford, H. Genuino, H.M. Galindo, J.B. Collins, G.E. Hoag, S.L. Suib, Biosynthesis of Iron and Silver Nanoparticles at Room Temperature Using Aqueous Sorghum Bran Extracts, Langmuir. 27 (2011) 264–271. https://doi.org/10.1021/la103190n

[122] T. Wang, J. Lin, Z. Chen, M. Megharaj, R. Naidu, Green synthesized iron nanoparticles by green tea and eucalyptus leaves extracts used for removal of nitrate in aqueous solution, J Clean Prod. 83 (2014) 413–419. https://doi.org/10.1016/j.jclepro.2014.07.006

[123] B. Ajitha, Y. Ashok Kumar Reddy, P. Sreedhara Reddy, Green synthesis and characterization of silver nanoparticles using Lantana camara leaf extract, Materials Science and Engineering: C. 49 (2015) 373–381. https://doi.org/10.1016/j.msec.2015.01.035

[124] A.R.J.A. de M. Lima, A.S. Siqueira, M.L.S. Möller, R.C. de Souza, J.N. Cruz, A.R.J.A. de M. Lima, R.C. da Silva, D.C.F. Aguiar, J.L. da S.G.V. Junior, E.C. Gonçalves, In silico improvement of the cyanobacterial lectin microvirin and mannose

interaction, J Biomol Struct Dyn. (2020). https://doi.org/10.1080/07391102.2020.1821782

[125] G. Singaravelu, J.S. Arockiamary, V.G. Kumar, K. Govindaraju, A novel extracellular synthesis of monodisperse gold nanoparticles using marine alga, Sargassum wightii Greville, Colloids Surf B Biointerfaces. 57 (2007) 97–101. https://doi.org/10.1016/j.colsurfb.2007.01.010

[126] K.N. Thakkar, S.S. Mhatre, R.Y. Parikh, Biological synthesis of metallic nanoparticles, Nanomedicine. 6 (2010) 257–262. https://doi.org/10.1016/j.nano.2009.07.002

[127] M.S. Mauter, M. Elimelech, Environmental Applications of Carbon-Based Nanomaterials, Environ Sci Technol. 42 (2008) 5843–5859. https://doi.org/10.1021/es8006904

[128] F.S. Alves, J. de A. Rodrigues Do Rego, M.L. da Costa, L.F. Lobato Da Silva, R.A. da Costa, J.N. Cruz, D.D.S.B. Brasil, Spectroscopic methods and in silico analyses using density functional theory to characterize and identify piperine alkaloid crystals isolated from pepper (Piper Nigrum L.), J Biomol Struct Dyn. 38 (2020) 2792–2799. https://doi.org/10.1080/07391102.2019.1639547

[129] E. Asadian, M. Ghalkhani, S. Shahrokhian, Electrochemical sensing based on carbon nanoparticles: A review, Sens Actuators B Chem. 293 (2019) 183–209. https://doi.org/10.1016/j.snb.2019.04.075

[130] Y. Wang, A. Hu, Carbon quantum dots: synthesis, properties and applications, J Mater Chem C Mater. 2 (2014) 6921. https://doi.org/10.1039/C4TC00988F

[131] J. Deng, M. Li, Y. Wang, Biomass-derived carbon: synthesis and applications in energy storage and conversion, Green Chemistry. 18 (2016) 4824–4854. https://doi.org/10.1039/C6GC01172A

[132] O. Zaytseva, G. Neumann, Carbon nanomaterials: production, impact on plant development, agricultural and environmental applications, Chemical and Biological Technologies in Agriculture. 3 (2016) 17. https://doi.org/10.1186/s40538-016-0070-8

[133] A. Saha, C. Jiang, A.A. Martí, Carbon nanotube networks on different platforms, Carbon N Y. 79 (2014) 1–18. https://doi.org/10.1016/j.carbon.2014.07.060

[134] A. Moisala, A.G. Nasibulin, E.I. Kauppinen, The role of metal nanoparticles in the catalytic production of single-walled carbon nanotubes—a review, Journal of Physics: Condensed Matter. 15 (2003) S3011–S3035. https://doi.org/10.1088/0953-8984/15/42/003

[135] V.I. Sokolov, I. v Stankevich, The fullerenes — new allotropic forms of carbon: molecular and electronic structure, and chemical properties, Russian Chemical Reviews. 62 (1993) 419–435. https://doi.org/10.1070/RC1993v062n05ABEH000025

[136] N.B. Singh, P. Jain, A. De, R. Tomar, Green Synthesis and Applications of Nanomaterials, Curr Pharm Biotechnol. 22 (2021) 1705–1747. https://doi.org/10.2174/1389201022666210412142734

[137] A. Rana, K. Yadav, S. Jagadevan, A comprehensive review on green synthesis of nature-inspired metal nanoparticles: Mechanism, application and toxicity, J Clean Prod. 272 (2020) 122880. https://doi.org/10.1016/j.jclepro.2020.122880

[138] S. Pal, Y.K. Tak, J.M. Song, Does the antibacterial activity of silver nanoparticles depend on the shape of the nanoparticle? A study of the gram-negative bacterium Escherichia coli, Appl Environ Microbiol. 73 (2007) 1712–1720. https://doi.org/10.1128/AEM.02218-06

[139] C.M.A. Rego, A.F. Francisco, C.N. Boeno, M. v Paloschi, J.A. Lopes, M.D.S. Silva, H.M. Santana, S.N. Serrath, J.E. Rodrigues, C.T.L. Lemos, R.S.S. Dutra, J.N. da Cruz, C.B.R. dos Santos, S. da S. Setúbal, M.R.M. Fontes, A.M. Soares, W.L. Pires, J.P. Zuliani, Inflammasome NLRP3 activation induced by Convulxin, a C-type lectin-like isolated from Crotalus durissus terrificus snake venom, Sci Rep. 12 (2022) 1–17. https://doi.org/10.1038/s41598-022-08735-7

[140] M.N. Nadagouda, A.B. Castle, R.C. Murdock, S.M. Hussain, R.S. Varma, In vitro biocompatibility of nanoscale zerovalent iron particles (NZVI) synthesized using tea polyphenols, Green Chem. 12 (2010) 114–122. https://doi.org/10.1039/B921203P

[141] Q. Shou, C. Guo, L. Yang, L. Jia, C. Liu, H. Liu, Effect of pH on the single-step synthesis of gold nanoparticles using PEO–PPO–PEO triblock copolymers in aqueous media, J Colloid Interface Sci. 363 (2011) 481–489. https://doi.org/10.1016/j.jcis.2011.07.021

Nanobiomaterials
Materials Research Forum LLC
Materials Research Foundations 145 (2023) 131-161
https://doi.org/10.21741/9781644902370-5

Chapter 5

Processes of Synthesis and Characterization of Silver Nanoparticles with Antimicrobial Action and their Future Prospective

Vishnu Vardhan Palem[1,*], Gokul Paramasivam[1], Nibedita Dey[1], Anu Swedha Ananthan[2]

[1]A1Department of Biotechnology, Saveetha School of Engineering, Saveetha Institute of Medical And Technical Sciences (SIMATS), Saveetha Nagar, Thandalam, Chennai - 602 105, India

[2]Department of Microbiology, Justice Basheer Ahmed Sayeed College for Women, Chennai, India

vishnuvardhanp.sse@saveetha.com

Abstract

The discovery of novel therapies is required due to the stark rise in microbial resistance to currently available conventional antibiotics, which poses a significant obstacle to the effective management of infectious diseases. Nanomaterials between 1 and 100 nm in size have recently become effective antibacterial agents. In particular, several classes of antimicrobial nanomaterials and nanosized carriers for antibiotic delivery have demonstrated their efficacy for treating infectious diseases, including antibiotic-resistant ones, in vitro and in animal models. Because of their high surface area-to-volume ratios, these materials can provide better therapy than conventional drugs and have new mechanical, chemical, electrical, optical, magnetic, electro-optical, and magneto-optical properties. So, nanoparticles have been proven to be fascinating in the fight against bacteria. In this chapter, we will go into detail about the various characteristics of microorganisms and how they differ across each strain. The toxicity mechanisms change depending on the stain. Even the effectiveness of nanomaterials to treat different bacteria and their defence mechanisms varies depending on strains, particularly the composition of cell walls, the makeup of the enzymes, and other factors. As a result, a perspective on nanomaterials in the microbial world, a method to combat drug resistance by tagging antibiotics in nanomaterials, as well as predictions for their future in science.

Keywords

Nanoparticles, Antibacterial Action, Microbial Resistance, NP-Assisted Drug Delivery, Nanoparticle-Assisted Therapy

Materials Research Foundations 145 (2023) 131-161 https://doi.org/10.21741/9781644902370-5

Contents

1. Introduction

The frequent occurrence of infectious disease outbreaks and illnesses brought on by disease-carrying microorganisms has always been a major concern. These outbreaks effects include substantial remediation expenses, interruption of everyday life, regional and national economic downturns, and in the worst cases, fatalities. Some of these contagious diseases have been wiped off. However, certain diseases that were believed to have completely wiped out have made a comeback due to drug-resistant variants [1]. Antibiotics and antibiotic-resistant microbes (also known as "superbugs") are caused by the overuse, improper dosage, and widespread misuse of antimicrobials like antibiotics, antifungals, antivirals, and antiparasitic, which causes these pathogens to develop resistance to the antimicrobials. By boosting people's immunity to certain diseases, vaccinations have been designed to stop the spread of diseases. Bacterial infections can cause serious illnesses including endocarditis, meningitis, and pneumonia that are a serious threat to human health globally [2-5]. One of the main obstacles to treating diseases caused by bacteria is antimicrobial resistance (AMR), particularly from biofilm formation. According to a World Health Organization (WHO) report, antibiotic-resistant (MDR) bacteria kill about 7.5 million people annually worldwide and are expected to kill 12 million people by the year 2050 [6-8]. Antibacterial activity is defined as a substance's capacity to either kill or slow down bacterial growth. Currently available antibacterial products are primarily either chemically produced or naturally derived [9]. Both organic substances, like aminoglycosides, and wholly synthetic antibiotics, such sulfonamides, are frequently used. Broad spectrum drugs can have mediator molecules that are either bacteriostatic (or) bactericidal.

However, a substantial portion of the populace either has a negative reaction to these vaccinations or refuses to take them because of personal beliefs and/or preconceptions. Therefore, decreasing widespread use puts the general public and efficiency at risk [10,11]. This has developed into a significant issue for public health, resulting in expensive pharmaceuticals that do not work. The spread of these infections is also widespread, particularly in view of the inadequate antimicrobials available to address the issue. An avenue for the potential spread of microbes through goods and products contaminated especially at manufacturing facilities can be problematic currently of widespread advanced industrialization, manufacturing, and commercialization of all kinds of products with the potential to reach all corners of the globe. To curb the development of diseases and bacterial strains resistant to antibiotics, researchers and pharmaceutical corporations have been motivated to look for new, powerful antibacterial medicines.

Recently, the use of antimicrobial substances like nanoparticles to control bacteria has attracted the interest of researchers all around the world. These materials' morphological and physicochemical characteristics, such as their high surface area to volume ratio and other physical and chemical characteristics that have been effectively applied in other fields, have generated interest in their utilization [12,13]. Additionally, the surface charge of these nanoparticles can make it easier for them to bind to bacteria's opposing surface charges, resulting in efficient antibacterial actions [14]. Moreover, due to their insolubility and intimate interaction with microbial membranes, the lifespan and endurance of these antimicrobial nanoparticles when used in antimicrobial applications seem promising [15.16]. Antibiotics can be directed at an infection site and nanoparticle carriers can reduce systemic adverse effects.

Materials Research Foundations 145 (2023) 131-161 | https://doi.org/10.21741/9781644902370-5

By using a carrier, we can promote high-dose drug absorption at the targeted region while reducing side effects, such as drug toxicity. Systems for delivering antibacterial medications based on nanoparticles reach the drug's target site, minimizing side effects. Active or passive targeting are involved in targeted nanoparticle-based medication delivery. Passive targeting is accomplished through better penetration and retention at the infection site, whereas active targeting is accomplished through nanoparticle surface modification, enabling the nanoparticle-based drug delivery system to selectively identify exact ligands on the cells at the site of infection. Receptor, temperature, and magnetic targeting are all included in active targeting [17].

Researchers are shifting their attention to investigating alternate antibacterial techniques with reduced risk of developing antimicrobial resistance due to the rising expense and difficulties in generating new antibiotics. One of the most well-known techniques in the antibacterial therapeutic field is the use of nanomaterials. Examples of this approach include the use of antibacterial polymers, photothermal therapy (PTT), photodynamic therapy (PDT), stimuli-triggered antibiotic release using nanomaterials, catalytic bacterial killing using nanozymes, and anti-virulence therapy [18]. The transmission of diseases caused by pathogens, particularly bacteria and viruses, and their prevention are two of the biggest issues facing the health sector. When pathogenic contamination is discovered, it can sometimes be lethal and difficult to stop. Due to the disease-causing bacteria' size in the micron and nanoscales, contamination or disease transmission through various routes is always a possibility, whether it be in medical facilities like hospitals, laboratories, or pharmaceutical businesses. Due to the nature of the environment, hospitals and laboratories are more at risk of contamination. As a result, there is a considerable risk of blood sample exposure in laboratories and cross contamination of numerous disease-causing, drug-resistant organisms due to the high patient traffic. It is unavoidable for diseases to spread through contact with people, objects, and equipment as well as through liquids, vapours, or airborne mists. Therefore, it is essential to create antimicrobial technologies to stop the progression and spread of illnesses right where they start in the healthcare industry. Recently, nanoparticles with antimicrobial properties have gained popularity and are vigorously being explored as an effective substance against a wide spectrum of infections. These particles can impair the growth and, in some cases, eradicate the pathogens. Due to their extensive antibacterial properties against a broad range of diseases, silver (Ag) and its compounds are among the nanoparticles that have been studied [19-21].

Nanomaterials may quickly cross cell membranes compared to bulk materials, which has a toxic effect on bacterial cells. To boost the antibacterial effects of the agents, many medicines can be combined within a single nanoparticle or with help from additional constructions. Due to the co-ordinated action of numerous mechanisms, the simultaneous combination of medications with different effects helps to develop efficiency. On the other hand, combining two or more nanoparticle types can enhance their antibacterial activities and prevent the development of resistance [22]. Recent research has shown that the size and shape of nanoparticles significantly affects their bio and antimicrobial activities [23,

Materials Research Foundations 145 (2023) 131-161 https://doi.org/10.21741/9781644902370-5

24]. Roughness [25], doping modification [26], and environmental issues are all the factors that displayed significant differences in antimicrobial activities.

2. Silver nanoparticle synthesis:

Researchers have focused on silver nanoparticles (NPs) because of their distinctive features, including size- and shape-dependent, optical, antibacterial, and electrical capabilities. For the synthesis of silver NPs, a variety of preparation methods have been documented; significant examples include laser ablation, gamma radiation, electron radiation, chemical reduction, photochemical procedures, microwave processing, and biological synthetic approaches. This chapter gives a general overview of the physical, chemical, and biological production of silver nanoparticles, therefore, to consider the present situation and probable outcomes, particularly the advantages and disadvantages of the industrial practices.

2.1 Physical methods

The most significant physical methods are laser ablation and evaporation-condensation. The produced thin films were free of solvent contamination, in comparison to chemical processes, the uniformity of NPs dispersion is a benefit of physical synthesis techniques. Atmospheric pressure physical synthesis of silver NPs has some drawbacks. For instance, the tube furnace takes up a lot of room, uses a lot of energy when raising the temperature around the source material, and takes a long time to reach thermal stability. According to Lee and Kang's research, monodispersed silver nano crystallites are produced as a result of the thermal decomposition of Ag^+ -oleate complexes [27]. A tiny ceramic heater was employed in a work by Jung et al. to create metal nanoparticles by evaporation/condensation processes. It was discovered that polydisperse nanoparticles were produced over time as the heater surface maintained a steady temperature. These silver nanoparticles were round and not clumped together [28]. To vaporize the raw materials, a little ceramic heater was used. As a result of the temperature gradient being far greater at the heater surface than it would be in a tube furnace, the evaporated vapor can cool at a suitable rapid rate. Because of this, highly concentrated tiny NPs can arise. Due to the heater surface's constant temperature, the particle creation is extremely stable. This physical technique can serve as a calibration device for nanoparticle measurement equipment as well as a nanoparticle generator for long-term investigations for inhalation toxicity studies [28]. Silver nanoparticles' geometric mean diameter and geometric standard deviation, respectively, ranged from 6.2-21.5 nm and 1.23-1.88 nm. Recent research has shown that the polyol method yields spherical nanoparticles of various sizes when laser ablation is used [29, 30]. Silver nanoparticles were created using laser ablation using various wavelengths to explore the effects of the wavelength on particle size. It was discovered that the average particle diameter decreased from ~ 29 to ~12 nm as the laser wavelength decreased [31]. Through a direct physical deposition of metal into the glycerol, Seigal et.al. investigated the creation of silver nanoparticles. It was discovered that this strategy works well in place of laborious chemical methods. Furthermore, the consequent nanoparticles had a restricted size distribution and were resistant to aggregation [32]. The benefits of physical methods of manufacturing silver NPs include speed, the lack of hazardous reagents, and the use of radiation as a reducing agent. Physical methods' drawbacks include solvent contamination, low yield, uneven distribution, and significant energy consumption (Table 1).

Table 1. Physical methods used in the synthesis of silver nanoparticles.

Silver nanoparticle type	Reducing agent	Biological activity	Method	Size (nm)	Ref.
Polydiallyldimethylammonium chloride and polymethacrylic acid capped silver nanoparticles	Methacrylic acid polymers	Antimicrobial	Laser ablation	10-50	[33]
$AgNO_3$	Sodium citrate	Antimicrobial	Electrical arc discharge	14-27	[34]

2.2 Chemical methods

Chemical reduction using reducing chemicals that are both organic and inorganic is the most used method for creating silver nanoparticles. Chemical procedures are advantageous because the necessary equipment is more practical and straightforward than that employed in biological methods. For the reduction of silver ions (Ag+) in aqueous or non-aqueous solutions, a variety of reducing agents are typically used, including sodium citrate, ascorbate, sodium borohydride (NaBH4), elemental hydrogen, polyol process, Tollen's reagent, N, N-dimethylformamide (DMF), and poly (ethylene glycol)-block copolymers. $AgNO_3$ is one of the most often utilized silver salts in chemical production of silver nanoparticles because of features like low cost. Monodispersed silver nanocubes were created by Sun and Xia by reducing nitrate [35]. These reducing chemicals cause the reduction of Ag^+ to metallic silver (Ag^0), which then aggregates into oligomeric clusters. Eventually, these clusters cause the emergence of metallic colloidal silver particles [36 - 38]. $AgNO_3$ was used as a precursor, sodium borohydride and trisodium citrate as stabilizing agents, and silver nanoparticles were synthesized by Mukherji and Agnihotri. For the synthesis of silver nanoparticles with a size range of 5–20 nm, $NaBH_4$ has reportedly been found to be an effective reducing agent. Comparatively, trisodium citrate is the most efficient reducing agent for the synthesis of silver nanoparticles in the 60-100 nm size range [39]. According to reports, silver nanoparticles with an average size of less than 10 nm can be produced using ethylene glycol as a solvent and a reducing agent together with polyvinylpyrrolidone (PVP) as a size controller and capping agent [40].

By employing polyvinyl alcohol as the stabilizing agent and hydrazine hydrate as the reducing agent, Patil et al. were able to successfully create silver nanoparticles. According to their findings, the resulting nanoparticles had a spherical form and had important uses in biotechnology and biomedical science [41]. When compared to physical procedures, the main benefit of chemical approaches is high yield. To synthesize silver nanoparticles chemically, substances like borohydride, 2-mercaptoethanol, citrate, and thiol-glycerol are noxious, and chemical approaches are very expensive. Obtaining silver nanoparticles of a specific size is highly challenging, and a further step is needed to prevent particle aggregation (Table 2) [42].

Table 2. Chemical methods used in the synthesis of silver nanoparticles.

Silver nanoparticle type	**Reducing agent**	**Biological activity**	**Method**	**Size (nm)**	**Ref.**
$AgNO_3$	$NaBH_4$	Antimicrobial	Chemical	5-20	[39]
$AgNO_3$	DMF	Antimicrobial	Chemical	< 20	[43]
$AgNO_3$	Trisodium citrate	Antimicrobial	Chemical	< 50	[39]
$AgNO_3$	PVP/Ethylene glycol	Antimicrobial	Chemical	30-50	[40]
$AgNO_3$	Hydrazine hydrate	Antimicrobial	Chemical	2-5	[41]

3. Biological Methods

Physical and chemical methods for synthesizing silver nanoparticles are costly, time-consuming, and environmentally harmful. Therefore, it is essential to develop a system that is both inexpensive and environmentally benign, eliminates the use of harmful substances [44] and other issues related to both physical and chemical means of production. By controlling a variety of biological activities, biological approaches close these gaps and have numerous applications in the management of health. Fungi, bacteria, and yeasts are used in biological production techniques in addition to plant sources. These sources make this method highly well-liked for using silver nanoparticles in medicinal applications.

When important factors, such as the types of organisms, inheritable and genetic characteristics of organisms, ideal conditions for cell growth and enzyme activity, ideal reaction conditions, and selection of the biocatalyst state, have been taken into consideration, bio-based protocols could be used to synthesize highly stable and well-characterized silver NPs. Changes to several crucial factors, such as substrate concentration, pH, light, temperature, buffer strength, an electron donor (such as glucose or fructose), biomass and substrate concentration, mixing speed, and exposure time, can affect the sizes and morphologies of the NPs.

3.1 Bacteria

It has been investigated how bacteria can produce silver nanoparticles. According to a study, Bacillus licheniformis, a non-pathogenic bacterium, might be used to bio reduce aqueous silver ions into very stable silver nanoparticles of size ~ 40 nm [45]. Additionally, utilizing the bacteria B. licheniformis, well-dispersed silver nanocrystals of size 50 nm were synthesized [46]. By combining B. subtilis culture supernatant with microwave irradiation in water, Saifuddin and colleagues have developed a unique combinational synthesis technique to produce silver nanoparticles of size ranging from 5-50 nm. [47]. Like-wise variety of silver nanoparticles have been demonstrated using Aeromonas sp. SH10, Klebsiella pneumonia, Lactobacillus strains, Pseudomonas stutzeri AG259, Corynebacterium sp. SH09 and Enterobacter cloacae.

3.2 Plants

Because plant-based synthesis of NPs is so inexpensive, it can be employed as a practical and profitable alternative to the large-scale manufacturing of NPs [48]. As a reducing and stabilizing agent to produce silver NPs, camellia sinensis (green tea) and black tea leaf extracts has been utilized [49,50]. These NPs' production appeared to be mediated by polyphenols and flavonoids. Alfalfa (Medicago sativa), lemongrass (Cymbopogon flexuosus), and geranium (Pelargonium graveolens) plant extracts have been used as green reactants in the manufacture of silver nanoparticles. Additionally, by subjecting silver ions to Datura metel leaf extract, a large density of exceptionally stable silver NPs (16–40 nm) was quickly synthesized [51]. The leaf broths of Pinus desiflora, Diospyros kaki, Ginkgo biloba, Magnolia kobus, and Platanus orientalis all produced stable silver NPs with average particle sizes ranging from 15 to 500 nm extracellularly, according to research by Song et.al. When the reaction temperature was raised in the case of M. kobus and D. kaki leaf broths, the rate of synthesis and ultimate conversion to silver NPs was faster. However, when the temperature was raised from 25°C to 95°C, the average particle sizes produced by the D. kaki leaf broth fell from 50 nm to 16 nm [52]. Likewise, a variety of silver NPs have been synthesized using various plant extracts such as Aloe vera, Azadirachta indica, Cinnamomum camphora, Emblica Officinalis, Pelargonium graveolens, Pelargonium graveolens, Pinus eldarica leaf extracts.

3.3 Algae

There are a few studies on the accumulation of gold employing algae species, especially cyanobacteria, as a biological reagent. For the reduction of silver ions and subsequent production of Ag NPs, marine algae such as Chaetoceros calcitrans, Chlorella salina, Isochrysis galbana, and Tetraselmis gracilis can also be employed [53]. Oscillatoria willei

(NTDM01), a marine cyanobacterium, was used to manufacture silver nanoparticles of size 100–200 nm. After 72 hours of incubation with washed marine cyanobacteria, the colour of the silver nitrate solution changed to yellow, signifying the production of silver nanoparticles (NPs). Others like Spirulina platensis, Oscillatoria willei, and Gelidiella acerosa were also used to synthesize silver NPs. It was demonstrated that extracellular nanoparticles were created when Humicola sp. interacted with Ag^+ ions to reduce the precursor solution.

3.4 Fungi

There have been reports that several fungi are involved in the synthesis of silver nanoparticles [54]. It has been discovered that fungi may produce silver nanoparticles quite quickly. Numerous researchers have thoroughly investigated how fungus produce silver nanoparticles [55]. According to one study, Fusarium solani and silver nitrate interact to produce spherical silver nanoparticles outside of cells [56]. It was demonstrated that extracellular nanoparticles were synthesized when Humicola sp. interacted with Ag^+ ions to decrease a precursor solution [57]. According to Owaid et. al., the extract of Pleurotus cornucopiae caused the bioreduction of silver nitrate, which led to the creation of silver nanoparticles [58]. According to reports, silver nanoparticles accumulated on the surface of the Aspergillus flavus fungus's cell wall as a result of silver nitrate solution's interaction with it [59]. Bhainsa and D'Souza also looked at Aspergillus fumigatus's role in the extracellular production of silver nanoparticles [60]. The outcomes showed that silver nanoparticles were produced quickly by the interaction of silver ions with the cell filtrate. Silver nanoparticles between 5 to 50 nm in size are produced extracellularly when Fusarium oxysporum is used [61]. Furthermore, Phanerochaete chrysosporium mycelium is incubated with a silver nitrate solution to yield silver nanoparticles [62]. Fusarium oxysporum were employed by Korbekandi and colleagues to demonstrate the bio reductive synthesis of silver nanoparticles [63].

4. Characterization of silver nanoparticles

Nanoparticles need to be characterized to identify and determine their behaviour, efficacy, and safety. Characterization of silver nanoparticles will evaluate the functional attributes and efficiency of the particle. Various analytical techniques like UV-Vis spectroscopy, Fourier transform infrared spectroscopy (FTIR), dynamic light scattering (DLS), Transmission electron microscopy (TEM), High resolution transmission electron microscopy (HR-TEM), Field emission scanning electron microscopy (FESEM), scanning electron microscopy, atomic force microscopy (AFM), Gas chromatography mass spectroscopy (GC-MS), etc. Many renowned articles and books have given extensive reviews on the techniques used to characterize their principles. Some of the inevitable techniques that need to be considered or characterization of silver nanoparticles are listed and discussed below.

4.1 Ultra- violet spectroscopy

This is a primary, simple, selective, quick, and sensitive technique that is quite reliable and used to analyse the stability of silver nanoparticles [64]. Optical properties of silver interact quite strongly with wavelengths of light [65]. Conduction and valence bands lie very adjacent to each other. Hence movement of electrons seems to be quite free. This gives rise to surface resonance (SPR). Oscillations at the conduction band are seen when exposed to light waves [66,67]. Green synthesis of silver from natural sources is generally preferred by many scientists and the nanoparticles yielded in the process have shown stability for about a year with SPR peaks at ranges around 400-500 nm.

4.2 X-ray diffraction (XRD)

XRD is a non-destructive technique opted to determine the orientation, purity, imperfections as well as size of crystalline and metallic structures [64]. This can identify catalysts, compounds, superconductors, chemical groups, isomorphous structures, etc. [68]. X-ray exposed on crystals gets reflected and patterns of diffraction are generated onto a film. These patterns are found to be characteristic for a given metal at a particular orientation. Patterns are compared to a reference database from the joint committee on powder diffraction also known as JCPDS. This test is one of the main tests for confirming a nanoparticle's size and orientation [69]. It works on the principle of bragg's law. The cons associated with this technique is its ability to assess only single state and low diffraction intensity when compared to electron probing [70]. General orientation and peaks that confirm the presence of silver nanoparticles are 38°, 54° and 46°. This relates to JCPDS file number 04-0783 [71]. Indices of these peaks are (200), (311), (111) and (220). Mostly crystalline face centred cubic lattice is observed in silver nanoparticles [72].

4.3 Dynamic light scattering (DLS)

Radiation scattering ability of nanoparticles determine their biological activities and size distribution. This technique gives a narrow size distribution of particles ranging from 2nm to 500 nm. DLS analysis non-destructively the laser light scattering ability of particles as they pass through a colloidal suspension. It relies on Rayleigh scattering [73]. Modulation in the intensity of scattered light is calculated in terms of time and hydrodynamic size is calibrated [74]. Toxicity of silver nanoparticles is quite dependent on the size of the particle. Size devised by DLS (due to brownian motion) is larger when compared to TEM. DLS probes large particles in liquid suspension but limitation of sample specific interaction that influences the result [75].

4.4 Fourier transform Infrared spectroscopy (FTIR)

FTIR provides reproducible, non-invasive, economical, accurate, and favourable signals that can detect small changes in absorbance due interaction with light and particles, especially to analyse the reduction of silver to its nanoparticle form. Changes in the range of 0.001 can be distinguished with ease using FTIR. Biomolecules used in the fabrication of nanoparticles are detected well using FTIR. They study the conformation of molecules

Materials Research Foundations 145 (2023) 131-161 https://doi.org/10.21741/9781644902370-5

that provide additional functionality to silver molecules [76]. FTIR collects data rapidly with less sample damage as well as strong signal and large ratio between signal and ratio [77]. Functional groups found in silver nanoparticles are amines (both primary and secondary), ketones, alkynes, aldehydes, and carboxylic acid [78].

4.5 X-ray photoelectron dpectroscopy (XPS)

XPS is a surface analysis electron spectroscopic method that can be used to quantify the chemical composition of a metal nanoparticle and devise its empirical formulae [79]. It is performed at vacuum conditions where interaction of the sample with laser emits electrons from the sample. The kinetic energy of the emitted electrons is measured in comparison to the number of electrons released from the surface. Combined data generates spectra specific for the given metal nanoparticle [80].

4.6 Scanning electron microscopy (SEM)

High energy beams of electrons have been subjected to samples under study to generate high resolution images [81]. Its raster scans the surface to generate three dimensional images of living and non-living samples. Non-conductive samples need to be sputtered using gold to make the scanning resolution more accurate. It can resolve particles of various sizes, shapes, morphology, and scales [82]. Combination of SEM with energy-dispersive X-ray spectroscopy (EDX) has been used to infer silver structures along with compositional analysis. Internal structures cannot be identified and analysed using SEM, but the morphology images generated with purity check are quite high in SEM.

4.7 Transmission electron microscope (TEM)

TEM is used to measure quantitatively the size of the grain, particle, morphology as well as internal components of a nanoparticle. The ratio of the distances between sample as well as objective lens and to its image plane determines the magnification of TEM [83]. Along with advanced spatial resolution, TEM provides extra measurements on the sample analytically [84]. The cons involved in using TEM is the tedious sample preparation, sectioning and working conditions like high vacuum [85].

4.8 Atomic force microscope (AFM)

Aggregation and dispersion of particles on the surface, size, sorption, shape, and topology of the samples have been analysed at different rates of scanning by AFM. Three general modes are used namely contact, tapping and non-contact using Vander Waals attraction as the main mode of interaction between the probe and sample surface. Interaction of nanoparticles with layers of biomolecules can be easily studied by AFM. Specifications of the sample to be oxide free, conductive and abrasion free are not required to be analysed using AFM. Probe used in AFM does not damage the sample but measures even the nanometre scale structures in liquids [86]. But the cantilever dimension influences the lateral size of the sample [87].

Materials Research Foundations 145 (2023) 131-161
https://doi.org/10.21741/9781644902370-5

4.9 Localized surface plasmon resonance (LSPR)

Collective oscillations of spatial electrons on a metallic nanoparticle are known as LSPR. It generally occurs in the visible light region. LSPR depends on shape, size, temperature, and dielectric nature of the particle. Refractive index of the nanoparticle influences the frequency of spectra generated by LSPR [88]. This technique forms the basis to evaluate molecular, thermodynamic, kinetic as well as imaging properties of nanoparticles. Table 3 depicts few of the recent studies done for biological source-based silver nanoparticle synthesis and the characterization used by authors to confirm the fabrication of silver in them.

Table 3. Characterization used by researchers recently to confirm the presence of silver nanoparticles

Source	**Synthesis technique**	**Techniques used for characterization**	**Size**	**UV-Vis spectroscopy peak**	**Ref.**
Acacia nilotica	Reduction	SEM, EDX, XRD, UV-Vis spectroscopy	50 nm	380-420 nm	[89]
Pomelo peel waste	Ultrasound	SEM, DLS, Zeta, EDX, FTIR. XRD, UV-Vis spectroscopy	40 nm	420 nm	[90]
Tricholoma ustale/Agaricus arvensis	microwave	STEM, DLS, FTIR. XRD, UV-Vis spectroscopy	20 nm	400 nm	[91]
Bryophyllum pinnatum	Reduction	FESEM, DLS, FTIR. XRD, UV-Vis spectroscopy	35 nm	465 nm	[92]
Acremonium borodinense	Reduction	FESEM, DLS, FTIR. XRD, HR-TEM, UV-Vis spectroscopy	0.19 nm	420-450 nm	[93]
Psidium guajava	Reduction	Zeta, FTIR, UV-Vis spectroscopy, SEM	65 nm	420 nm	[94]

Curcumin	Reduction	MTT, UV-Vis spectroscopy, XRD, SEM, FTT	51 nm	435 nm	[95]
Cyanobacterium Pseudanabaena	Reduction	UV-Vis spectroscopy, XRD, TGA, FTIR, XPS, AFM, MTT	50 nm	560 nm	[96]
Heteroderimia leucomela	Reduction	GC-MS, FTIR, UV-Vis spectroscopy, XRD	20 nm	450 nm	[97]
Silver nitrate	Reduction	FESEM, EDX, FTIR. XRD, UV-Vis spectroscopy	12.4 nm	428 nm	[98]

5. Mechanism of Action in silver nanoparticles

The exact mechanism opted by silver nanoparticles on cells is still under study. But literature holds many prospective suggestions as to how silver works on cells and pathogens. These nanoparticles can interact with cells either physically or chemically. Silver seems to physically accumulate on surfaces of the pathogen especially their cell membranes. Internalizations by porins on the cell membranes lead to penetration of silver into the cytoplasm of the microorganisms. Gram negative bacteria often internalize silver nanoparticles much faster than their gram-positive counterparts [99]. Porins enable hydrophilic molecules of varying sizes as well as charges to move across the membrane. Gram positive strains have thicker peptidoglycan layers that make penetration quite tedious in comparison to gram negative bacteria [99]. Lipopolysaccharide making up the cellular integrity of gram-negative bacteria makes it more sensitive to silver nanoparticles thus causing easy internalization [14]. Interaction between the phosphate, amino and carboxyl groups present on the bacterial membrane and positively charged silver leads to successful penetration into the pathogen. These electrostatic interactions lead to changes in the structural components of cell membrane resulting in enhanced permeability and dissipation of hydrogen ions from the surface. Thus, the membrane disrupts and kills the pathogen [100]. As the hydrogen ion gets released in the form of reduced proton motive force, the pH decreases and releases silver ion concentration from the nanoparticle [102].

Chemically silver nanoparticles can generate free radicals on contact with microbes and damage cellular membranes by creating pores [103]. Interaction with thiol moieties, proteins and complexes in the membrane can lead to dissolution of the structural layer [104]. These reactions are possible as donors of electrons from molecules of nitrogen, oxygen, sulphur, and phosphorus [105]. Membrane bound enzymes and proteins are inactivated as silver nanoparticles interact with their di-sulphide bonds and block their

Materials Research Foundations 145 (2023) 131-161
https://doi.org/10.21741/9781644902370-5

active sites. Orientations of lipids are altered from cis to trans and vice versa by silver nanoparticles resulting in alterations in membrane fluidity and thus the overall composition of the layer. Biofilm components are neutralized by adhesion of silver nanoparticles on them [23]. Cellular apoptosis is also initiated in microbes as actin cytoskeleton (MreB) is disrupted and alters the fluidity as well as integrity of the outer membrane by these particles [106]. Reactive oxygen species (ROS) or oxidation of cellular components of pathogens by silver nanoparticles has been previously reported in *Pseudomonas aeruginosa* [107]. Double bonds in lipid bilayers are oxidized to generate radical species that cause subsequent damage to organelles and components of the microbial cell [96].

Enzymes are affected structurally and functionally by silver nanoparticles as it alters the shape and morphology of enzymes. 900 parts per billion administration of silver nanoparticles seem to affect expressions of some important cellular enzymes like maltose transporter, ribosomal units (30S), fructose bisphosphate aldolase and succinyl coenzyme A synthetase. These nanoparticles deactivate the enzyme or protein complexes and hinder the functions of the respective cell component. Cellular metabolism is disrupted by silver nanoparticle's interaction with an important citric acid cycle enzyme called succinyl coenzyme A synthetase [108]. Suppression or manipulation of cellular metabolism creates oxidative stress and hinders gene expression, nutrition, energy production and blocking enzymes of the electron transport chain like cytochrome oxidase.

DNA damage by silver nanoparticles is also reported by many articles in past few years [109]. Replication process is interrupted by complex formation by silver nanoparticles by breaking hydrogen bonds between the complementary strands. Change in the relaxed to condensed state of the DNA decreases its ability to multiply, subsequently suppressing transcription and eventually translation [110]. Modulation of tyrosine residues by dephosphorylation is supported by silver nanoparticles, thus inhibiting pathogen multiplication [111]. Size plays a major role in determining the overall effect of the nanoparticle on the pathogen. Smaller the nanoparticle, easier is its penetration in the cytoplasm through the membranes [112]. Hence the link between synthesis and mode of action in silver nanoparticles on target cells for better practical applications is indisputable. Table 4 gives a summary of the perspective mode of action by silver nanoparticles on microorganisms.

Materials Research Forum LLC
https://doi.org/10.21741/9781644902370-5

Table 4. Overview of the prospective mode of action by silver nanoparticles on pathogens

Interaction	**Mode**	**Cells/ Cellular component affected**	**Ref.**
Physical	Adhesion	Gram negative Bacteria	[98]
	Penetration	Gram negative Bacteria, Fungi	[98]
	Disruption of organelles	Bacteria	[99]
	Disruption of biomolecules	Bacteria	[99]
	Electrostatic attraction	Gram negative bacteria	[14,100,101]
	Reduced Proton motive force	Gram negative bacteria	]102]
	Apoptosis/ MreB disruption	Bacteria	[106]
Chemical	Oxidative stress	Enzymes	[108]
	Signal modulation	Dephosphorylation of cellular signals	[113]
	Free radicals	Bacteria	[103]
	Inactivation of disulfide bonds	Proteins	[105]
	Alter Cis/trans orientation and fluidity	Lipids	[105]
	Neutralization	Biofilms	[23]
	Alter relaxed state to condensed state by breaking hydrogen bond	DNA	[110]

Materials Research Foundations 145 (2023) 131-161 https://doi.org/10.21741/9781644902370-5

6. Role of silver nanoparticles in antimicrobial action

The biggest challenge of the century in the healthcare sector is the global rise of antimicrobial resistance. Though development of antimicrobial resistance is an evolutionary mechanism, it has accelerated immensely due to the indiscriminate use and misuse of antimicrobials. This has resulted in the development of drug resistance amongst the microbial pathogens that is potentially irreversible and untreatable with the drugs currently in use. Prevalence of such disease pathogens may result in increased morbidity and mortality amongst the patient population, increased hospital stays and thereby higher healthcare costs. This may ultimately impact the country's economic burden. Irrational use of antimicrobials further escalated during the COVID-19 pandemic which has resulted in adverse effects. Antibiotic resistance superbugs are strains of microbial pathogens that are resistant to most of the antimicrobial agents used for treatment of infectious diseases that they cause. Some of the important superbugs complicating treatment strategies include methicillin resistant staphylococcus aureus (MRSA), carbapenem resistant enterobacteriaceae including escherichia coli and klebsiella pneumoniae, vancomycin resistant enterococci (VRE) and the hospital bug, acinetobacter baumannii. In the current scenario, it becomes a necessity to look for alternative approaches to combat the silent pandemic of antimicrobial resistance.

Nanotechnology provides huge potential applications in the field of infection biology and its processes. Nanomedicine is one of the important applications of nanotechnology that involves the use of nanomaterials for therapeutics, drug delivery, vaccine development, implants, diagnostics, and imaging tools and in screening platforms. Nanomaterials that are used for therapeutic antimicrobial action have distinctive properties over other chemical components which increases the efficacy of being used in disease control [114]. The antimicrobial action of metals, metallic salts and metallic oxides have been known since time immemorial for the treatment of various bacterial and fungal infections. Silver nanoparticles are one of the most promising antimicrobials amongst the metal and metal oxides. Apart from silver nanoparticles, gold and other metallic oxide nanoparticles like copper oxide, zinc oxide, iron oxide, magnesium oxide and titanium dioxide have been extensively studied for antimicrobial efficacies. Gold nanoparticles possess excellent photothermal properties that produces heat to efficiently disrupt biofilm and kill microbes. The metallic oxides are highly biocompatible and are effective against both gram positive and gram-negative bacteria including drug resistant strains. Their mode of action involves generation of ROS production in bacteria thereby killing by creating oxidative stress [115].

Silver has been used since ancient times for curing various ailments. Silver nanoparticles in the dimension of 1-100 nm have strong capacity and higher surface area to volume ratio that enables it to be used as a promising therapeutic agent. The unique properties of AgNPs are its ability to kill almost all the drug resistant bacteria and potentiality in targeting different sites of a single bacteria in any given time. AgNPs have exhibited antimicrobial, antibiofilm and wound healing properties that makes it the most extensively researched metal oxide nanoparticle [116].

The action mechanisms of AgNPs on bacteria are multiple, the most common being direct interaction of AgNPs with the plasma membrane to create pores causing bacterial cell lysis or inhibition of cell wall synthesis. Silver ions are released on silver oxidation which binds to thiol containing key enzymes thereby preventing their functions. AgNPs can also trigger the formation of ROS that kills bacteria by creating oxidative stress [117]. The action mechanism of AgNPs is attributed to the release of Ag^+ ions which is responsible for its antibacterial activity. AgNPs may release the Ag^+ ions which interact with the nucleosides of the nucleic acids thereby inhibiting DNA replication. Due to the electrostatic attraction and affinity with the sulphur proteins, Ag^+ ions bind to the cytoplasmic membrane and cell wall causing increased permeability leading to cell lysis. Cytoplasmic membrane can also be denatured due to accumulation of Ag^+ ions by modification of the cell membrane arrangements. Ag^+ ions can also cause protein synthesis inhibition by denaturation of the cell's ribosomal components. Another mechanism is the disruption of microbial signal transduction by phosphorylation of protein substrates like tyrosine which may ultimately cause apoptosis of the cell and inhibition of cell proliferation. AgNPs also deactivates the respiratory enzymes leading to the formation of ROS and interruptions in ATP release. Gram negative bacteria is significantly affected by AgNPs than gram positive bacteria owing to its cell wall components. Studies suggest that as the gram-positive cell wall is made up of a thick layer of peptidoglycan when compared to the thin layered peptidoglycan in the gram-negative bacteria, the AgNPs are not be to penetrate and therefore produce the desired effect [118].

The main limitation of using AgNPs for microbial inhibition is the cytotoxic effects. Research studies in mammals, especially in rabbits and rats have shown that use of AgNPs have a direct effect on the organs of the animals thereby resulting in irreversible damage to the reproduction and growth processes. The cytotoxicity of the AgNps depends on its physical and chemical properties and it is now convenient to synthesize silver nanoparticles using green technology which produces nanoparticles which are less toxic compared to other synthesis methods [118].

7. Applications of silver nanoparticles with antimicrobial action in healthcare

The antimicrobial properties of silver nanoparticles are widely exploited in the health care sector and medicine because of its many advantages such as wide functionality, biocompatibility, high infusibility across the tissue barriers, solubility, and multiple antimicrobial mechanisms. Face masks have become an indispensable tool in the prevention of airborne infections especially during times of the COVID-19 pandemic. Facemasks incorporated with AgNPs have been widely studied to prevent infection spread in hospital settings where the most drug resistant microbes exist. About 100% reduction of E. coli and staphylococcus aureus CFU were observed in facemasks which were incorporated with silver nitrate and titanium dioxide. In another similar study, commercial masks treated with AgNPs were shown to inhibit E. coli and staphylococcus aureus at 50 and 100ppm concentrations. AgNPs were also used along with disinfectants to decontaminate surgical masks with an high potential of antibacterial activity against E.coli, klebsiella pneumoniae and staphylococcus aureus [116].

Catheters associated with urinary tract infection can cause serious complications like urosepsis and septicaemia. Therefore, AgNPs are coated in catheters to significantly reduce the growth and prevent biofilm formation in catheters. AgNPs catheters are shown to prevent and inhibit biofilm producing bacteria such as E.coli, staphylococcus aureus, pseudomonas aeruginosa, enterococcus and also the opportunistic yeast, candida albicans. Catheters coated with AgNPs also did not induce toxic effects or inflammation in animal models making it an efficient and safe for long term use. The concentrations of AgNPs coated and the release rate needs to be validated for commercial applications of such catheters. However, commercially available catheters coated with AgNPs in different countries include ON-Q Silver Soaker™, SilverlineR, and AgTive.

The healing and antimicrobial characteristics of AgNPs have progressed its use in wound dressings for faster curing and skin regeneration. It is assumed that use of AgNPs has an immunomodulatory effect on promodulatory cytokines by decreasing the inflammation period and hence allow the regeneration to happen quickly.AgNPs when combined with organic molecules show enhanced wound healing properties, thereby making its applications as hydrogels in topical applications for better wound repair. Silver ions also destroy the pathogenic bacteria found in wound exudates. Today, many drugs containing AgNPs in combination with other biopolymeric compounds are used commercially as wound dressings. Acticoat™ and Bactigras™, Aquacel™ , PolyMem Silver™ , and Tegaderm™ are some bio composites modified with ionic silver and approved by the FDA for applications as wound dressings [117].

Silver nanoparticle technology finds extensive use in the field of orthopaedics for its antimicrobial action. Contamination of implants from opportunistic pathogens may result in loss of implants and may hinder the restoration and repair of bone function. AgNPs with antimicrobial activity is extensively used in trauma implants, tumour prosthesis, bone cement, and in combination with hydroxyapatite coatings in implants. Silver nanoparticles have a unique ability to improve the differentiation process of MC3T3-1 pre-osteoblast cells and subsequently bone-like tissue mineralization. Silver nanoparticles combined with hydroxyapatite coatings in trauma implants aids osseo-integration. Human bone, dentin and dental enamel made with hydroxyapatite and nano silver is evidenced to be effective against both gram positive and gram-negative bacteria. Electrically generated silver ions have been investigated to be effective in treatment of osteomyelitis and in non-union of bones. Nano silver loaded bone cement had shown good antimicrobial activity against Staphylococcus.

The antimicrobial and biofilm inhibiting properties of AgNPs are exploited in the dental field by incorporating it in dental materials and implants. Silver nanoparticles are either used alone or in combination with other biomaterials as major components in adhesive resins, orthodontic cements, antimicrobial filling agents, dental composites, and as biocidal coatings in titanium-based implants. AgNPs based nano systems are evaluated as performance enhanced drug delivery systems for therapeutic molecules based on their biocompatibility and functionality with tremendous antioxidant, antimicrobial and anti-inflammatory properties in current healthcare practices. Because of its intrinsic anticancer properties, AgNP based nanocarriers for anticancer drugs are extensively investigated as efficient antitumor based drug delivery systems. The broad-spectrum bioactivity of AgNPs makes them potential and promising agents for widespread applications in biomedicine and healthcare settings [118].

8. Limitations

Worldwide, the usage of silver nanoparticles is expanding quickly across various industries, including health care. However, it is crucial to reduce the risk of silver nanoparticles' negative effects on both human patients and the environment. It should be noted that many of NPs' antibacterial processes are still mostly unknown, which may surprise some. For instance, a lot of research links oxidative stress and ROS for antibacterial activity, while for other NPs, like MgO NPs, the antibacterial mechanism might not be connected to the control of bacterial strains' metabolism. Therefore, future research should pay close attention to the antibacterial processes of NPs.

Nanoparticles have shown significant potential for antibacterial activities and applications due to their enormous surface area and size, which increases contact with bacteria. The research that has been done so far on the antimicrobial mechanisms of NPs are constrained by the absence of unifying criteria. Because each type of NP has a unique microbial impact, no single method can be used to collect evidence about the antimicrobial processes of NPs. To assess the toxicity of silver nanoparticles and their impact on physiology and tissue architecture, numerous research using animal models have been carried out. The mitochondrial inner membrane's permeability increases non classically as a result of Ag^{+}. Additionally, there was a higher degree of permeability in the rat liver mitochondria, which led to swelling in the mitochondria, aberrant metabolism, and ultimately cellular death [119]. The smaller silver nanoparticles (10 nm) caused the highest level of congestion, single cell necrosis, and focal necrosis in the liver and congestion in the spleen in a study on female mice exposed to different sizes of silver nanoparticles (10, 60, and 100 nm). This suggests that the smaller-sized particles caused greater acute toxicity in mice [120]. An additional investigation discovered a large reduction in glutathione levels, a reduction in mitochondrial membrane potential, and an increase in reactive oxygen species. These findings imply that oxidative stress likely facilitates the cytotoxicity of Ag particles between 15 and 100 nm in liver cells [121].

The development of an adaptive NP medication for antibacterial therapy is extremely challenging, as was already mentioned in above reports. In order to alleviate the multidrug resistance (MDR) and associated negative effects, researchers came up with the idea of antibiotic-tagged NPs. The synergistic effects of silver NPs with eight antibiotics against harmful microorganisms were therefore examined by Kumar et. al. in 2016. Apart from B. cereus, which experienced a 6.1-fold increase, the synergistic interaction of AgNPs with streptomycin resulted in a minute increase in the inhibitory zone against seven harmful bacteria in the range of 0.1 to 0.9. Additionally, this research offers valuable information on the development of innovative antimicrobial drugs. A new formulation of NPs that work in synergy with antibiotics can be developed since the combination of antibiotics and NPs will make it more difficult for pathogenic bacteria to acquire resistance, which would otherwise render the present antibiotics ineffective.

Future prospective

Multiple scientific domains, including electronics, probes, illness diagnosis and therapy, cleanup, imaging, and cellular transportation, have numerous uses for silver nanoparticles. Also, due to their numerous uses in food packaging, agriculture, the healthcare industry, and as antibacterial and antitumor agents, silver nanoparticles have a substantial impact on health management. Moreover, it is generally recognised that most empirical antibiotic use results in resistance, which renders the drugs ineffective. Alternative therapy approaches are receiving more attention globally as a means of combating the issue of antibiotic resistance. Among these alternate procedures is the possible application of silver nanoparticles as antioxidant and antimicrobial agents, as well as chemical surface or copulation of nanomaterials. Due to their potential use in the fight against multidrug-resistant pathogens, nanomaterials can be associated to various cell processes as opposed to antibiotics, which may only have one mechanism of action. When fully understood, it will transform both laboratory and industrial microbiologyThe entire manuscript must be in English. Please use the Time New Roman font with 13 font size.

References

[1] B. Wiley, Y. Sun, B. Mayers, Y. Xia, Shape-controlled synthesis of metal nanostructures: The case of silver, Chem. - A Eur. J. 11 (2005) 454–463. https://doi.org/10.1002/chem.200400927

[2] J.R. Anacona, J. Santaella, R.K.R. Al-shemary, J. Amenta, A. Otero, C. Ramos, F. Celis, Ceftriaxone-based Schiff base transition metal(II) complexes. Synthesis, characterization, bacterial toxicity, and DFT calculations. Enhanced antibacterial activity of a novel Zn(II) complex against S. aureus and E. coli, J. Inorg. Biochem. 223 (2021) 111519. https://doi.org/10.1016/j.jinorgbio.2021.111519

[3] K. Kalimuthu, R. Suresh Babu, D. Venkataraman, M. Bilal, S. Gurunathan, Biosynthesis of silver nanocrystals by Bacillus licheniformis, Colloids Surfaces B Biointerfaces. 65 (2008) 150–153. https://doi.org/10.1016/j.colsurfb.2008.02.018

[4] K. Kalishwaralal, V. Deepak, S. Ramkumarpandian, H. Nellaiah, G. Sangiliyandi, Extracellular biosynthesis of silver nanoparticles by the culture supernatant of Bacillus licheniformis, Mater. Lett. 62 (2008) 4411–4413. https://doi.org/10.1016/j.matlet.2008.06.051

[5] S.S. Nath, D. Chakdar, G. Gope, D.K. Avasthi, Effect of 100 MeV nickel ions on silica coated ZnS quantum dots, J. Nanoelectron. Optoelectron. 3 (2008) 180–183. https://doi.org/10.1166/jno.2008.212

[6] S. Iravani, Green synthesis of metal nanoparticles using plants, Green Chem. 13 (2011) 2638–2650. https://doi.org/10.1039/c1gc15386b

[7] S. Nagaraja, S.S. Ahmed, D.R. Bharathi, P. Goudanavar, K.M. Rupesh, S. Fattepur, G. Meravanige, A. Shariff, P.N. Shiroorkar, M. Habeebuddin, M. Telsang, Green Synthesis and Characterization of Silver Nanoparticles of Psidium guajava Leaf Extract and Evaluation for Its Antidiabetic Activity, Molecules. 27 (2022) 4336. https://doi.org/10.3390/molecules27144336

[8] H. Müller, Optical Properties of Metal Clusters, Zeitschrift Für Phys. Chemie. 194 (1996) 278–279. https://doi.org/10.1524/zpch.1996.194.part_2.278

[9] N. Chauhan, A.K. Tyagi, P. Kumar, A. Malik, Antibacterial potential of Jatropha curcas synthesized silver nanoparticles against food borne pathogens, Front. Microbiol. 7 (2016) 1748. https://doi.org/10.3389/fmicb.2016.01748

[10] M.A. Hossain, B. Paul, K.A. Khan, M. Paul, M.A. Mamun, M.E. Quayum, Green synthesis and characterization of silver nanoparticles by using Bryophyllum pinnatum and the evaluation of its power generation activities on bio-electrochemical cell, Mater. Chem. Phys. 282 (2022) 125943. https://doi.org/10.1016/j.matchemphys.2022.125943

[11] H. Bedford, D. Elliman, Concerns about immunisation, Br. Med. J. 320 (2000) 240–243. https://doi.org/10.1136/bmj.320.7229.240

[12] D.C. Tien, K.H. Tseng, C.Y. Liao, J.C. Huang, T.T. Tsung, Discovery of ionic silver in silver nanoparticle suspension fabricated by arc discharge method, J. Alloys Compd. 463 (2008) 408–411. https://doi.org/10.1016/j.jallcom.2007.09.048

[13] P. Panchal, E. Ogunsona, T. Mekonnen, Trends in advanced functional material applications of nanocellulose, Processes. 7 (2019) 10. https://doi.org/10.3390/pr7010010

[14] N. Vigneshwaran, A.A. Kathe, P. V. Varadarajan, R.P. Nachane, R.H. Balasubramanya, Biomimetics of silver nanoparticles by white rot fungus, Phaenerochaete chrysosporium, Colloids Surfaces B Biointerfaces. 53 (2006) 55–59. https://doi.org/10.1016/j.colsurfb.2006.07.014

[15] P. Hinterdorfer, M.F. Garcia-Parajo, Y.F. Dufrêne, Single-molecule imaging of cell surfaces using near-field nanoscopy, Acc. Chem. Res. 45 (2012) 327–336. https://doi.org/10.1021/ar2001167

[16] V. Sambhy, M.M. MacBride, B.R. Peterson, A. Sen, Silver bromide nanoparticle/polymer composites: Dual action tunable antimicrobial materials, J. Am. Chem. Soc. 128 (2006) 9798–9808. https://doi.org/10.1021/ja061442z

[17] Y. Sun, Y. Xia, Shape-controlled synthesis of gold and silver nanoparticles, Science (80-.). 298 (2002) 2176–2179. https://doi.org/10.1126/science.1077229

[18] W. Russin, Scanning Electron Microscopy for the Life Sciences. Heide Schatten (Ed.). Cambridge University Press, Cambridge, UK, 2013, 261 pages. ISBN: 978-0-521-19599-7 (Hardcover), Microsc. Microanal. 20 (2014) 313–313. https://doi.org/10.1017/s1431927613014062

[19] S. Iravani, Bacteria in Nanoparticle Synthesis: Current Status and Future Prospects, Int. Sch. Res. Not. 2014 (2014) 1–18. https://doi.org/10.1155/2014/359316

[20] H. Cao, Toward selectively toxic silver nanoparticles, CRC Press, 2017. https://doi.org/10.1201/9781315370569

[21] S. Agnihotri, S. Mukherji, S. Mukherji, Size-controlled silver nanoparticles synthesized over the range 5-100 nm using the same protocol and their antibacterial efficacy, RSC Adv. 4 (2014) 3974–3983. https://doi.org/10.1039/c3ra44507k

[22] C.W. Hall, T.F. Mah, Molecular mechanisms of biofilm-based antibiotic resistance and tolerance in pathogenic bacteria, FEMS Microbiol. Rev. 41 (2017) 276–301. https://doi.org/10.1093/femsre/fux010

[23] S. Anees Ahmad, S. Sachi Das, A. Khatoon, M. Tahir Ansari, M. Afzal, M. Saquib Hasnain, A. Kumar Nayak, Bactericidal activity of silver nanoparticles: A mechanistic review, Mater. Sci. Energy Technol. 3 (2020) 756–769. https://doi.org/10.1016/j.mset.2020.09.002

[24] A.R. Vilchis-Nestor, V. Sánchez-Mendieta, M.A. Camacho-López, R.M. Gómez-Espinosa, M.A. Camacho-López, J.A. Arenas-Alatorre, Solventless synthesis and optical properties of Au and Ag nanoparticles using Camellia sinensis extract, Mater. Lett. 62 (2008) 3103–3105. https://doi.org/10.1016/j.matlet.2008.01.138

[25] C. Willyard, The drug-resistant bacteria that pose the greatest health threats, Nature. 543 (2017) 15. https://doi.org/10.1038/nature.2017.21550

[26] D. Kim, S. Jeong, J. Moon, Synthesis of silver nanoparticles using the polyol process and the influence of precursor injection, Nanotechnology. 17 (2006) 4019–4024. https://doi.org/10.1088/0957-4484/17/16/004

[27] E. Weir, A. Lawlor, A. Whelan, F. Regan, The use of nanoparticles in anti-microbial materials and their characterization, Analyst. 133 (2008) 835–845. https://doi.org/10.1039/b715532h

[28] L. Nelsonjoseph, B. Vishnupriya, Ramasamy, D. Bharathi, S. Thangabalu, P. Rehna, Synthesis and characterization of silver nanoparticles using Acremonium borodinense and their anti-bacterial and hemolytic activity, Biocatal. Agric. Biotechnol. 39 (2022) 102222. https://doi.org/10.1016/j.bcab.2021.102222

[29] P. Jena, M. Bhattacharya, G. Bhattacharjee, B. Satpati, P. Mukherjee, D. Senapati, R. Srinivasan, Bimetallic gold-silver nanoparticles mediate bacterial killing by

disrupting the actin cytoskeleton MreB, Nanoscale. 12 (2020) 3731–3749. https://doi.org/10.1039/c9nr10700b

[30] D. Karageorgou, P. Zygouri, T. Tsakiridis, M.A. Hammami, N. Chalmpes, M. Subrati, I. Sainis, K. Spyrou, P. Katapodis, D. Gournis, H. Stamatis, Green Synthesis and Characterization of Silver Nanoparticles with High Antibacterial Activity Using Cell Extracts of Cyanobacterium Pseudanabaena/Limnothrix sp., Nanomaterials. 12 (2022) 2296. https://doi.org/10.3390/nano12132296

[31] D. Kim, S. Jeong, J. Moon, Synthesis of silver nanoparticles using the polyol process and the influence of precursor injection, Nanotechnology. 17 (2006) 4019–4024. https://doi.org/10.1088/0957-4484/17/16/004

[32] A. Ahmad, P. Mukherjee, S. Senapati, D. Mandal, M.I. Khan, R. Kumar, M. Sastry, Extracellular biosynthesis of silver nanoparticles using the fungus Fusarium oxysporum, Colloids Surfaces B Biointerfaces. 28 (2003) 313–318. https://doi.org/10.1016/S0927-7765(02)00174-1

[33] P. Hobson-West, Understanding vaccination resistance: Moving beyond risk, Heal. Risk Soc. 5 (2003) 273–283. https://doi.org/10.1080/13698570310001606978

[34] M. Sastry, K.S. Mayya, K. Bandyopadhyay, pH Dependent changes in the optical properties of carboxylic acid derivatized silver colloidal particles, Colloids Surfaces A Physicochem. Eng. Asp. 127 (1997) 221–228. https://doi.org/10.1016/S0927-7757(97)00087-3

[35] J.S. McQuillan, H. Groenaga Infante, E. Stokes, A.M. Shaw, Silver nanoparticle enhanced silver ion stress response in Escherichia coli K12, Nanotoxicology. 6 (2012) 857–866. https://doi.org/10.3109/17435390.2011.626532

[36] S. Thomas, P. McCubbin, A comparison of the antimicrobial effects of four silver-containing dressings on three organisms., J. Wound Care. 12 (2003) 101–107. https://doi.org/10.12968/jowc.2003.12.3.26477

[37] M.M.O. Rashid, K.N. Akhter, J.A. Chowdhury, F. Hossen, M.S. Hussain, M.T. Hossain, Characterization of phytoconstituents and evaluation of antimicrobial activity of silver-extract nanoparticles synthesized from Momordica charantia fruit extract, BMC Complement. Altern. Med. 17 (2017) 336. https://doi.org/10.1186/s12906-017-1843-8

[38] A. Heidari, Small-Angle X-Ray Scattering (SAXS), Ultra-Small Angle X-Ray Scattering (USAXS), Fluctuation X-Ray Scattering (FXS), Wide-Angle X-Ray Scattering (WAXS), Grazing-Incidence Small-Angle X-Ray Scattering (GISAXS), Grazing-Incidence Wide-Angle X-Ray Scattering (GIWAXS), Small-Angle Neutron Scattering (SANS), Grazing-Incidence Small-Angle Neutron Scattering (GISANS), X-Ray Diffraction (XRD), Powder X-Ray Diffraction (PXRD), Wide-Angle X-Ray Diffraction (WAXD), Grazing- Incidence X-Ray Diffraction (GIXD) and Energy-Dispersive X-Ray Diffraction (EDXRD) Comparative Study on Malignant and Benign

Human Cancer Cells and Tissues under Synchrotron Radiation, Oncol. Res. Rev. 1 (2017). https://doi.org/10.15761/orr.1000104

[39] T. Bruna, F. Maldonado-Bravo, P. Jara, N. Caro, Silver nanoparticles and their antibacterial applications, Int. J. Mol. Sci. 22 (2021) 7202. https://doi.org/10.3390/ijms22137202

[40] S.T. Dubas, P. Kumlangdudsana, P. Potiyaraj, Layer-by-layer deposition of antimicrobial silver nanoparticles on textile fibers, Colloids Surfaces A Physicochem. Eng. Asp. 289 (2006) 105–109. https://doi.org/10.1016/j.colsurfa.2006.04.012

[41] M. Rawat, a Review on Green Synthesis and Characterization of Silver Nanoparticles and Their Applications: a Green Nanoworld., World J. Pharm. Pharm. Sci. (2016) 730–762. https://doi.org/10.20959/wjpps20167-7227

[42] F. von Nussbaum, M. Brands, B. Hinzen, S. Weigand, D. Häbich, Antibakterielle Naturstoffe in der medizinischen Chemie – Exodus oder Renaissance?, Angew. Chemie. 118 (2006) 5194–5254. https://doi.org/10.1002/ange.200600350

[43] C. Gopu, P. Chirumamilla, S. Kagithoju, S. Taduri, Green synthesis of silver nanoparticles using Momordica cymbalaria aqueous leaf extracts and screening of their antimicrobial activity: AgNPs studies in Momordica cymbalaria, Proc. Natl. Acad. Sci. India Sect. B - Biol. Sci. 92 (2022) 771–782. https://doi.org/10.1007/s40011-022-01367-x

[44] A. Syed, S. Saraswati, G.C. Kundu, A. Ahmad, Biological synthesis of silver nanoparticles using the fungus Humicola sp. And evaluation of their cytoxicity using normal and cancer cell lines, Spectrochim. Acta - Part A Mol. Biomol. Spectrosc. 114 (2013) 144–147. https://doi.org/10.1016/j.saa.2013.05.030

[45] N. Vigneshwaran, N.M. Ashtaputre, P. V. Varadarajan, R.P. Nachane, K.M. Paralikar, R.H. Balasubramanya, Biological synthesis of silver nanoparticles using the fungus Aspergillus flavus, Mater. Lett. 61 (2007) 1413–1418. https://doi.org/10.1016/j.matlet.2006.07.042

[46] Silver Nanoparticles - Fabrication, Characterization and Applications, Silver Nanoparticles - Fabr. Charact. Appl. (2018). https://doi.org/10.5772/intechopen.71247

[47] S. V. Kyriacou, W.J. Brownlow, X.H.N. Xu, Using Nanoparticle Optics Assay for Direct Observation of the Function of Antimicrobial Agents in Single Live Bacterial Cells, Biochemistry. 43 (2004) 140–147. https://doi.org/10.1021/bi0351110

[48] W. He, H.K. Kim, W.G. Wamer, D. Melka, J.H. Callahan, J.J. Yin, Photogenerated charge carriers and reactive oxygen species in ZnO/Au hybrid nanostructures with enhanced photocatalytic and antibacterial activity, J. Am. Chem. Soc. 136 (2014) 750–757. https://doi.org/10.1021/ja410800y

[49] M. Farré, D. Barceló, Introduction to the analysis and risk of nanomaterials in environmental and food samples, Compr. Anal. Chem. 59 (2012) 1–32. https://doi.org/10.1016/B978-0-444-56328-6.00001-3

[50] N.A. Begum, S. Mondal, S. Basu, R.A. Laskar, D. Mandal, Biogenic synthesis of Au and Ag nanoparticles using aqueous solutions of Black Tea leaf extracts, Colloids Surfaces B Biointerfaces. 71 (2009) 113–118. https://doi.org/10.1016/j.colsurfb.2009.01.012

[51] S. Gurunathan, J.W. Han, D.N. Kwon, J.H. Kim, Enhanced antibacterial and anti-biofilm activities of silver nanoparticles against Gram-negative and Gram-positive bacteria, Nanoscale Res. Lett. 9 (2014) 1–17. https://doi.org/10.1186/1556-276X-9-373

[52] Pr. Shankar, Book review: Tackling drug-resistant infections globally, Arch. Pharm. Pract. 7 (2016) 110. https://doi.org/10.4103/2045-080x.186181

[53] X. Pan, Y. Wang, Z. Chen, D. Pan, Y. Cheng, Z. Liu, Z. Lin, X. Guan, Investigation of antibacterial activity and related mechanism of a series of nano-Mg(OH)2, ACS Appl. Mater. Interfaces. 5 (2013) 1137–1142. https://doi.org/10.1021/am302910q

[54] J.M. V. Makabenta, A. Nabawy, C.H. Li, S. Schmidt-Malan, R. Patel, V.M. Rotello, Nanomaterial-based therapeutics for antibiotic-resistant bacterial infections, Nat. Rev. Microbiol. 19 (2021) 23–36. https://doi.org/10.1038/s41579-020-0420-1

[55] S.S. Alharthi, T. Gomathi, J.J. Joseph, J. Rakshavi, J.A.K. Florence, P.N. Sudha, G. Rajakumar, M. Thiruvengadam, Biological activities of chitosan-salicylaldehyde schiff base assisted silver nanoparticles, J. King Saud Univ. - Sci. 34 (2022) 102177. https://doi.org/10.1016/j.jksus.2022.102177

[56] P. Kesharwani, K.K. Singh, Preface, Nanoparticle Ther. Prod. Technol. Types Nanoparticles, Regul. Asp. (2022) xvii–xix. https://doi.org/10.1016/B978-0-12-820757-4.09991-9

[57] Y.M. Cho, Y. Mizuta, J.I. Akagi, T. Toyoda, M. Sone, K. Ogawa, Size-dependent acute toxicity of silver nanoparticles in mice, J. Toxicol. Pathol. 31 (2018) 73–80. https://doi.org/10.1293/tox.2017-0043

[58] N. Saifuddin, C.W. Wong, A.A.N. Yasumira, Rapid biosynthesis of silver nanoparticles using culture supernatant of bacteria with microwave irradiation, E-Journal Chem. 6 (2009) 61–70. https://doi.org/10.1155/2009/734264

[59] J.P. Kratohvil, Light Scattering, Anal. Chem. 36 (1964) 458–472. https://doi.org/10.1021/ac60211a039

[60] Y. Dong, L. Wang, D.P. Burgner, J.E. Miller, Y. Song, X. Ren, Z. Li, Y. Xing, J. Ma, S.M. Sawyer, G.C. Patton, Infectious diseases in children and adolescents in

China: Analysis of national surveillance data from 2008 to 2017, BMJ. 369 (2020) m1043–m1043. https://doi.org/10.1136/bmj.m1043

[61] R.P. Allaker, Critical review in oral biology & medicine: The use of nanoparticles to control oral biofilm formation, J. Dent. Res. 89 (2010) 1175–1186. https://doi.org/10.1177/0022034510377794

[62] M. Garland, S. Loscher, M. Bogyo, Chemical Strategies To Target Bacterial Virulence, Chem. Rev. 117 (2017) 4422–4461. https://doi.org/10.1021/acs.chemrev.6b00676

[63] J. Rello, L. Campogiani, V.K. Eshwara, Understanding resistance in enterococcal infections, Intensive Care Med. 46 (2020) 353–356. https://doi.org/10.1007/s00134-019-05875-9

[64] D.K. Carpenter, Dynamic Light Scattering with Applications to Chemistry, Biology, and Physics (Berne, Bruce J.; Pecora, Robert), J. Chem. Educ. 54 (1977) A430. https://doi.org/10.1021/ed054pa430.1

[65] R. Das, S.S. Nath, D. Chakdar, G. Gope, R. Bhattacharjee, Synthesis of silver nanoparticles and their optical properties, J. Exp. Nanosci. 5 (2010) 357–362. https://doi.org/10.1080/17458080903583915

[66] A. Karatutlu, A. Barhoum, A. Sapelkin, Liquid-phase synthesis of nanoparticles and nanostructured materials, Emerg. Appl. Nanoparticles Archit. Nanostructures Curr. Prospect. Futur. Trends. (2018) 1–28. https://doi.org/10.1016/B978-0-323-51254-1.00001-4

[67] S. Saint, J.G. Elmore, S.D. Sullivan, S.S. Emerson, T.D. Koepsell, The Efficacy of Silver Alloy-Coated Urinary Catheters in Preventing Urinary Tract Infection: A Meta-Analysis, J. Urol. 161 (1999) 1422–1422. https://doi.org/10.1016/s0022-5347(01)61753-1

[68] U. Klueh, V. Wagner, S. Kelly, A. Johnson, J.D. Bryers, Efficacy of silver-coated fabric to prevent bacterial colonization and subsequent device-based biofilm formation, J. Biomed. Mater. Res. 53 (2000) 621–631. https://doi.org/10.1002/1097-4636(2000)53:6<621::AID-JBM2>3.0.CO;2-Q

[69] S.M. Hussain, K.L. Hess, J.M. Gearhart, K.T. Geiss, J.J. Schlager, In vitro toxicity of nanoparticles in BRL 3A rat liver cells, Toxicol. Vitr. 19 (2005) 975–983. https://doi.org/10.1016/j.tiv.2005.06.034

[70] P.E. Champness, Diffraction and the electron microscope, Electron Diffr. Transm. Electron Microsc. (2020) 1–23. https://doi.org/10.1201/9781003076872-3

[71] J.Y. Song, B.S. Kim, Rapid biological synthesis of silver nanoparticles using plant leaf extracts, Bioprocess Biosyst. Eng. 32 (2009) 79–84. https://doi.org/10.1007/s00449-008-0224-6

[72] D.D. Evanoff, G. Chumanov, Size-controlled synthesis of nanoparticles. 2. Measurement of extinction, scattering, and absorption cross sections, J. Phys. Chem. B. 108 (2004) 13957–13962. https://doi.org/10.1021/jp0475640

[73] P.I. Haris, Andrew J. Macnab – An innovator and pioneer in the field of Biomedical Near Infrared Spectroscopy, Biomed. Spectrosc. Imaging. 3 (2014) 307–309. https://doi.org/10.3233/bsi-140096

[74] M. Vallet-Regí, B. González, I. Izquierdo-Barba, Nanomaterials as promising alternative in the infection treatment, Int. J. Mol. Sci. 20 (2019) 3806. https://doi.org/10.3390/ijms20153806

[75] Noble Metal Nanoparticles: Preparation, Composite Nanostructures, Biodecoration and Collective Properties, Focus Catal. 2018 (2018) 7. https://doi.org/10.1016/j.focat.2018.11.106

[76] M.R. Almofti, T. Ichikawa, K. Yamashita, H. Terada, Y. Shinohara, Silver ion induces a cyclosporine A-insensitive permeability transition in rat liver mitochondria and release of apoptogenic cytochrome c, J. Biochem. 134 (2003) 43–49. https://doi.org/10.1093/jb/mvg111

[77] B.D. Ratner, A.S. Hoffman, F.J. Schoen, J.E. Lemons, Biomaterials Science, Biomater. Sci. (1996) 1–9. https://doi.org/10.1016/b978-012582460-6/50002-5

[78] Publishers note, Vib. Spectrosc. 43 (2007) 1. https://doi.org/10.1016/j.vibspec.2006.11.001

[79] A. Abbaszadegan, Y. Ghahramani, A. Gholami, B. Hemmateenejad, S. Dorostkar, M. Nabavizadeh, H. Sharghi, The effect of charge at the surface of silver nanoparticles on antimicrobial activity against gram-positive and gram-negative bacteria: A preliminary study, J. Nanomater. 2015 (2015) 1–8. https://doi.org/10.1155/2015/720654

[80] G. Merga, R. Wilson, G. Lynn, B.H. Milosavljevic, D. Meisel, Redox catalysis on "naked" silver nanoparticles, J. Phys. Chem. C. 111 (2007) 12220–12226. https://doi.org/10.1021/jp074257w

[81] X. Yan, B. He, L. Liu, G. Qu, J. Shi, L. Hu, G. Jiang, Antibacterial mechanism of silver nanoparticles in: Pseudomonas aeruginosa: Proteomics approach, Metallomics. 10 (2018) 557–564. https://doi.org/10.1039/c7mt00328e

[82] M. Zubair, M. Azeem, R. Mumtaz, M. Younas, M. Adrees, E. Zubair, A. Khalid, F. Hafeez, M. Rizwan, S. Ali, Green synthesis and characterization of silver nanoparticles from Acacia nilotica and their anticancer, antidiabetic and antioxidant efficacy, Environ. Pollut. 304 (2022) 119249. https://doi.org/10.1016/j.envpol.2022.119249

[83] A.C. Burduşel, O. Gherasim, A.M. Grumezescu, L. Mogoantă, A. Ficai, E. Andronescu, Biomedical applications of silver nanoparticles: An up-to-date overview, Nanomaterials. 8 (2018) 681. https://doi.org/10.3390/nano8090681

[84] K.P. Rumbaugh, K. Sauer, Biofilm dispersion, Nat. Rev. Microbiol. 18 (2020) 571–586. https://doi.org/10.1038/s41579-020-0385-0

[85] R.S. Patil, M.R. Kokate, C.L. Jambhale, S.M. Pawar, S.H. Han, S.S. Kolekar, One-pot synthesis of PVA-capped silver nanoparticles their characterization and biomedical application, Adv. Nat. Sci. Nanosci. Nanotechnol. 3 (2012) 15013. https://doi.org/10.1088/2043-6262/3/1/015013

[86] M. Torras, A. Roig, From Silver Plates to Spherical Nanoparticles: Snapshots of Microwave-Assisted Polyol Synthesis, ACS Omega. 5 (2020) 5731–5738. https://doi.org/10.1021/acsomega.9b03748

[87] K.C. Bhainsa, S.F. D'Souza, Extracellular biosynthesis of silver nanoparticles using the fungus Aspergillus fumigatus, Colloids Surfaces B Biointerfaces. 47 (2006) 160–164. https://doi.org/10.1016/j.colsurfb.2005.11.026

[88] K. Tiede, A.B.A. Boxall, S.P. Tear, J. Lewis, H. David, M. Hassellöv, Detection and characterization of engineered nanoparticles in food and the environment, Food Addit. Contam. - Part A Chem. Anal. Control. Expo. Risk Assess. 25 (2008) 795–821. https://doi.org/10.1080/02652030802007553

[89] A.C. Burduşel, O. Gherasim, A.M. Grumezescu, L. Mogoantă, A. Ficai, E. Andronescu, Biomedical applications of silver nanoparticles: An up-to-date overview, Nanomaterials. 8 (2018) 681. https://doi.org/10.3390/nano8090681

[90] Woodhead Publishing Series in Biomaterials, Bioresorbable Polym. Biomed. Appl. (2017) xix–xxiv. https://doi.org/10.1016/b978-0-08-100262-9.09002-9

[91] R.I. Barbhuiya, P. Singha, N. Asaithambi, S.K. Singh, Ultrasound-assisted rapid biological synthesis and characterization of silver nanoparticles using pomelo peel waste, Food Chem. 385 (2022) 132602. https://doi.org/10.1016/j.foodchem.2022.132602

[92] T. Sannomiya, C. Hafner, J. Voros, In situ sensing of single binding events by localized surface plasmon resonance, Nano Lett. 8 (2008) 3450–3455. https://doi.org/10.1021/nl802317d

[93] S.K. Sharma, D.S. Verma, L.U. Khan, S. Kumar, S.B. Khan, Handbook of Materials Characterization, Handb. Mater. Charact. (2018) 1–613. https://doi.org/10.1007/978-3-319-92955-2

[94] B. Galeano, E. Korff, W.L. Nicholson, Inactivation of vegetative cells, but not spores, of Bacillus anthracis, B. cereus, and B. subtilis on Stainless steel surfaces coated with an antimicrobial silver- and zinc-containing zeolite formulation, Appl.

Environ. Microbiol. 69 (2003) 4329–4331. https://doi.org/10.1128/AEM.69.7.4329-4331.2003

[95] Ö. Kaplan, N. Gökşen Tosun, R. İmamoğlu, İ. Türkekul, İ. Gökçe, A. Özgür, Biosynthesis and characterization of silver nanoparticles from Tricholoma ustale and Agaricus arvensis extracts and investigation of their antimicrobial, cytotoxic, and apoptotic potentials, J. Drug Deliv. Sci. Technol. 69 (2022) 103178. https://doi.org/10.1016/j.jddst.2022.103178

[96] K. Deplanche, I. Caldelari, I.P. Mikheenko, F. Sargent, L.E. Macaskie, Involvement of hydrogenases in the formation of highly catalytic Pd(0) nanoparticles by bioreduction of Pd(II) using Escherichia coli mutant strains, Microbiology. 156 (2010) 2630–2640. https://doi.org/10.1099/mic.0.036681-0

[97] C. Shanmugam, G. Sivasubramanian, B. Parthasarathi, K. Baskaran, R. Balachander, V.R. Parameswaran, Antimicrobial, free radical scavenging activities and catalytic oxidation of benzyl alcohol by nano-silver synthesized from the leaf extract of Aristolochia indica L.: a promenade towards sustainability, Appl. Nanosci. 6 (2016) 711–723. https://doi.org/10.1007/s13204-015-0477-8

[98] T. Karan, R. Erenler, B. Moran Bozer, Synthesis and characterization of silver nanoparticles using curcumin: Cytotoxic, apoptotic, and necrotic effects on various cell lines, Zeitschrift Fur Naturforsch. - Sect. C J. Biosci. 77 (2022) 343–350. https://doi.org/10.1515/znc-2021-0298

[99] S. Mukherjee, B.L. Bassler, Bacterial quorum sensing in complex and dynamically changing environments, Nat. Rev. Microbiol. 17 (2019) 371–382. https://doi.org/10.1038/s41579-019-0186-5

[100] V.R. Netala, V.S. Kotakadi, V. Nagam, P. Bobbu, S.B. Ghosh, V. Tartte, First report of biomimetic synthesis of silver nanoparticles using aqueous callus extract of Centella asiatica and their antimicrobial activity, Appl. Nanosci. 5 (2015) 801–807. https://doi.org/10.1007/s13204-014-0374-6

[101] M.H. Xiong, Y.J. Li, Y. Bao, X.Z. Yang, B. Hu, J. Wang, Bacteria-responsive multifunctional nanogel for targeted antibiotic delivery, Adv. Mater. 24 (2012) 6175–6180. https://doi.org/10.1002/adma.201202847

[102] K.A. Linder, P.N. Malani, Meningococcal Meningitis, JAMA - J. Am. Med. Assoc. 321 (2019) 1014. https://doi.org/10.1001/jama.2019.0772

[103] X.F. Zhang, Z.G. Liu, W. Shen, S. Gurunathan, Silver nanoparticles: Synthesis, characterization, properties, applications, and therapeutic approaches, Int. J. Mol. Sci. 17 (2016) 1534. https://doi.org/10.3390/ijms17091534

[104] A. Ingle, M. Rai, A. Gade, M. Bawaskar, Fusarium solani: A novel biological agent for the extracellular synthesis of silver nanoparticles, J. Nanoparticle Res. 11 (2009) 2079–2085. https://doi.org/10.1007/s11051-008-9573-y

[105] S. Pal, Y.K. Tak, J.M. Song, Does the antibacterial activity of silver nanoparticles depend on the shape of the nanoparticle? A study of the gram-negative bacterium Escherichia coli, Appl. Environ. Microbiol. 73 (2007) 1712–1720. https://doi.org/10.1128/AEM.02218-06

[106] J. Kesharwani, K.Y. Yoon, J. Hwang, M. Rai, Phytofabrication of silver nanoparticles by leaf extract of Datura metel: Hypothetical mechanism involved in synthesis, J. Bionanoscience. 3 (2009) 39–44. https://doi.org/10.1166/jbns.2009.1008

[107] E. Ogunsona, E. Ojogbo, T. Mekonnen, Advanced material applications of starch and its derivatives, Eur. Polym. J. 108 (2018) 570–581. https://doi.org/10.1016/j.eurpolymj.2018.09.039

[108] M. Ovais, A.T. Khalil, A. Raza, M.A. Khan, I. Ahmad, N.U. Islam, M. Saravanan, M.F. Ubaid, M. Ali, Z.K. Shinwari, Green synthesis of silver nanoparticles via plant extracts: Beginning a new era in cancer theranostics, Nanomedicine. 12 (2016) 3157–3177. https://doi.org/10.2217/nnm-2016-0279

[109] M.N. Owaid, J. Raman, H. Lakshmanan, S.S.S. Al-Saeedi, V. Sabaratnam, I. Ali Abed, Mycosynthesis of silver nanoparticles by Pleurotus cornucopiae var. citrinopileatus and its inhibitory effects against Candida sp., Mater. Lett. 153 (2015) 186–190. https://doi.org/10.1016/j.matlet.2015.04.023

[110] V.U.M. Nallal, K.N. Devi, M. Razia, Biogenic fabrication and characterization of Silver nanoparticles using high altitude lichen Heteroderimia leucomela extract and its potential applications, Mater. Today Proc. 50 (2021) 365–370. https://doi.org/10.1016/j.matpr.2021.10.017

[111] H. Korbekandi, S. Iravani, S. Abbasi, Optimization of biological synthesis of silver nanoparticles using Lactobacillus casei subsp. casei, J. Chem. Technol. Biotechnol. 87 (2012) 932–937. https://doi.org/10.1002/jctb.3702

[112] A.E.F. Oliveira, A.C. Pereira, M.A.C. de Resende, L.F. Ferreira, Synthesis of a silver nanoparticle ink for fabrication of reference electrodes, Talanta Open. 5 (2022) 100085. https://doi.org/10.1016/j.talo.2022.100085

[113] R.G. Wunderink, G. Waterer, Advances in the causes and management of community acquired pneumonia in adults, BMJ. 358 (2017) j2471. https://doi.org/10.1136/bmj.j2471

[114] M. Guilger-Casagrande, R. de Lima, Synthesis of Silver Nanoparticles Mediated by Fungi: A Review, Front. Bioeng. Biotechnol. 7 (2019) 287. https://doi.org/10.3389/fbioe.2019.00287

[115] J.M. V. Makabenta, A. Nabawy, C.H. Li, S. Schmidt-Malan, R. Patel, V.M. Rotello, Nanomaterial-based therapeutics for antibiotic-resistant bacterial infections, Nat. Rev. Microbiol. 19 (2021) 23–36. https://doi.org/10.1038/s41579-020-0420-1

Materials Research Foundations 145 (2023) 131-161 https://doi.org/10.21741/9781644902370-5

[116] A.R. Shahverdi, S. Minaeian, H.R. Shahverdi, H. Jamalifar, A.A. Nohi, Rapid synthesis of silver nanoparticles using culture supernatants of Enterobacteria: A novel biological approach, Process Biochem. 42 (2007) 919–923. https://doi.org/10.1016/j.procbio.2007.02.005

[117] E. Roeven, L. Scheres, M.M.J. Smulders, H. Zuilhof, Design, Synthesis, and Characterization of Fully Zwitterionic, Functionalized Dendrimers, ACS Omega. 4 (2019) 3000–3011. https://doi.org/10.1021/acsomega.8b03521

[118] T.M.D. Dang, T.T.T. Le, E. Fribourg-Blanc, M.C. Dang, Influence of surfactant on the preparation of silver nanoparticles by polyol method, Adv. Nat. Sci. Nanosci. Nanotechnol. 3 (2012) 35004. https://doi.org/10.1088/2043-6262/3/3/035004

[119] N. Nasri, A. Rusli, N. Teramoto, M. Jaafar, K.M. Ku Ishak, M.D. Shafiq, Z.A. Abdul Hamid, Green synthesis and characterization of silver nanoparticles by using turmeric extract and chitosan mixture, Mater. Today Proc. 66 (2022) 3044–3048. https://doi.org/10.1016/j.matpr.2022.07.335

[120] K.B. Holt, A.J. Bard, Interaction of silver(I) ions with the respiratory chain of Escherichia coli: An electrochemical and scanning electrochemical microscopy study of the antimicrobial mechanism of micromolar Ag, Biochemistry. 44 (2005) 13214–13223. https://doi.org/10.1021/bi0508542

Nanobiomaterials Materials Research Forum LLC
Materials Research Foundations 145 (2023) 162-176 https://doi.org/10.21741/9781644902370-6

Chapter 6

Biology Method and Mechanism of Antimicrobial Action of Silver Nanoparticles

Nahid Ahmadi[1,*], Ali Ramazani[2,*]

[1]A1School of Mahdieh Shahed, Education of Zanjan, Zanjan 45186-17981, Iran

[2]Department of Chemistry, Faculty of Science, University of Zanjan, Zanjan 45371-38791, Iran

Nahid Ahmadi: nahid.ahmadi0041@gmail.com and Ali Ramazani: aliramazani@gmail.com

Abstract

Silver nanoparticles are synthesized by various processes such as chemical, physical, and biology. However, biology method is the best method between all of them due to safe, eco-friendly, eco-environmental, and sometimes available. The obtained nanoparticle was characterized by TEM, SEM, XRD, and UV. The size and shape of nanoparticles play a main key in enter to bacteria cell. Silver nanoparticle destroyed the cell membrane and entered to the cells. With change silver atoms to silver ions and interaction ions with proteins, enzymes, DNA molecular, ribosomes, and mitochondria lead to inactive them and as result cells die.

Keywords

Nanoparticles, Characterization, TEM, SEM, XRD

Contents

Materials Research Foundations 145 (2023) 162-176 https://doi.org/10.21741/9781644902370-6

1. Introduction

Silver has many applications at different industrials. However, its medical property has been identified for over 2000 years. Silver has been employed due to its antimicrobial activity since 19 centuries. The activity can be against microorganisms on the surface of infected tissue or hard objects. It is observed Silver ions and silver -based particles have high toxicity to bacteria. Therefore, silver will be an excellent suitable option for medicinal field. They are adsorbed by some microorganisms. Besides of silver particles, silver nanoparticles have also extensive applications in biomedicine and high antibacterial property. Nanoparticles of silver have high area to silver particles to be involve with microbes. No exact mechanism has been known for reaction between silver particles and microbes. Nevertheless, some literature introduced a little mechanism. However, it has been well averred silver nanoparticles has anti-microbial activity such as anti-viral, anti-platelet, anti-inflammatory and anti-fungal activity [1].

There are usually different methods for synthesis silver nanoparticles containing chemical, physical, and biological (figure 1). Synthesis can be created as extracellular and intracellular. Each method has advantage and disadvantages. But chemical method is toxic and hazardous process for generating harmful by products besides of using expensive chemicals which are main factors for choosing a suitable synthesis process. A valuable and flexible method for synthesis of silver nanoparticles is extract of plant by biological.

Nanoparticles have various application but the main application of silver nanoparticles is antimicrobial property. That is why today silver nanoparticles are used at socks, composites (dental resins), water filters, toothpaste, wet wipe, soap, detergents, sprays, pillows, and many products. In spite of nanoparticles of silver have many benefits, some of particles may be entered water and environmental and led to damage the organism's health.

Materials Research Foundations 145 (2023) 162-176 https://doi.org/10.21741/9781644902370-6

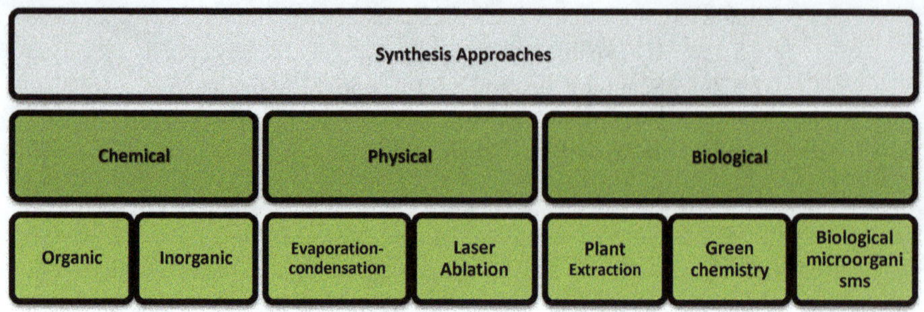

Figure 1. Synthesis approaches of silver nanoparticles

2. Chemical method

The nanoparticles of silver are prepared by two main methods, top-down and bottom-up. The most common chemical approach is silver reduction of its derivatives (figure 2A). In this case, there are two type of reduction agents, organic and inorganic. Inorganic reductants including hydrazine hydrate solution ($N_2H_4.H_2O$) [2], $NaBH_4$, and hydrogen. Ascorbic acid ($C_6H_8O_6$), dimethyl formamide, dimethyl sulfoxide, ethylene glycol, formaldehyde, tannic acid, and sodium citrate ($C_6H_5O_7Na_3$) were employed as an organic reduced agent for reduction of Ag ions in the silver salts. The another chemical method was researchers interesting is electrochemical and sonodecomposition. In electrochemical process (figure 2B), the aqueous solution of $AgNO_3$ with a polymer (long polyethylene glycol, sodium poly acrylate and etc.) were added in the reaction vessel. Then, reduction of Ag ions occurred by electron transferring. Synthesis of silver nanoparticles performance via ultrasonic waves. An aqueous solution of $AgNO_3$ faced under ultrasonic waves and produced bubbles of silver nitrate. After that, silver ions reduced in the presence of argon-hydrogen atmosphere. Aqueous foam informed through pouring silver ions to anionic surfactant aerosol led to information silver complexes as electrostatically. The complex can reduce with sodium borohydride and yield silver nanoparticles of foam[3]. Silver nanoparticles was also synthesized under microwave radiation. This method is a fast reaction and at the high temperature. Microwave irradiation is led to form uniformly monodispersed particles which is because of faster heating and uniform heating. In addition, explained approaches, a mixed method is for preparing of silver colloidal nanoparticle consist cryochemical synthesis, cryochemical reduction, vacuum evaporation, the use of pulsed lasers, sol-gel etc.

Materials Research Forum LLC
https://doi.org/10.21741/9781644902370-6

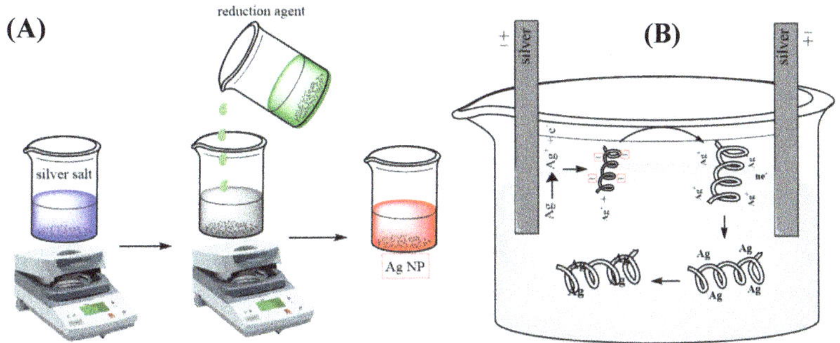

Figure 2. Synthesis of Ag Nanoparticle (A) by metal reduction. (B) by electrochemical

3. Physical method

Metal nanoparticles are synthesized by different physical methods including high energy ball milling, pulsed electron evaporation, CO_2 laser radiation, arc discharge, desiccation condensate, lase ablation, spray drying, electrical explosion wire, and microwave. The chemical reagent used less than chemical method and usually generated no byproduct at this method. From all methods were named conventional evaporation-condensation is the cheapest and laser ablation makes low impurities and nanoparticles with irregular shape and size. Several parameters can be role at shape and size of obtained nanoparticles, the wavelength of laser, the laser pulse duration and fluency, and impact surfactant or reagents. When the wavelength of laser is low the size of particles decreased.

Basically, the physical synthesis of silver nanoparticles usually employs the physical energies to make silver nanoparticles with proximately limited size dispensation. Also, may deterrent notably abundant amounts of sampling of silver nanoparticles in one procedure, which is the most superior secondary method for producing silver nanoparticles powder.

4. Biological process

The most important problems of the physical and chemical process for synthesis silver nanoparticles are expensive, toxic, and hazardous which can be dangerous for the environment and biologic systems. Using biological processes can help to delete prior issues. Biosynthesis processes were performed as a natural reduction technique of particles following; biological microorganisms, polysaccharides, fungi, bacteria, and plant extraction. The biosynthesis of metal nanoparticles is an environmentally and economically feasible approach. Therefore, it is a fact synthesis of silver nanoparticle via biologic is cheap and development.

silver nanoparticles were synthesized by reduction/oxidation reactions that usually called a bottom-up approach. In this case the microbial enzymes or the plant phytochemicals react with the compounds inside reaction vessel and produce the desired nanoparticles. Using biological process in the preparation nanoparticles involved The three main components, solvent, reducing agent, and stabilizing agent.

4.1 Bacteria method

Using bacteria is common for producing inorganic compounds intracellularly or extracellularly[4,5]. Bacteria employed for generating of silver nanoparticles are Anabaena variabilis[6], Gluconacetobacter xylinus[7], Ocimum basilicum L. var. thyrsiflorum[8], Penicillium sps.[9,10], Aneurinibacillus migulanus[11], Fuzarium Oxysporum[10], etc. Silver nanoparticles were synthesized using the Pseudomonas stutzeri AG259 strain for the first time.

some microorganisms can survive and grow in the metal ion concentrations due to their resistance to that metal. The parameters involved in the resistance are efflux systems, alteration of solubility and toxicity via reduction or oxidation, biosorption, bioaccumulation, extracellular complex formation or precipitation of metals, and lack of specific metal transport systems.

Biosynthesis of silver nanoparticles leads to acting nitrate reductase enzyme which nitrate into nitrite and in reaction during electron transferred to silver ions. In hence, this mechanism is more acceptable. There are some methods that biosynthesis of silver nanoparticles needs no enzyme. For example, cells of Lactobacillus sp. A09 reduces silver ions via the microbial cell wall.

4.2 Fungi method

Nanoparticle is obtained through fungi more than bacteria since they can exude proteins that help to synthesize nanoparticles. The steps of producing nanoparticles by fungi are the following: at first, silver ions are trapped on the surface of fungal cells, and subsequently, enzymes on the cells reduce those ions (figure 3). There are extracellular and intracellular enzymes which extracellular enzymes like naphthoquinones and anthraquinones simplify the reduction. Although no exact mechanism was discovered in silver nanoparticle production by fungi, it is believed that nicotine adenosine diphosphate (NADPH)-dependent nitrate reductase and a shuttle quinine extracellular process are responsible for the process. in comparison with plant extraction, the fungi methodology is a very slow process for preparing of silver nanoparticles [12,13].

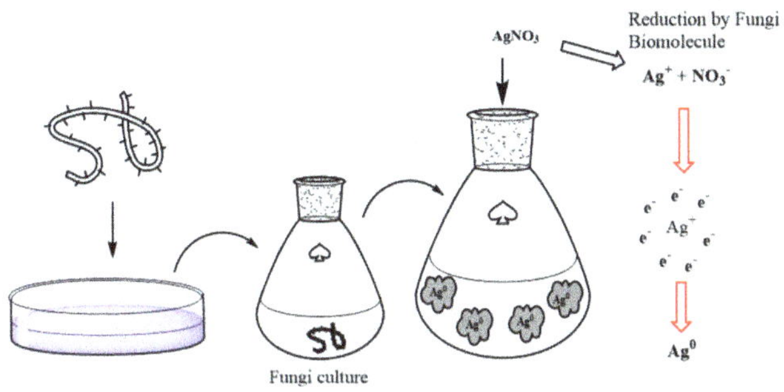

Figure 3. Synthesis of silver nanoparticles by fungi method.

4.3 Plants extraction

The several main advantages of plant extracts include being available, safe, and non-toxic which lead to being used for silver nanoparticle synthesis (figure 4). They possess a broad diversity of metabolites that can support the reduction of silver ions and are quicker than bacteria and fungi in the synthesis. Phytochemical is a process that is commonly used for the reduction. The phytochemical process implicates many substances containing terpenoids, flavones, ketones, aldehydes, amides, and carboxylic acids. Some of them such as flavones, organic acids, and quinones are water-soluble and conducted quick reduction of silver ions. Studies reported biomolecules such as proteins, flavonoids, phenols, and carbohydrates not only handle capping of nanoparticles but also play the main role in the reduction of silver ions and the preparation of nanoparticles [8,14].

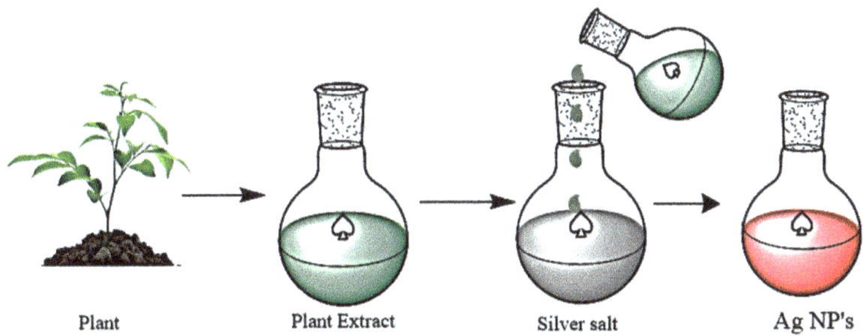

Figure 4. Synthesis of silver nanoparticles by plant extract.

Materials Research Foundations 145 (2023) 162-176
https://doi.org/10.21741/9781644902370-6

The plant extract is performance by different plants such as G. amansii algal[15]thiol, Ocimum basilicum L. var.thyrsiflorum plant[8], honey solution[16], Cannonball leaves[17], seaweed[8], Lantana camara leaf[18], Moringa oleifera flower[19], Alcea rosea flower[20], fritillaria flower [21], Caesalpinia pulcherrima flower, Terminalia chebula leaf[22,23], Plantago major L. leaf[24], Arabinoxylan from P. ovata husk[25], Leaves of Tinospora cordifolia[26], Capparis zeylanica leaves[27], Aquilegia pubiflora leaves[8], Costus pictus leaf, Catharanthus roseus[28], curcumin[29], Spondias mombin[30], Elaeagnus angustifolia bark[31], Eclipta alba[32], Ferocactus echidne[33], Momordica cymbalaria[34], and Boerhaavia diffusa plant[35]. The obtained nanoparticles from all of above plants has no the same size and the particle size are different from 13 to 60nm.

5. Characterization

Obtained nanoparticles were characterized by variety technique, TEM, SEM, DLS, XRD, and UV. DLS, SEM, and TEM are employed for measuring of nanoparticle size which it is said TEM, SEM, XRD, and UV are the most usable.

5.1 UV-visible spectroscopy

The UV-visible spectrum analysis is a useful technique for considering of nanoparticles. The UV spectrum of silver nanoparticles revealed an absorbance in the range of wavelength 390 to 470nm. The range is determined by the shape and size of nanoparticles. It maybe the lower absorbance obtained the smaller size (figure 5) however, the shape of nanoparticles isn't unaffected. Of course the size of nanoparticles depends on the synthesis method. For example, cell extract lead to have the large nanoparticles(100nm) so absorbance of them will be at high wavelength(440nm)[6]. The parameters include pH, concentration of $AgNO_3$ solution, reaction temperature, reaction time, and the ratio of cell extract and $AgNO_3$ never effect on wavelength but it can be changed intensity of peak[6][4][9].

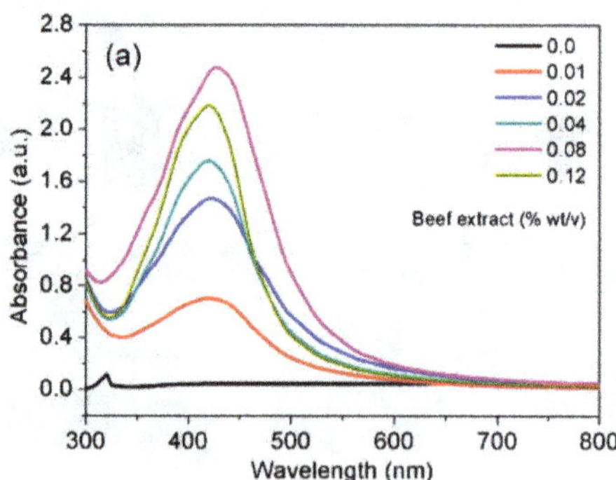

Figure 5. UV-vis spectra of Ag NPs obtained with different concentrations of cell-free beef extract at 420 nm.

Materials Research Foundations 145 (2023) 162-176 https://doi.org/10.21741/9781644902370-6

5.2 Scanning electron microscope (SEM) and transmission electron microscopy (TEM)

SEM and TEM are a microscopy technique employed to consider the morphology of particles especial metals nanoparticles. Both of them are used to measure the size of nanoparticles. The image of TEM displayed the silver nanoparticles have the size of 2 to 100 nm is various based on solution concentration of stabilizing [36] [37], synthesis method and temperature, and reducing agent[8]. Figure 6 show with increasing the concentration of silver nitrate from 3M to 22M, decreasing the size of nanoparticles from 25nm to 14nm[37].

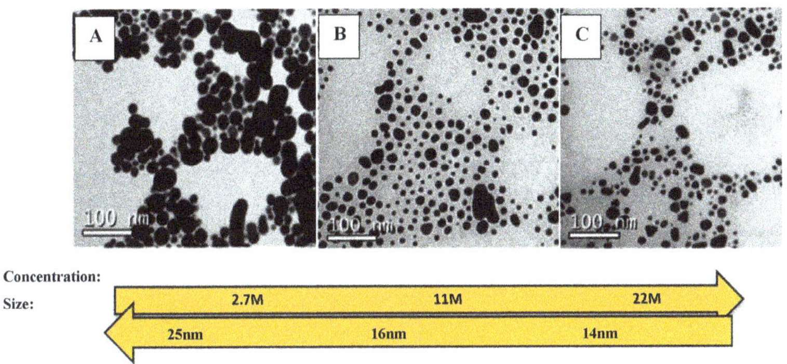

Figure 6. TEM image of silver nanoparticles that changed the size base on concentration of stabilizing, A)2.7M(25nm) B) 11M(16nm) C)22M(14nm)

The TEM image of Ag NP's display the spherical or oval morphology of these particles. Albite the shape of nanoparticles can changed for the type of coating, polymers, composites, and metals. According to TEM image, some nanoparticles have the different size that depends on the synthesized method like the nanoparticles of cell-free extract extracellular synthesis have 4 to 17nm[4], bacteria synthesis 20 to 60nm[11], biosynthesis by Fuzarium Oxysporum (fungi culture) 30-45nm[10], the sol-gel method 11nm[2], and Ag NP's coating Co 30-80nm,

5.3 X-ray Diffraction (XRD)

XRD is a characterization technique used for determination crystallite of particle structure. By XRD can measure the angles and intensities of diffracted beams and can generate a three-dimensional image of the density of electrons within the crystal [38,39]. The presence of sharp peaks indicates the bioactive compounds present on the surface of the NPs. The most common of lattice planes for silver nanoparticles is (111), (200), (220) and (311) have

Materials Research Foundations 145 (2023) 162-176 https://doi.org/10.21741/9781644902370-6

been gotten at 2θ indicating the formation of face centered cubic (fcc) crystalline structure (figure 7) [40][20].

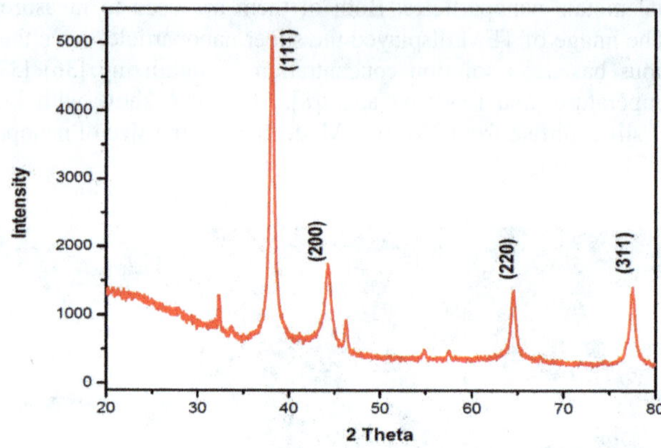

Figure 7. XRD pattern of AgNP.

6. Antimicrobial action

Antimicrobial activity was found to be a very complex process that kill microorganisms. As observed in figure 8, parts of microorganism can involve with antimicrobials contain cell membrane, DNA, ribosome, mitochondria, proteins, and enzymes in microorganism. The presence of toxic substance can have affected nutrition, metabolism, respiration, and breeding of organisms. One of toxic substance is employed silver nanoparticles. No exact mechanism is for antimicrobial activity of silver nanoparticles. Many mechanisms have been proposed for interaction between various categories of substances and various types of living organisms such as fungi, bacteria, viruses, termites, etc. However, some perspectives are not yet obvious, and it is necessary to be performed some progress for exact understanding in the future.

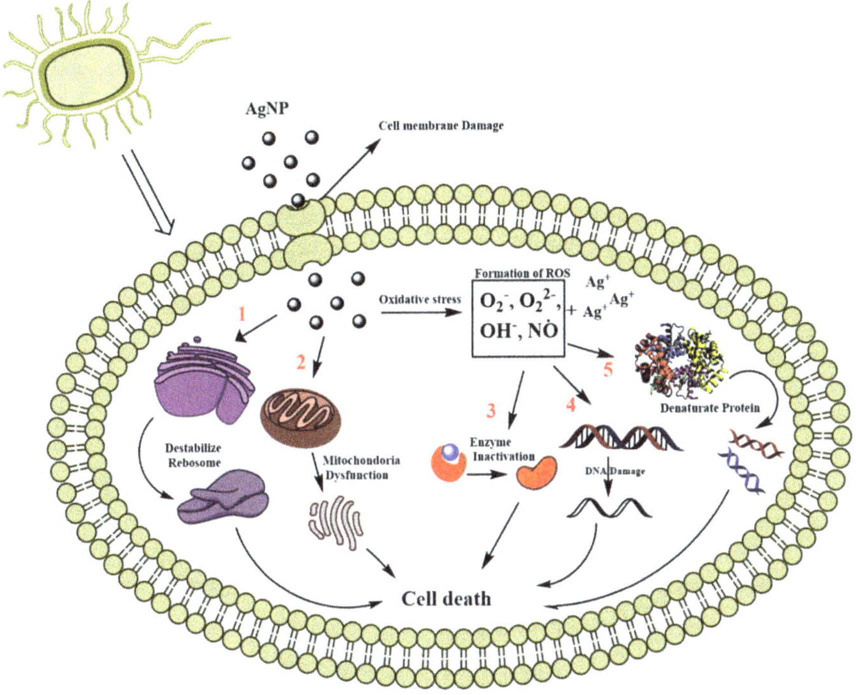

Figure 8. Various actions of silver nanoparticles on bacteria. Step1: effecting on ribosome Step 2: effecting on mitochodoria Step 3: effecting of silver ions on enzyme Step 4: effecting of silver ions on DNA Step 5: effecting of silver ions on protein

Silver nanoparticles have the capability to link to the bacterial cell wall and interpenetrate it, as result causing structural shifts in the cell membrane similar to the penetrance of the cell membrane and death of the cell. In continue, holes are created on the cell surface, and nanoparticles stuff on the cell surface.

There is another mechanism that silver nanoparticles produced radical species of oxygen which kill cells. The study of electron spin resonance spectroscopy indicates that the obtained free radicals of silver nanoparticles be capable of damage to the cell membrane of bacteria. After making it spongy can led to the death of the cell.

Antimicrobial properties of silver nanoparticles are demonstrated to be because of the oxidation of silver atoms and escape of silver ions from the surface of the silver nanoparticles. when silver nanoparticles contact with air leads to the release of silver ions which is why the antimicrobial properties of silver nanoparticles increased 2.3-fold. In fact, there is a mechanism that explains silver nanoparticles have a small size and large surface, therefore it can provide an expanded surface for attracting microorganisms. The treatment prevents several function of cells. On the other hand, silver nanoparticles are strong oxidants that can be taken down cell structures. Antioxidant properties of silver nanoparticles was confirmed using DPPH, H_2O_2, and ABTS as reported literature[6].

Silver nanoparticles after entering inside the cell of bacteria, and formation ions turn the DNA molecular into a compressed form, damaged the cells, and consequently cause the cell dies. The DNA molecular have phosphorous and sulfur atoms mainly. Silver ions joined them and depraved the DNA replication. Moreover, silver ions linked to proteins denaturants them. On the other hand, silver nanoparticles with the destruction of the cell membrane wall caused to leak of proteins and the reducing sugars, and inactive the respiratory chain dehydrogenases. Silver ions are also coupled with thiol groups[41] of vital enzymes and inactive the enzymes. it has been proved that silver nanoparticles can dampened the signal transition in bacteria which influence by phosphorylation of protein substrates. inhibition of signal transition and standing of bacteria growth are the treatments that taken place at the effect dephosphorylate of peptides on the tyrosine by silver nanoparticles.

Conclusions

Silver nanoparticles can have prepared as colloidal or powder. Both of them can't effected on the antibacterial activity. The nature of silver particles has antioxidant that kept antimicrobial property. Silver nanoparticles either through cell membrane or through organs in cells can be caused the death of cell.

References

[1] S. Sharma, H.I. Rasool, V. Palanisamy, C. Mathisen, M. Schmidt, D.T. Wong, J.K. Gimzewski, Nano core-A Review on 5G Mobile Communications, ACS Nano. 4 (2010) 1921–1926

[2] M. Shahjahan, Synthesis and Characterization of Silver Nanoparticles by Sol-Gel Technique, Nanosci. Nanometrology. 3 (2017) 34. https://doi.org/10.11648/j.nsnm.20170301.16

[3] Q. Bao, D. Zhang, P. Qi, Synthesis and characterization of silver nanoparticle and graphene oxide nanosheet composites as a bactericidal agent for water disinfection, J. Colloid Interface Sci. 360 (2011) 463–470. https://doi.org/10.1016/j.jcis.2011.05.009

[4] G. Ghodake, M. Kim, J.S. Sung, S. Shinde, J. Yang, K. Hwang, D.Y. Kim, Extracellular synthesis and characterization of silver nanoparticles—antibacterial activity against multidrug-resistant bacterial strains, Nanomaterials. 10 (2020) 360. https://doi.org/10.3390/nano10020360

[5] A.R.J.A. de M. Lima, A.S. Siqueira, M.L.S. Möller, R.C. de Souza, J.N. Cruz, A.R.J.A. de M. Lima, R.C. da Silva, D.C.F. Aguiar, J.L. da S.G.V. Junior, E.C. Gonçalves, In silico improvement of the cyanobacterial lectin microvirin and mannose interaction, J. Biomol. Struct. Dyn. (2020). https://doi.org/10.1080/07391102.2020.1821782

[6] I. Ahamad, N. Aziz, A. Zaki, T. Fatma, Synthesis and characterization of silver nanoparticles using Anabaena variabilis as a potential antimicrobial agent, J. Appl. Phycol. 33 (2021) 829–841. https://doi.org/10.1007/s10811-020-02323-w

[7] L.C. de Santa Maria, A.L.C. Santos, P.C. Oliveira, H.S. Barud, Y. Messaddeq, S.J.L. Ribeiro, Synthesis and characterization of silver nanoparticles impregnated into bacterial cellulose, Mater. Lett. 63 (2009) 797–799. https://doi.org/10.1016/j.matlet.2009.01.007

[8] H. Jan, G. Zaman, H. Usman, R. Ansir, S. Drouet, N. Gigliolo-Guivarc'h, C. Hano, B.H. Abbasi, Biogenically proficient synthesis and characterization of silver nanoparticles (Ag-NPs) employing aqueous extract of Aquilegia pubiflora along with their in vitro antimicrobial, anti-cancer and other biological applications, J. Mater. Res. Technol. 15 (2021) 950–968. https://doi.org/10.1016/j.jmrt.2021.08.048

[9] J.U. Shareef, M. Navya Rani, S. Anand, D. Rangappa, Synthesis and characterization of silver nanoparticles from Penicillium sps., Mater. Today Proc. 4 (2017) 11923–11932. https://doi.org/10.1016/j.matpr.2017.09.113

[10] S.J. Ukkund, M. Ashraf, A.B. Udupa, M. Gangadharan, A. Pattiyeri, Y.K. Marigowda, R. Patil, P. Puthiyllam, Synthesis and Characterization of Silver Nanoparticles from Fuzarium Oxysporum and Investigation of Their Antibacterial Activity, Mater. Today Proc. 9 (2019) 506–514. https://doi.org/10.1016/j.matpr.2018.10.369

[11] B. Syed, M.N. Nagendra Prasad, S. Satish, Synthesis and characterization of silver nanobactericides produced by Aneurinibacillus migulanus 141, a novel endophyte inhabiting Mimosa pudica L., Arab. J. Chem. 12 (2019) 3743–3752. https://doi.org/10.1016/j.arabjc.2016.01.005

[12] A.E.D.A. Gawad, M. Ibrahim, Computational Studies of the Interaction of Chitosan Nanoparticles and αB-Crystallin, Bionanoscience. 3 (2013) 302–311. https://doi.org/10.1007/s12668-013-0096-3

[13] C.M.A. Rego, A.F. Francisco, C.N. Boeno, M. V. Paloschi, J.A. Lopes, M.D.S. Silva, H.M. Santana, S.N. Serrath, J.E. Rodrigues, C.T.L. Lemos, R.S.S. Dutra, J.N. da Cruz, C.B.R. dos Santos, S. da S. Setúbal, M.R.M. Fontes, A.M. Soares, W.L. Pires, J.P. Zuliani, Inflammasome NLRP3 activation induced by Convulxin, a C-type lectin-like isolated from Crotalus durissus terrificus snake venom, Sci. Rep. 12 (2022) 1–17. https://doi.org/10.1038/s41598-022-08735-7

[14] N.C. da R. Galucio, D. de A. Moysés, J.R.S. Pina, P.S.B. Marinho, P.C. Gomes Júnior, J.N. Cruz, V.V. Vale, A.S. Khayat, A.M. do R. Marinho, Antiproliferative, genotoxic activities and quantification of extracts and cucurbitacin B obtained from Luffa operculata (L.) Cogn, Arab. J. Chem. 15 (2022) 103589. https://doi.org/10.1016/j.arabjc.2021.103589

[15] A. Pugazhendhi, D. Prabakar, J.M. Jacob, I. Karuppusamy, R.G. Saratale, Synthesis and characterization of silver nanoparticles using Gelidium amansii and its antimicrobial property against various pathogenic bacteria, Microb. Pathog. 114 (2018) 41–45. https://doi.org/10.1016/j.micpath.2017.11.013

[16] A.J. González Fá, A. Juan, M.S. Di Nezio, Synthesis and Characterization of Silver Nanoparticles Prepared with Honey: The Role of Carbohydrates, Anal. Lett. 50 (2017) 877–888. https://doi.org/10.1080/00032719.2016.1199558

[17] P. Devaraj, P. Kumari, C. Aarti, A. Renganathan, Synthesis and characterization of silver nanoparticles using cannonball leaves and their cytotoxic activity against MCF-7 cell line, J. Nanotechnol. 2013 (2013). https://doi.org/10.1155/2013/598328

[18] B. Ajitha, Y. Ashok Kumar Reddy, P. Sreedhara Reddy, Green synthesis and characterization of silver nanoparticles using Lantana camara leaf extract, Mater. Sci. Eng. C. 49 (2015) 373–381. https://doi.org/10.1016/j.msec.2015.01.035

[19] M.R. Bindhu, M. Umadevi, G.A. Esmail, N.A. Al-Dhabi, M.V. Arasu, Green synthesis and characterization of silver nanoparticles from Moringa oleifera flower and assessment of antimicrobial and sensing properties, J. Photochem. Photobiol. B Biol. 205 (2020) 111836. https://doi.org/10.1016/j.jphotobiol.2020.111836

[20] A. Ebrahiminezhad, Y. Barzegar, Y. Ghasemi, A. Berenjian, Zelena sinteza I karakterizacija nanočestica srebra pomoću ekstrakta cveta alcea rosea kao antimikrobnog sredstva nove generacije, Chem. Ind. Chem. Eng. Q. 23 (2017) 31–37. https://doi.org/10.2298/CICEQ150824002E

[21] S. Hemmati, A. Rashtiani, M.M. Zangeneh, P. Mohammadi, A. Zangeneh, H. Veisi, Green synthesis and characterization of silver nanoparticles using Fritillaria flower extract and their antibacterial activity against some human pathogens, Polyhedron. 158 (2019) 8–14. https://doi.org/10.1016/j.poly.2018.10.049

[22] C.S. Espenti, K.S.V.K. Rao, K.M. Rao, Bio-synthesis and characterization of silver nanoparticles using Terminalia chebula leaf extract and evaluation of its antimicrobial potential, Mater. Lett. 174 (2016) 129–133. https://doi.org/10.1016/j.matlet.2016.03.106

[23] B. Dheeba, PLANT MEDIATED SYNTHESIS AND CHARACTERIZATION OF SILVER NANOPARTICLES Original Article Climatic change View project PLANT MEDIATED SYNTHESIS AND CHARACTERIZATION OF SILVER NANOPARTICLES Original Article DHEEBA B., SIVAKUMAR R., SHEIK ABDULLA S, (n.d.). https://www.researchgate.net/publication/268812989

 Materials Research Forum LLC
https://doi.org/10.21741/9781644902370-6

[24] J. Sukweenadhi, K.I. Setiawan, C. Avanti, K. Kartini, E.J. Rupa, D.C. Yang, Scale-up of green synthesis and characterization of silver nanoparticles using ethanol extract of Plantago major L. leaf and its antibacterial potential, South African J. Chem. Eng. 38 (2021) 1–8. https://doi.org/10.1016/j.sajce.2021.06.008

[25] M.U.A. Khan, S.I. Abd Razak, H. Mehboob, M.R. Abdul Kadir, T.J.S. Anand, F. Inam, S.A. Shah, M.E.F. Abdel-Haliem, R. Amin, Synthesis and Characterization of Silver-Coated Polymeric Scaffolds for Bone Tissue Engineering: Antibacterial and in Vitro Evaluation of Cytotoxicity and Biocompatibility, ACS Omega. 6 (2021) 4335–4346. https://doi.org/10.1021/acsomega.0c05596

[26] J. Mittal, A. Singh, A. Batra, M.M. Sharma, Synthesis and characterization of silver nanoparticles and their antimicrobial efficacy, Part. Sci. Technol. 35 (2017) 338–345. https://doi.org/10.1080/02726351.2016.1158757

[27] M. Nilavukkarasi, S. Vijayakumar, S. Prathip Kumar, Biological synthesis and characterization of silver nanoparticles with Capparis zeylanica L. leaf extract for potent antimicrobial and anti proliferation efficiency, Mater. Sci. Energy Technol. 3 (2020) 371–376. https://doi.org/10.1016/j.mset.2020.02.008

[28] A. Verma, N. Bharadvaja, Plant-Mediated Synthesis and Characterization of Silver and Copper Oxide Nanoparticles: Antibacterial and Heavy Metal Removal Activity, J. Clust. Sci. 33 (2022) 1697–1712. https://doi.org/10.1007/s10876-021-02091-8

[29] M.J. Khan, K. Shameli, A.Q. Sazili, J. Selamat, S. Kumari, Rapid green synthesis and characterization of silver nanoparticles arbitrated by curcumin in an alkaline medium, Molecules. 24 (2019) 719. https://doi.org/10.3390/molecules24040719

[30] S. Samuggam, S. V. Chinni, P. Mutusamy, S.C.B. Gopinath, P. Anbu, V. Venugopal, L.V. Reddy, B. Enugutti, Green synthesis and characterization of silver nanoparticles using spondias mombin extract and their antimicrobial activity against biofilm-producing bacteria, Molecules. 26 (2021) 2681. https://doi.org/10.3390/molecules26092681

[31] S. Mortazavi-Derazkola, A. Yousefinia, A. Naghizadeh, S. Lashkari, M. Hosseinzadeh, Green Synthesis and Characterization of Silver Nanoparticles Using Elaeagnus angustifolia Bark Extract and Study of Its Antibacterial Effect, J. Polym. Environ. 29 (2021) 3539–3547. https://doi.org/10.1007/s10924-021-02122-5

[32] P. Premasudha, M. Venkataramana, M. Abirami, P. Vanathi, K. Krishna, R. Rajendran, Biological synthesis and characterization of silver nanoparticles using Eclipta alba leaf extract and evaluation of its cytotoxic and antimicrobial potential, Bull. Mater. Sci. 38 (2015) 965–973. https://doi.org/10.1007/s12034-015-0945-5

[33] A.T. Shah, M.I. Din, S. Bashir, M.A. Qadir, F. Rashid, Green Synthesis and Characterization of Silver Nanoparticles Using Ferocactus echidne Extract as a Reducing Agent, Anal. Lett. 48 (2015) 1180–1189. https://doi.org/10.1080/00032719.2014.974057

[34] M.K. Swamy, M.S. Akhtar, S.K. Mohanty, U.R. Sinniah, Synthesis and characterization of silver nanoparticles using fruit extract of Momordica cymbalaria and assessment of their in vitro antimicrobial, antioxidant and cytotoxicity activities, Spectrochim. Acta - Part A Mol. Biomol. Spectrosc. 151 (2015) 939–944. https://doi.org/10.1016/j.saa.2015.07.009

[35] P.P.N. Vijay Kumar, S.V.N. Pammi, P. Kollu, K.V.V. Satyanarayana, U. Shameem, Green synthesis and characterization of silver nanoparticles using Boerhaavia diffusa plant extract and their anti bacterial activity, Ind. Crops Prod. 52 (2014) 562–566. https://doi.org/10.1016/j.indcrop.2013.10.050

[36] W. Zhang, X. Qiao, J. Chen, Synthesis and characterization of silver nanoparticles in AOT microemulsion system, Chem. Phys. 330 (2006) 495–500. https://doi.org/10.1016/j.chemphys.2006.09.029

[37] A.L. Nogueira, R.A.F. Machado, A.Z. De Souza, F. Martinello, C. V. Franco, G.B. Dutra, Synthesis and characterization of silver nanoparticles produced with a bifunctional stabilizing agent, Ind. Eng. Chem. Res. 53 (2014) 3426–3434. https://doi.org/10.1021/ie4030903

[38] D. Psimadas, P. Georgoulias, V. Valotassiou, G. Loudos, Molecular Nanomedicine Towards Cancer :, J. Pharm. Sci. 101 (2012) 2271–2280. https://doi.org/10.1002/jps

[39] F.S. Alves, J. de A. Rodrigues Do Rego, M.L. Da Costa, L.F. Lobato Da Silva, R.A. Da Costa, J.N. Cruz, D.D.S.B. Brasil, Spectroscopic methods and in silico analyses using density functional theory to characterize and identify piperine alkaloid crystals isolated from pepper (Piper Nigrum L.), J. Biomol. Struct. Dyn. 38 (2020) 2792–2799. https://doi.org/10.1080/07391102.2019.1639547

[40] K. Ponsanti, B. Tangnorawich, N. Ngernyuang, C. Pechyen, A flower shape-green synthesis and characterization of silver nanoparticles (AgNPs) with different starch as a reducing agent, J. Mater. Res. Technol. 9 (2020) 11003–11012. https://doi.org/10.1016/j.jmrt.2020.07.077

[41] A.W. Orbaek, M.M. McHale, A.R. Barron, Synthesis and characterization of silver nanoparticles for an undergraduate laboratory, J. Chem. Educ. 92 (2015) 339–344. https://doi.org/10.1021/ed500036b

Nanobiomaterials Materials Research Forum LLC
Materials Research Foundations 145 (2023) 177-206 https://doi.org/10.21741/9781644902370-7

Chapter 7

Use of Nanomaterials in Bone Regeneration

N. Fattahi[1] and A. Ramazani[1,*]

[1] Department of Chemistry, Faculty of Science, University of Zanjan, Zanjan 45371-38791, Iran

aliramazani@gmail.com

Abstract

Over the past few decades, studies on bone tissue engineering have inspired innovation in novel materials, processing methods, performance evaluations, and applications. Nanomaterials have great promise for the creation of novel treatment solutions, such as bone regeneration and repair, as well as the replacement of organs and tissues. Bone tissue engineering is facilitated by nanomaterials that replicate bone characteristics and provide special functions. Nanomaterials can be employed in their natural state as drug delivery system carriers or as fillers to strengthen bone regeneration scaffolds. This chapter is focused on nanomaterials used in or being developed for bone regeneration. The present chapter aims to inspire readers to explore new avenues for designing and developing efficient and effective nanomaterials for bone regeneration applications.

Keywords

Bone Regeneration, Nanomaterials, Bone Biology, Biomaterials, Tissue Engineering

Contents

Materials Research Foundations 145 (2023) 177-206 https://doi.org/10.21741/9781644902370-7

1. Introduction

Bones are the main structural components of a skeleton in our body. They are biological composite materials with multiscale structures. The bone tissue is a specialized connective tissue adapted for protection and support, comprising organic and inorganic materials [1,2]. Millions of people worldwide have bone loss due to non-healing fractures after trauma, diseases like osteoporosis, and tumours, which are one of the leading causes of disability globally [3].

The growing need for innovative treatments to address severe tissue loss has made regenerative medicine as an attractive field of study. This area of research has created innovative therapies that use cells, growth factors, and biomaterials either separately or in combination with any type of tissue, including bone [4]. Fig. 1 shows bone fracture healing in four phases. As shown in Fig. 1, a hematoma forms at the fracture site during the first inflammatory phase, which begins 1-5 days after the fracture and is characterized by a hypoxic microenvironment that restricts oxygen and nutrient supply. Cell migration is induced by local inflammation. A fibrocartilaginous callus is formed in the second stage, and a bony callus replaces the fibrocartilaginous callus in the third stage. The fracture healing process will be completed when the bony callus is replaced by hard cortical bone in the fourth stage [5].

Numerous studies have focused on creating advanced biomaterials that promote and accelerate bone regrowth in defect areas [4]. Nanomaterials have an important function in proliferation and stimulating cell growth, which is crucial for bone regeneration. Because of their unique physical and chemical properties (e.g., electrical, magnetic) that are uniquely different from their bulk counterparts, nanostructured biomaterials have demonstrated superiority in promoting bone regeneration [6,7]. These differences are the result of the capability to precisely mimic the composition and nanoarchitecture of bone while enabling the recapitulation of essential elements of its metabolic environment at the nanoscale [8]. Ceramics, polymers, metals, and organic materials can all be utilized to prepare nanoparticles (NPs) [9,10]. Moreover, NPs can be incorporated into a different matrix, so nanocomposite scaffolds are produced [11,12].

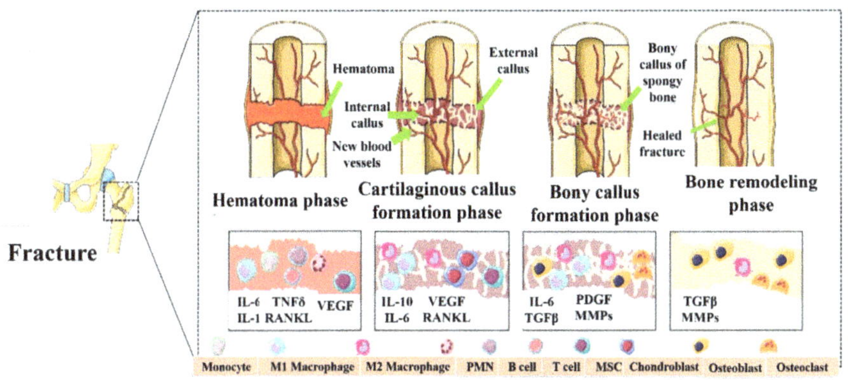

Figure 1. Schematic presentation of the four phases of fracture healing. IL, interleukin; RANKL, receptor activator of the nuclear factor-κB ligand; TGF-α, β, transforming growth factor-alpha, beta; PDGF, platelet-derived growth factor; MMP, matrix metalloproteinase; VEGF, vascular endothelial growth factor; PMN, polymorphonuclear leukocyte [5].

In this chapter, we will focus on the utilization of nanomaterials for bone regeneration. Following a short overview of bone biology, different nanomaterials in bone regeneration are reviewed. The approaches reported here serve as a guide for constructing novel and promising nanomaterials for future bone regeneration applications.

2. Bone biology

The bone skeleton protects internal organs, supports the entire body, and maintains mineral homeostasis and base/acid balance [13,14]. It supplies the environment for hematopoiesis inside the bone marrow and acts as a source of growth factors and cytokines. Organic type I collagen and the inorganic minerals phosphate (PO_4^{3-}) and calcium (Ca^{2+}) make up the majority of bone's structural elements and act as a platform to successfully stimulate cell proliferation and differentiation [15,16]. The bone also has endocrine capabilities that can influence other organs. As shown in Fig. 2, bone is generally composed of two parts (outer cortical bone and inner cancellous or spongy bone) and three significant cells (osteocytes, osteoclasts, and osteoblasts) [13]. Parallel bone lamellae make up the surface of compact bone, which is made up of a system of parallel-arranged osteons. The collagen fibers that make up the lamellae are parallelly arranged inside an amorphous mineralized matrix. On the surface of the bone, they either form a system of parallel mantle lamellae or a concentrated cluster around the central canal. Cancellous bone creates a network of trabeculae, which are thin bone beams with a significant interior surface. The optimal

Materials Research Foundations 145 (2023) 177-206 https://doi.org/10.21741/9781644902370-7

directional force transmission is reflected in the trabecular networks' orientation. It is essential for the remodelling and recovery of bone tissue [17]. One cancellous layer at the center and two layers of compact bone at the outside create flat bones. Bone marrow containing pluripotent mesenchymal stem cells (MSCs) occupies the space in the cancellous bone between trabeculae. In the body, MSCs are able to develop into adipose, muscle, bone, and cartilage tissue. Osteocytes and osteoblasts play an important role in bone formation, while osteoclasts are responsible for the reabsorption of existing bone tissue [18].

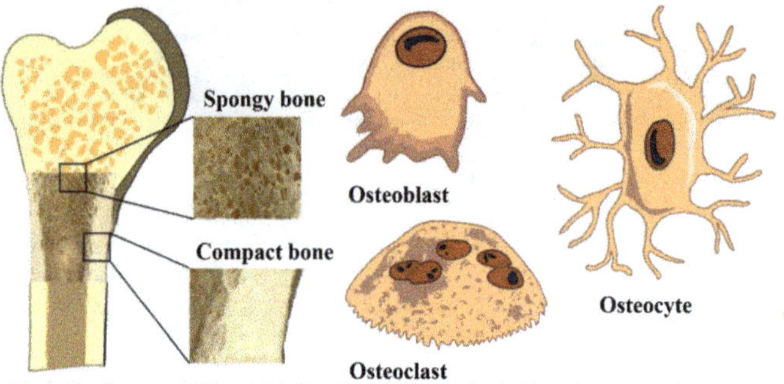

Figure 2. The schematic structure of the bone and the kinds of bone cells [19].

Despite the fact that bone tissue often has some degree of self-healing capacity, this capability is restricted when the bone defect is larger than a crucial value (2 cm) or when it results from an orthopedic disorder such as infection or tumour [20–22]. Failure of bone healing will result in nonunion of the bone owing to osteonecrosis, bone loss, and ischemia [23]. Additional therapies are necessary to promote bone regeneration in order to prevent these problems. In the following section, we focus on the various nanomaterials for bone regeneration.

3. Promising nanomaterials for bone regeneration

The ideal materials for bone should possess the following characteristics, such as good biocompatibility, appropriate biodegradability, optimal plasticity and mechanical properties, good osteoinductivity and osteoconductivity, a three-dimensional (3D) porous structure, and easily sterilized. In recent years, there has been a remarkable advancement in nanotechnology and the use of nanomaterials in regenerative medicine. Compared to their micron-sized equivalents, nanomaterials have demonstrated better bone cell capabilities [6,24]. This is because the hierarchical and nanoscale properties of bones and

Materials Research Forum LLC
https://doi.org/10.21741/9781644902370-7

nanomaterials may be exactly mimicked by nanomaterials, and the addition of magnetic NPs may deliver mechanical stimuli as needed or provide unique "smart" functions. When compared to related materials with conventional grain sizes, nanocrystalline materials exhibit unique mechanical, chemical, and physical properties due to their large surface areas and small grain sizes [25,26]. A material's beneficial qualities, such as increased toughness and strength, can be achieved by reducing the grain size. Metal and their oxide-based NPs have a number of benefits, including facile functionalization with different molecules, incorporation into hydrophilic and hydrophobic systems, excellent stability, no swelling changes, and simple preparation of the required size and shape [27]. In the following sections, we will discuss on various promising and important nanomaterials in bone regeneration applications.

3.1 Gold nanoparticles

The use of gold nanoparticles (Au-NPs) as the next-generation osteogenic agents for bone regeneration, has attracted considerable attention in recent years [28–30]. In an interesting study reported by Zhang et al. [31], the modulatory activities of Au-NPs on the osteogenesis of osteoblasts and macrophages were demonstrated. It was found that Au-NPs could promote the osteogenic differentiation of periodontal ligament stem cell sheets by upregulating bone-related protein expression and mineralization. The Au-NPs with a diameter of 45 nm were more effective than Au-NPs with a diameter of 13 nm. On the other hand, Au-NPs with the size of 5 nm remarkably decreased alkaline phosphatase (ALP) activity, the expression of osteogenic genes, and the preparation of mineralized nodules [31]. These findings support the hypothesis that smaller NPs do not provide sufficient free energy, leading to cluster formation where they can get the necessary driving force to enter cells [32]. In another study [33], Au-NPs functionalized with oligonucleotides demonstrated high stability, biocompatibility, and effective cell uptake. The TEM analysis of the Au-NPs indicated that these nanomaterials can interact with cytoplasmic proteins, downregulate adipogenic genes, and upregulate osteogenic genes [33]. In an interesting study, biogenic multifunctional Au-NPs were prepared using an immobilized crude enzyme of Bacillus licheniformis. Au-NPs exhibited good antioxidant activities, osteocompatibility, hemocompatibility and osteopromotive features. Moreover, Au-NPs showed increased calcium deposition and ALP activity through rat bone marrow stromal cells (BMSCs) and MC3T3-E1 murine preosteoblast cells [34]. In another interesting study, Heo et al. [35] synthesized hydrogel incorporating Au-NPs by photo-crosslinking. It was found that ALP activity, osteogenic differentiation of adipose-derived stromal cells, cell viability, and cell proliferation were promoted by the addition of Au-NPs [35]. Liang et al. [36] studied the modulatory activities of AuNPs-loaded mesoporous silica NPs (Au-MSNs) on osteoblastic lineage cells and macrophages. The direct osteogenic stimulation and immunomodulatory effects by Au-MSNs synergistically improved the osteogenic differentiation potential of MC3T3 cells as a result of crosstalk between Au-MSNs-conditioned macrophages and Au-MSNs-treated osteoblasts in a co-culture system. *In vivo* investigations further demonstrated that Au-MSNs could facilitate bone healing in rat cranial defects [36]. In an interesting study reported by Huang et al.

[37], multifunctional nanofibrous scaffolds were developed using the coaxial electrospinning technique with polyvinyl pyrrolidone (PVP), ethylcellulose (EC), and citrate-stabilized Au-NPs (Fig. 3). The obtained findings showed that citrate-stabilized Au-NPs were successfully encapsulated into electrospun scaffolds. *In vitro* investigations demonstrated that Au-NPs encapsulated coaxial electrospun scaffolds demonstrated high biocompatibility and osteogenic effects. Moreover, following the incorporation of Au-NPs, the mineralized nodule formation, osteogenic-related gene expression and ALP activity level were increased. Au-NPs-incorporated electrospun scaffolds were implanted into the defect site of the rat skull bone to investigate their bone-repairing potential *in vivo*. It was found that Au-NPs-incorporated scaffolds rapidly promoted bone regeneration [37]. In another study reported by Huang et al. [38], PEGylated Au-NPs were encapsulated into the bacterial cellulose (BC) hydrogel via in-situ fermentation as a promising scaffold for bone regeneration. The influence of Au-NPs on physicochemical characteristics of BC hydrogel, *in vitro* osteogenic differentiation and *in vivo* bone-forming capacity of Au/BC hydrogels were fully evaluated. The obtained findings demonstrated that the increased feeding levels of Au-NPs could remarkably improve the Au/BC hydrogels with better mechanical features, higher surface area, biocompatibility and, higher porosity. In particular, the incorporation of Au-NPs into hydrogels considerably improved their bone regeneration potential *in vivo*, which was found by the considerable new bone formation [38].

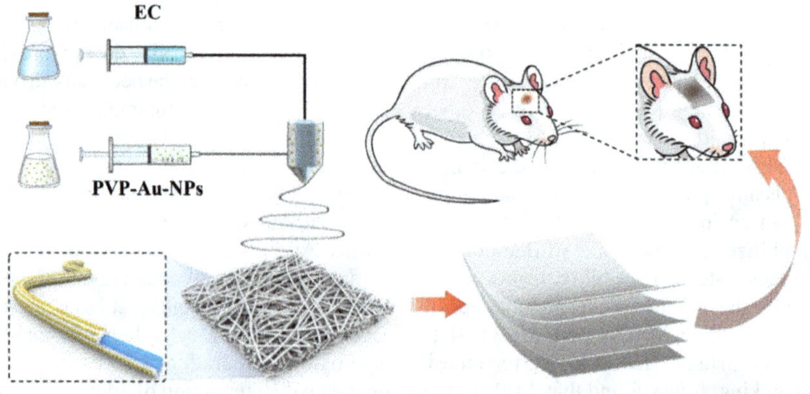

Figure 3. Schematic presentation of Au-NPs incorporated PVP/EC coaxial electrospun nanofibrous scaffolds and their potential in bone regeneration, Reprinted with permission from Ref. [37], Copyright 2021.

3.2 Palladium nanoparticles

Palladium nanoparticles (Pd-NPs) have recently been found to have considerable antimicrobial, catalytic, photosensitive, and electrical characteristics. In a study reported by Balaji et al. [39], reduced graphene oxide (rGO) was non-covalently functionalized with polypyrrole (PPy) and Pd-NPs, to improve the biological characteristics, including osteoproliferation, biocompatibility, and prevent bacterial infection (Fig. 4). It was found that the synthesized scaffold increased the proliferation of the Sarcoma osteogenic cells.

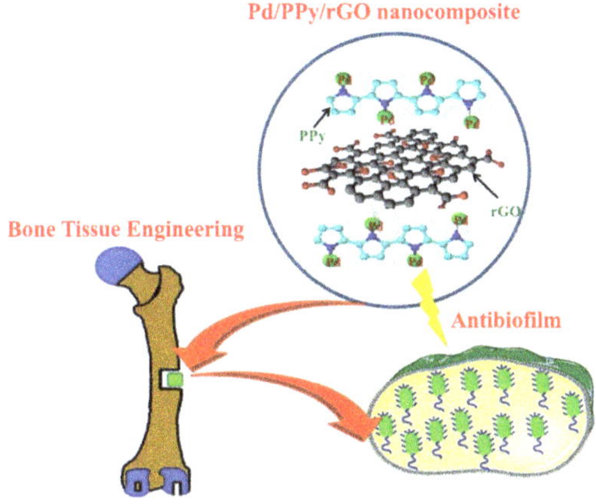

Figure 4. Antibiofilm associated orthopedic implantation, Reprinted with permission from Ref. [39], Copyright 2020.

Calabrese et al. [40] demonstrated that the growth of adipose-derived MSCs was inhibited up to 94% and 89% using Mg-hydroxyapatite-Collagen type I scaffolds modified with Pd-NPs and Au nanorods (Au-NRs), respectively. The OC and ON expression (specific for mature osteoblast and bone mineralization, respectively) and analysis of matrix mineralization density lead to comparable outcomes. Variants with Pd-NPs and Au-NPs also exhibited reduced cell differentiation than the control (20% lower) [40]. In another interesting study, Heidari et al. [40] synthesized and characterized the hydroxyapatite/ZnO/Pd nanocomposite scaffolds. Sol-gel and precipitation processes were used to prepare the initial materials. The biocompatibility of scaffolds was investigated in interaction with dental pulp stem cells and in simulated body fluid (SBF). The addition of Pd and ZnO-NPs improved the fracture toughness and compressive strength of scaffolds compared to scaffolds fabricated with free hydroxyapatite. Following 28 days of immersion in the SBF, the samples' surface roughness decreased, indicating that homogenous apatite was being deposited there. Moreover, the inclusion of Pd and ZnO made the scaffolds, showing antibacterial activities against P. aeruginosa. The findings demonstrated that the optimized hydroxyapatite/ZnO/Pd nanocomposite scaffold could be a potential bone graft substitute for bone regeneration applications [41]. Zhang et al. [42] synthesized porous Au/Pd alloy NPs as hyperthermia agent to develop mild localized heat (MLH) for rapid in situ bone regeneration under the irradiation of near-infrared laser (Fig. 5). After 6 weeks of photothermal therapy, about 97% of the cranial defect site was repaired by the newly-formed bone. RNA sequencing data was utilized to identify insight into the molecular mechanism of the MLH on bone formation and cell proliferation. The obtained findings showed that the WNT signalling pathway was involved in the MLH [42].

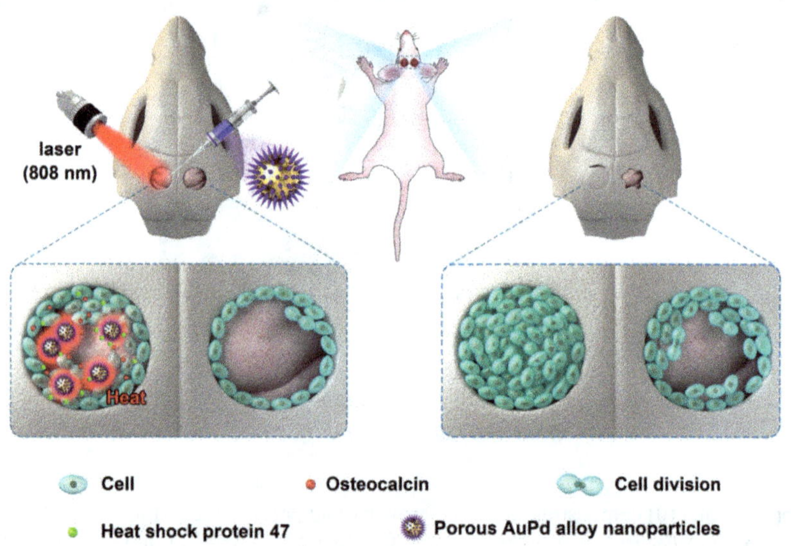

Figure 5. The schematic presentation of pAuPds for photothermal therapy of cranial defect. Reprinted with permission from Ref. [42] Copyright 2019.

3.3 Platinum nanoparticles

Platinum is a heavy and valuable metal utilized in the glass, chemical, and pharmaceutical industries [43]. Platinum Nanoparticles (Pt-NPs) are able to provide anti-inflammatory and anti-oxidant effects, protecting the bone by reducing the production of osteoclasts [44]. In an interesting study, Eid et al. [45] reported a better degradation rate and morphology for calcium phosphate scaffolds loaded with Pt-NPs than calcium phosphate scaffolds free of Pt-NPs. Moreover, it was found that the proliferation and adhesion of osteoblast cells on scaffolds incorporated with Pt-NPs were superior [45]. Radwan-Pragłowska et al. [46] developed 3D bioactive scaffolds with hierarchical structures. By electrospinning poly-L-lactic acid, microwave-assisted chitosan crosslinking, and doping with three different kinds of metallic NPs (TiO_2, Au, and Pt), the matrixes were successfully created. In contrast to scaffolds enriched with TiO_2-NPs, superior biomineralization, proliferation, cell adhesion, and differentiation were found in the presence of Au-NPs and Pt-NPs [46].

3.4 Iron oxide nanoparticles

Maghemite (γ-Fe_2O_3) and magnetite (Fe_3O_4) are two main types of iron oxide nanoparticles (Fe-NPs), and they both exhibit excellent biocompatibility and unique magnetic characteristics [7,47,48]. Fe-NPs and scaffolds comprising these materials are considerably utilized in bone tissue regeneration because magnetic field can exert suitable forces on magnetic particles, which are internalized into cells, presenting a strong tool to control cell behaviours [49–51]. In order to use Fe-NPs, their magnetic characteristics are essential. When nanoparticle-labeled cells can still be detected by magnetic resonance imaging after 7 days, this allows for *in vivo* observation and monitoring [52]. Therefore, it is possible to study cell behaviour, such as adhesion, proliferation, and differentiation. Further bone regeneration depends on the osteogenic potential of stem cells, and Fe-NPs can provide the proteins required to regulate the differentiation and growth of these cells. In the presence of the magnetic field, these magnetic nanomaterials have a positive influence on osteogenic differentiation, as shown in a trial with bone marrow-derived MSCs (BM-MSCs), where the expression of type I collagen was improved [53] and in s study with bovine serum albumin enriched with Fe-NPs, was also observed increased ALP activity, expression of osteocalcin, and calcium deposition [54]. Biopolymer scaffolds containing Fe-NPs demonstrated remarkable osteogenic differentiation and stimulation of cell adhesion in both *in vitro and in vivo* bone repair models, which was demonstrated by higher gene expression of main transcription factors for osteoblast (Osx and RUNX2) as well as ALP activity. Taken together, magnetic scaffolds are beneficial for bone regeneration, particularly in adjunctive treatment with foreign magnetic stimulation [55].

3.5 Copper nanoparticles

Copper (Cu) is involved in a broad range of biological processes, including bone metabolism control, blood clotting, Cu-dependent enzyme activity, and vision process. The presence of Cu ions in scaffolds can improve bone regeneration [56]. In a study [57], the effect of Cu-doped chitosan scaffolds was investigated on bone regeneration. After 4 weeks of implantation of synthesized scaffolds in calvarial defects of rats, the increase in bone volume was observed. It was found that the presence of Cu ions in scaffolds could improve bone tissue engineering [57]. In another study, Lin et al. [58] synthesized a kind of Cu-

Materials Research Foundations 145 (2023) 177-206
https://doi.org/10.21741/9781644902370-7

containing calcium phosphate cement (Cu-CPC) by incorporating it with Cu phosphate NPs through a photothermal anti-tumour effect. At different concentrations, the effect of these particles on angiogenesis and osteogenesis was evaluated on human umbilical vein endothelial cells and mouse BM-MSCs. It was found that Cu-CPC containing 0.05 and 0.01 wt% of Cu phosphate NPs could promote bone formation around cancerous bone defects. The differentiation was confirmed by elevated expression of osteogenesis-related genes runt-related transcription factor 2, osteocalcin, type I collagen, and by angiogenesis-related genes vascular endothelial growth factor, fibroblast growth factor, and endothelial nitric oxide synthase [58].

3.6 Zinc nanoparticles

For the human body to function normally, zinc is a necessary element. This element takes part in numerous physiological and enzymatic activities. Zinc oxide nanoparticles (ZnO-NPs) have antibacterial and anti-diabetic activities and can decrease oxidative stress by removing free radicals [59]. These NPs are employed as drug delivery vehicles, nanosensors, or contrast materials in biomedical imaging. They have been incorporated into the bioactive glass and bioactive ceramics because of their biocompatibility [60]. Due to the release of Zn^{2+} ions, Zn-NPs have osteogenic characteristics. These ions can promote bone formation, mineralization, and growth. Zn^{2+} ions promote the expression of the osteocalcin gene area by activating the mitogen-activated protein kinase pathway. By promoting Runx2 expression and increasing Zn^{2+} carriers like zinc transporter protein ZIP1, Zn^{2+} ions also help in bone formation [59]. Khader et al. [61] demonstrated the influence of Zn-NPs composite scaffolds in bone regeneration. In this work, fibrous ZnO-NPs composite scaffolds were developed using the electrospinning method. Polycaprolactone (PCL) incorporated with Zn-NPs showed low dissolution and degradation of Zn^{2+}. The highest expression of cartilage-specific genes and collagen type II production were induced by little amounts of Zn-NPs, which increased chondrogenic differentiation. On the other hand, Zn-NPs in high concentrations promoted the osteogenic differentiation of human MSCs. The highest expression of bone-specific genes, ALP activity, and collagen production were evidence of this differentiation [61].

3.7 Magnesium oxide nanoparticles

In the human body, magnesium plays a significant biogenic role. It functions as a cofactor in over 300 enzymatic processes and is crucial for the production of RNA, DNA, and proteins. It also participates in the metabolism of calcium and vitamin D [62]. Therefore, scaffolds made from diverse materials and combined with magnesium regulate cellular behaviour as well as promote bone growth and healing. By releasing magnesium, these substances promote proliferation, cell adhesion, and osteogenic differentiation of MSCs [63]. Mg-based scaffolds and magnesium oxide nanoparticles (MgO-NPs) are easily biodegradable. However, high degradability can lead to scaffold failure. Using graphene NPs in the Mg-Al-Cu matrix can reduce the level of degradation and bactericidal potential of the synthesized scaffolds up to five times [64]. The release of acidic products during the degradation of synthetic polymer scaffolds is a drawback because it is responsible for initiating inflammation. The pH-neutralizing properties of the polymer containing magnesium hydroxide NPs significantly reduce the release of pro-inflammatory cytokines [65]. To improve the mechanical characteristics and biological reactivity of the polymeric scaffolds, Mg-doped bioactive glass can be used as reinforcement [66]. Making a scaffold that effectively releases magnesium cations is still a challenge. For instance, a self-assembled bisphosphonate-magnesium NPs hydrogel based on hyaluronic acid exhibits increased mineralization abilities and regulated release of Mg^{2+} ions [63]. MgO-NPs provide antibacterial activities with comparable outcomes to conventional sodium hypochlorite (5.25%), which is utilized for endodontic root canal cleaning. They cause oxidative stress and damage the cell membrane, which results in cell death [67]. For cell growth on the scaffold, the optimal concentration of MgO-NPs is significant. In a study reported by Wetteland et al. [68], it was found that when BM-MSCs were cultivated in 200 g/ml of MgO-NPs, their density was increased considerably in comparison to the control group. The reduction of bacterial adhesion is an additional advantage of MgO-NPs. Thus, MgO-NPs could be utilized as a part of a composite or as a coating material for medical implant applications [68].

3.8 Titanium dioxide nanoparticles

Titanium dioxide nanoparticles (TiO_2) are considered natural materials and are broadly utilized in medicine and industry [69]. TiO_2 can be utilized as a nanofiller for fabricating silk fibroin-based scaffolds. In a study reported by Johari et al. [70], SF/TiO_2 scaffolds containing different concentrations of TiO_2 (5, 10, 15, 20 wt%) were synthesized. Porosity in scaffolds was reduced by increasing the content of TiO_2-NPs. This is because of NPs agglomeration on the pore walls. In addition, the synthesized scaffold exhibited enhanced mechanical and directional stability owing to the increase in the content of TiO_2-NPs. Moreover, the biocompatibility tests with SaOS-2 cells revealed high proliferation and adhesion. With the growing content of SF, the viability of cells was increased [70]. TiO_2-NPs are also used in the synthesis of polymer-inorganic nanoparticle composite membranes. In an interesting study reported by Nechifor et al. [71], polysulfone-silica microfiber attached with TiO_2 NPs (PSf-SiO_2-TiO_2) composite membranes were fabricated. In this work, microspheres of TiO_2-NPs with a diameter of 1-3 μm were bonded to silica microfiber. In comparison with membranes without TiO_2 NPs, PSf-SiO_2-TiO_2 nanocomposite membranes provided better mechanical and flux characteristics, chemical oxidation resistance, and retention. The antimicrobial properties of synthesized scaffolds were investigated, and it was found that they have an inhibitory influence on five Gram (-) microorganisms and inhibited the proliferation of the *Candida albicans* strain. The obtained findings indicated that PSf-SiO_2-TiO_2 composite membranes are suitable as a barrier in guided bone regeneration [71]. It is also necessary to mention that in the human body, the degradation of Ti-based implants causes the release of solid debris and ions of TiO_2-NPs, leading to peri-implant inflammatory diseases. In a study reported by Bressan et al. [72], the effect of TiO_2-NPs on fibroblast and MSCs was investigated by the production of reactive oxygen species (ROS) and MTT test. It was found that TiO_2-NPs

Materials Research Foundations 145 (2023) 177-206 https://doi.org/10.21741/9781644902370-7

caused high production of ROS and a time-dependent decrease in cell viability. Molecular and histological studies of peri-implant tissue indicated the high levels of Ti particles, the occurrence of ROS, and improved bone turnover [72].

3.9 Calcium oxide nanoparticles

Calcium is a significant component in many geologic and biologic systems. In the human body, about 99% of the calcium is found in the skeleton and plays an important role as a mediator in a number of regulatory and metabolic processes [73]. Calcium is broadly utilized and investigated for bone regenerative applications. Calcium oxide nanoparticles (CaO-NPs) are excellent materials for medicinal use since they are low-toxic, biocompatible, and have remarkable antibacterial properties. It is necessary to mention that, CaO-NPs are well known for their antifungal qualities and their antibacterial activity. Trichophyton-induced dermatophytosis can be treated with CaO in heated scallop-shell powder [74]. The encapsulation of CaO-NPs in various polymers can provide antimicrobial effects and increase their mechanical performance. In a study, Silva et al. [75] investigated the biocidal and mechanical properties of polyethylene composites containing CaO-NPs with diameters of 25 and 55 nm. It was observed that Young's modulus of polymer modified with CaO-NPs was higher than that of pure polymer. More than 99.99% of the bacteria were killed using nanomaterial modified with CaO-NPs. The fact that CaO-NPs' antibacterial properties increase with size reduction has been highlighted. CaO-NPs have the significant benefit of not requiring external stimuli, such as any type of irradiation, to produce an antibacterial effect [75]. In another study, Münchow et al. [76] evaluated PCL/gelatin and PCL electrospun matrices enriched with CaO-NPs. It was found that the electrospun matrices containing CaO-NPs improved cell viability, adhesion, osteogenic differentiation, and proliferation [76].

3.10 Aluminum oxide nanoparticles

Aluminum oxide nanoparticles (Al_2O_3-NPs) integrated with ceramic coatings exhibit low cytotoxicity, good osseointegration, high biodegradability, and resistant to long-term implantation [77]. In a study reported by Yu et al. [78], the 3D porous Ca_2SiO_4 scaffold and its modified form by Al ions were compared. Rat BM-MSCs on the scaffolds modified with aluminum showed significantly higher proliferation and differentiation, as well as a slower rate of degeneration. Increased expression of osteogenic markers such as osteocalcin, osteopontin, and collagen type I has demonstrated that enrichment with Al_2O_3 improved bone regeneration [78]. In another study reported by Toloue et al. [78], reinforcement of polyhydroxy butyrate-chitosan scaffolds with Al_2O_3 nanowires improved mechanical performance with a tensile strength of 10 times higher, in comparison with the same basic scaffold without Al nanowires [79].

3.11 Cerium oxide nanoparticles

Cerium oxide (CeO_2-NPs) are free radical scavengers because of their multifunctionality and unique atomic lattice architectures with oxygen vacancies [80]. *In vitro* studies have demonstrated that CeO_2-NPs improved osteogenic differentiation and proliferation of MSCs [81]. CeO_2-NPs can be utilized to modify the surface of titanium implants. Studies demonstrate that these coatings have anti-inflammatory and antibacterial characteristics [82]. The use of CeO_2-NPs to reinforce scaffolds is another area of research. Purohit et al. [83] synthesized and investigated the CeO_2-alginate-gelatin nanocomposite scaffold. It was found that adding CeO_2-NPs to the scaffold matrix significantly altered its properties by improving cell adhesion, proliferation, and viability, as well as its capacity to scavenge free radicals [83]. In another study, Lukin et al. [84] investigated the effects of nanocrystalline CeO_2 in the early stages of bone regeneration in rabbit models. In the lower jaw, the formed defects were inserted with a formulation comprising nanocrystalline CeO_2, barium sulphate, and calcium hydroxide. This study demonstrated the ability of nanocrystalline CeO_2 to increase osteogenesis and decrease inflammatory response [84].

3.12 Carbon nanoparticles

Carbon-based nanomaterials are very promising and valuable materials owing to their biocompatibility and mechanical strength. Carbon can be found in a variety of structurally unique forms, including carbon nanotubes (CNTs), graphene, nanodiamonds (NDs), fullerenes, carbon dots (CDs), and their derivatives [85,86]. NDs are carbon-based NPs having various functional groups on their surface. They typically have a diameter of 4 to 10 nm and an sp^3 hybridized crystal lattice structure. NDs are employed in the bone-anchoring stems of articular prostheses or other permanent bone implants, with the aim of increase their interaction with the surrounding bone in the body. In addition to their excellent chemical and mechanical resistance, this advantageous property can also be attributed to their surface nanotopography and nanoroughness. These features could induce the differentiation of bone marrow mesenchymal stem cells into osteoblastic lineage and stimulate the growth, maturation of osteoblasts, and adhesion [87]. The surface of NDs can act as carriers for bioactive chemicals (e.g., fibroblast growth factor or bone morphogenetic protein) because of their capacity to bind proteins that promote bone formation [88]. In an interesting study, Choi et al. [89] designed icariin (ICA)-functionalized NDs (ICA-NDs) and showed that the synthesized ICA-NDs could promote osteogenic ability. The ICA-NDs showed a prolonged ICA release for up to 4 weeks. *In vitro* cell studies revealed that MC3T3 osteoblastic cells treated with ICA-NDs produced significantly more osteogenic markers, including runt-related transcript factor 2-E1, collagen type I alpha 1, osteopontin, and alkaline phosphatase, and even calcium content, in comparison with free NDs treated cells [89].

CNTs are cylindrical macromolecules composed of carbon atoms having remarkable electrical, mechanical, thermal, and chemical properties. Due to their distinctive characteristics and potential to have a significant impact on many fields of technology and science, CNTs are currently the subject of intense research. Two reputable categories of

CNTs are multi-walled (MW-CNTs) and single-walled (SW-CNTs) carbon nanotubes [90]. Nanocomposite-based scaffolds in which CNTs act as a filler of hydroxyapatite exhibit a remarkable enhancement in tensile strength and fracture toughness. SW-CTNs are most often employed to create scaffolds that promote the formation and nucleation of hydroxyapatite because they have a particle size ranging from 0.7 to 1.5 nm, close to the length of the collagen triple helix. In an interesting study, Zhao et al. [91] reported chemically modified SW-CNTs as a scaffold for the growth of artificial bone material. They demonstrated that SW-CNTs/hydroxyapatite scaffolds with various functional moieties have a significant influence on hydroxyapatite formation [91]. In another study reported by Meng et al. [92], a nanocomposite scaffold with an appropriate roughness was synthesized for cell attachment by adding MW-CNTs to the porous bioglass foam. Owing to the presence of MW-CNTs, it was found that the interaction among cells was remarkably improved during their adhesion, the proliferation of osteoblasts was promoted, and osteoclastogenesis was suppressed [92]. Because CNTs have insufficient solubility in water and highly hydrophobic character, their dispersion into the chitosan solution can overcome this drawback and improve the mechanical characteristics of the chitosan itself [93]. The synthesized scaffold exhibited high biodegradability and biocompatibility. The prepared pores were appropriate for the proliferation, differentiation, and penetration of MSCs. Therefore, the synthesized nanomaterials can be desirable for bone regeneration and other biomedical purposes [93]. In another study, a cryogenic technique was utilized to synthesize 3D biomimetic chitosan/MW-CNTs/nano-hydroxyapatite scaffolds suitable for bone regeneration and determine their viability for bone tissue engineering applications. It was found that, in combination with MSCs, this kind of scaffold exhibited significant mechanical properties and better biocompatibility [94]. One of the promising approaches for improving the mechanical characteristics of bone is the use of CNTs instead of collagen fibers. Jamilpour et al. [95] indicated that the use of CNTs instead of collagen fibers could provide a remarkable reduction in strain energy density. The findings also showed that this replacement might alter the strain energy distribution within the bone. Based on a semi-mechanistic bone remodelling theory, it is speculated that this alteration in strain energy distribution can destabilize the normal bone remodelling processes in artificial bone. Thus it is possible to cause some abnormalities in bone's biological and mechanical properties [95].

Conductive biomaterials can provide significant signal transmission from a foreign source to the seeded cells. In an interesting study, the surface and mechanical properties of PLA were remarkably increased by the addition of CNTs, thereby improving osteoblast proliferation and adhesion. Scaffolds containing PLA/MW-CNTs with uniformly dispersed nano-hydroxyapatite indicated a high degree of osteoblast differentiation [96]. Other nanomaterials fabricated by combining CNTs with various synthetic polymers, such as polypropylene, poly (lactic-co-glycolic acid), polymethacrylate, polyurethane, etc., also exhibited remarkably better proliferation, adhesion, mineralization, and ALP levels [85,97].

Materials Research Foundations 145 (2023) 177-206 https://doi.org/10.21741/9781644902370-7

Graphene, an sp^2-bonded carbon sheet with a single atom thickness, is also utilized in biomedical applications owing to its particular nanoscale and electronic properties. Graphene films can be incorporated into polymers to increase their physical qualities. In a study, Crowder et al. [98] utilized three-dimensional (3D) graphene foams for differentiating and culturing human MSCs. The obtained findings demonstrated not only promoted cell proliferation in the scaffolds but also accelerated differentiation into bone cells. It was found that the synthesized 3D graphene foams can promote osteogenic differentiation and maintain stem cell viability [98]. In an interesting study reported by Chen et al. [99], graphene oxide (GO)/ultrahigh molecular weight polyethylene (PE) composites were fabricated by liquid-phase ultrasonication dispersion followed by hot-pressing. Owing to the presence of GO sheets, the synthesized composites showed a considerable combination of improved mechanical properties and good biocompatibility, making the composites promising for the potential candidates as artificial joints [99]. In another study reported by Das et al. [100], polyvinyl alcohol (PVA)/pristine graphene nanofibers were fabricated by electrospinning an aqueous solution of polyvinylpyrrolidone-stabilized PVA and graphene. The synthesized fibers showed a remarkable enhancement in crystallinity and thermal stability, even at low graphene loading [100].

Fullerenes are carbon allotropes with a spherical structure comprising more than sixty carbon atoms linked via pentagonal and hexagonal rings. Thanks to their particular physicochemical properties, fullerene materials are expected to have great potential in various industries [101]. The carbon nanofibers with a fullerene surface layer increase the proliferation and improve the adhesion of MG-63 osteoblast bone cells [102]. As another class of carbon nanomaterial family, carbon dots (CDs) are promising diagnostic and therapeutic biomaterial with tunable and unique physicochemical properties. They are widely utilized in targeted drug delivery, bone regeneration materials, and cell imaging [103]. In an interesting study, Ren et al. [104] demonstrated a promising pharmacological target of extracellular signal-regulated kinases (ERK)/AMPK pathway for bone loss treatment. In this work, metformin carbon dots (MCDs) were fabricated using citric acid and metformin hydrochloride via a hydrothermal technique. The MCDs showed remarkable biocompatibility. In comparison with free metformin, MCDs could more significantly increase the osteogenesis of rat BM-MSCs (rBMSCs) under both normal and inflammatory conditions *in vitro*. Furthermore, the MCDs could activate the extracellular signal-regulated kinases (ERK)/AMP-activated protein kinase (AMPK) pathway to increase the osteogenic potential of rBMSCs (Fig. 6) [104].

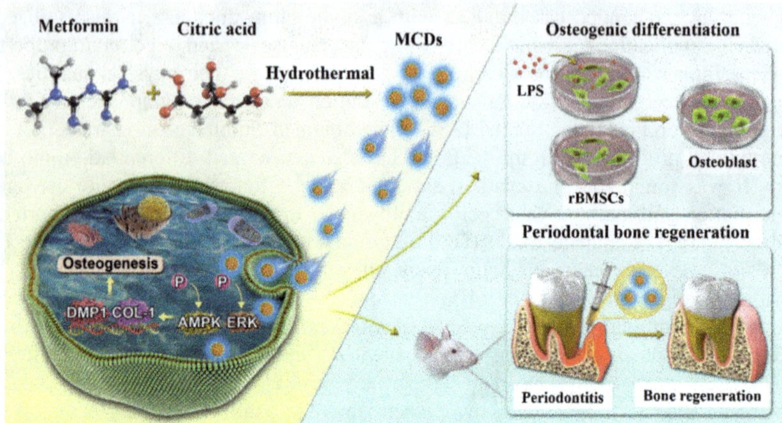

Figure 6. Schematic presentation of the synthetic of MCDs and the working mechanism for promoting periodontal bone regeneration. Reprinted with permission from Ref. [104], Copyright 2021.

In another study reported by Jin et al. [105], CDs remarkably promoted osteogenic differentiation, enhanced matrix mineralization *in vitro*, and promoted new bone formation in the skull defect model *in vivo* [105].

3.13 Silicon dioxide nanoparticles

Silicon dioxide nanoparticles (SiO_2-NPs) are one of the most common NPs owing to their commercial availability and low cost. Silica-based mesostructured nanomaterials can be regarded as a full family of biomaterials with significant potential applications in bone regeneration and tissue engineering [106,107]. Decomposition tests of mesoporous silica-based nanomaterials at 37 °C in simulated body fluid have demonstrated their gradual decomposition. Moreover, studies in mice have indicated their biodegradability and biocompatibility [108]. Recently, injectable hydrogels endowed with osteogenic potential can fill irregular bone defect sites by minimally invasive techniques and have thus been attracting increasing attention [109]. In a study reported by Gaihre et al. [110], an injectable material composed of chitosan and SiO_2-NPs was developed. It was observed that the differentiation and proliferation of OB-6 cells along the surface were increased in the presence of SiO_2-NPs [110]. In another study reported by Shi et al. [111], Gelatin-methacryloyl pre-polymer was prepared. In this work, stromal cell-derived factor-1 alpha and nano silicate (SN) were introduced into the pre-polymer to achieve controlled release property, injectability, efficient stem cell homing, and excellent osteogenic ability. The osteogenic differentiation was induced by SN without the requirement for any other osteoinductive factors. The synthesized hydrogel exhibited improved cell proliferation, migration, matrix mineralization, and expression of osteogenic biomarkers. Moreover, the significant benefit was it's crosslinking by UV light in situ and excellent injectability via a 17-G needle at room temperature [111]. In an interesting study, Maleki et al. [112] developed a multiscale porous hybrid silicasilk fibroin aerogel-based scaffold in an entirely aqueous-based sol−gel reaction via in situ processing of SF biopolymer with the tetraethyl orthosilicate (TEOS), followed by unidirectional freeze-casting and then by CO_2 supercritical drying (scCO2) (Fig. 7a). The nanostructured silica provides mechanical stiffness like filler or inorganic reinforcement as found in the bone. Cellular studies demonstrated that the synthesized scaffold promotes osteoblast growth and proliferation. The new bone formation was shown inside and around the implants (Fig. 7b). In the rat model, the biocompatibility of the scaffold was demonstrated since macroscopic degenerative changes and necrosis were not found in the defect area and around multichannel silica-based aerogel hybrid as a peripheral nerve repair scaffold. Moreover, the scaffold was well demonstrated by the local bone tissue without any significant side effects, such as bleeding, inflammation, or infection. μ-CT imaging also showed the favourable progression of bone regeneration in the scaffold-inserted animals (Fig. 7c). In the animals receiving scaffold implantation, the periosteum formation was complete, but in the defect area of the scaffold-free group, very poor growth of the periosteum was observed. Moreover, the X-ray radiography experiments indicated a considerable difference in the radiodensity of the defect area in surgery/scaffold animal groups and surgery/no scaffold, which demonstrates the favourable host-material interaction and successful scaffold-mediated bone formation [112,113].

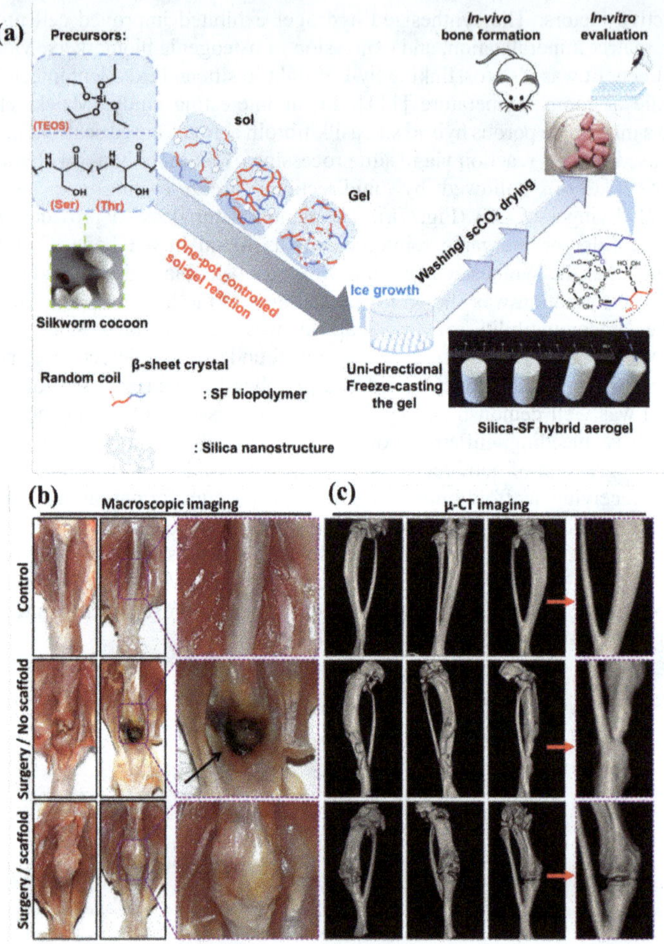

Figure 7. (a) Schematic presentation of the synthesis of silica-SF aerogel hybrid scaffold (b) Macroscopic image of the scaffold-implanted site after 25 days of placement in the region of the rat femur. (c) μ-CT images of the bone in the defect areas with and without scaffold and the control group without creating any defect after 25 days of surgery and implantation. The yellow area displays the sites without bone formation in the animals with bone defects but no scaffold implantation. Reprinted with permission from Ref. [112], Copyright 2019.

 Materials Research Forum LLC
https://doi.org/10.21741/9781644902370-7

Conclusions

This chapter aims to provide a comprehensive overview on nanostructured materials for bone tissue regeneration. Although in many cases, the bone itself can remodel and repair, these procedures are not adequate, and further process is needed. Due to the close similarity of nanostructure to native bone architecture, nanomaterials offer ideal platforms to produce the organization of natural extracellular matrix for the production of functional bone tissues. Thanks to the nanoscale size, physicochemical characteristics are remarkably improved, and nanomaterials regulate and support proliferation, migration, and cellular function. A wide variety of nanomaterials are currently being researched and developed. Different NPs are being investigated to modify the polymers in order to improve their properties. In the form of a hydrogel, the benefit of NPs as a segment of composites is that they can be injected into the defect sites by a minimally invasive intervention. However, further progress may provide new treatment options for bone regeneration in the field of smart nanomaterials in the future

References

[1] P. Kumar, M. Saini, B.S. Dehiya, A. Sindhu, V. Kumar, R. Kumar, L. Lamberti, C.I. Pruncu, R. Thakur, Comprehensive survey on nanobiomaterials for bone tissue engineering applications, Nanomaterials. 10 (2020) 1–60. https://doi.org/10.3390/nano10102019

[2] G.G. Walmsley, A. McArdle, R. Tevlin, A. Momeni, D. Atashroo, M.S. Hu, A.H. Feroze, V.W. Wong, P.H. Lorenz, M.T. Longaker, D.C. Wan, Nanotechnology in bone tissue engineering, Nanomedicine Nanotechnology, Biol. Med. 11 (2015) 1253–1263. https://doi.org/10.1016/j.nano.2015.02.013

[3] G.T. STRICKLAND, Hunter's Tropical Medicine and emerging infectious diseases, Elsevier Health Sciences, 2001. https://doi.org/10.1590/s0036-46652001000200018

[4] A. Atala, R. Lanza, J.A. Thomson, R.M. Nerem, Principles of Regenerative Medicine, Academic press, 2008. https://doi.org/10.1016/B978-0-12-369410-2.X5001-3

[5] M. Pfeiffenberger, A. Damerau, A. Lang, F. Buttgereit, P. Hoff, T. Gaber, Fracture healing research—shift towards in vitro modeling?, Biomedicines. 9 (2021) 748. https://doi.org/10.3390/biomedicines9070748

[6] R.A. Pérez, J.E. Won, J.C. Knowles, H.W. Kim, Naturally and synthetic smart composite biomaterials for tissue regeneration, Adv. Drug Deliv. Rev. 65 (2013) 471–496. https://doi.org/10.1016/j.addr.2012.03.009

[7] H. Ahankar, A. Ramazani, N. Fattahi, K. Ślepokura, T. Lis, P.A. Asiabi, V. Kinzhybalo, Y. Hanifehpour, S.W. Joo, Tetramethylguanidine-functionalized silica-coated iron oxide magnetic nanoparticles catalyzed one-pot three-component synthesis

of furanone derivatives, J. Chem. Sci. 130 (2018) 1–13. https://doi.org/10.1007/s12039-018-1572-7

[8] S. Minardi, F. Taraballi, L. Pandolfi, E. Tasciotti, Patterning biomaterials for the spatiotemporal delivery of bioactive molecules, Front. Bioeng. Biotechnol. 4 (2016) 45. https://doi.org/10.3389/fbioe.2016.00045

[9] F. Karkeh-Abadi, H. Safardoust-Hojaghan, L.S. Jasim, W.K. Abdulsahib, M.A. Mahdi, M. Salavati-Niasari, Synthesis and characterization of Cu2Zn1.75Mo3O12 ceramic nanoparticles with excellent antibacterial property, J. Mol. Liq. 356 (2022) 119035. https://doi.org/10.1016/j.molliq.2022.119035

[10] N. Yu, L. Zhao, D. Cheng, M. Ding, Y. Lyu, J. Zhao, J. Li, Radioactive organic semiconducting polymer nanoparticles for multimodal cancer theranostics, J. Colloid Interface Sci. 619 (2022) 219–228. https://doi.org/10.1016/j.jcis.2022.03.107

[11] N. Wang, D. Qi, L. Liu, Y. Zhu, H. Liu, S. Zhu, Fabrication of In Situ Grown Hydroxyapatite Nanoparticles Modified Porous Polyetheretherketone Matrix Composites to Promote Osteointegration and Enhance Bone Repair, Front. Bioeng. Biotechnol. 10 (2022). https://doi.org/10.3389/fbioe.2022.831288

[12] C. Covarrubias, J. Bejarano, M. Maureira, C. Tapia, M. Díaz, J.P. Rodríguez, H. Palza, F. Lund, A. Von Marttens, P. Caviedes, M. Yazdani-Pedram, Preparation of osteoinductive – Antimicrobial nanocomposite scaffolds based on poly (D,L-lactide-co-glycolide) modified with copper – Doped bioactive glass nanoparticles, Polym. Polym. Compos. 30 (2022) 09673911221098231. https://doi.org/10.1177/09673911221098231

[13] B. Clarke, Normal bone anatomy and physiology., Clin. J. Am. Soc. Nephrol. 3 Suppl 3 (2008) S131–S139. https://doi.org/10.2215/CJN.04151206

[14] I. Makhoul, C.O. Montgomery, D. Gaddy, L.J. Suva, The best of both worlds-managing the cancer, saving the bone, Nat. Rev. Endocrinol. 12 (2016) 29–42. https://doi.org/10.1038/nrendo.2015.185

[15] J.H. Jang, O. Castano, H.W. Kim, Electrospun materials as potential platforms for bone tissue engineering, Adv. Drug Deliv. Rev. 61 (2009) 1065–1083. https://doi.org/10.1016/j.addr.2009.07.008

[16] A.R.J.A. de M. Lima, A.S. Siqueira, M.L.S. Möller, R.C. de Souza, J.N. Cruz, A.R.J.A. de M. Lima, R.C. da Silva, D.C.F. Aguiar, J.L. da S.G.V. Junior, E.C. Gonçalves, In silico improvement of the cyanobacterial lectin microvirin and mannose interaction, J. Biomol. Struct. Dyn. (2020). https://doi.org/10.1080/07391102.2020.1821782

[17] J.S. Kenkre, J.H.D. Bassett, The bone remodelling cycle, Ann. Clin. Biochem. 55 (2018) 308–327. https://doi.org/10.1177/0004563218759371

[18] S. Torgbo, P. Sukyai, Bacterial cellulose-based scaffold materials for bone tissue engineering, Appl. Mater. Today. 11 (2018) 34–49. https://doi.org/10.1016/j.apmt.2018.01.004

[19] V. Babuska, P.B. Kasi, P. Chocholata, L. Wiesnerova, J. Dvorakova, R. Vrzakova, A. Nekleionova, L. Landsmann, V. Kulda, Nanomaterials in Bone Regeneration, Appl. Sci. 12 (2022) 6793. https://doi.org/10.3390/app12136793

[20] J.M. Wagner, F. Reinkemeier, C. Wallner, M. Dadras, J. Huber, S.V. Schmidt, M. Drysch, S. Dittfeld, H. Jaurich, M. Becerikli, K. Becker, N. Rauch, V. Duhan, M. Lehnhardt, B. Behr, Adipose-Derived Stromal Cells Are Capable of Restoring Bone Regeneration After Post-Traumatic Osteomyelitis and Modulate B-Cell Response, Stem Cells Transl. Med. 8 (2019) 1084–1091. https://doi.org/10.1002/sctm.18-0266

[21] H. Ma, C. Jiang, D. Zhai, Y. Luo, Y. Chen, F. Lv, Z. Yi, Y. Deng, J. Wang, J. Chang, C. Wu, A Bifunctional Biomaterial with Photothermal Effect for Tumor Therapy and Bone Regeneration, Adv. Funct. Mater. 26 (2016) 1197–1208. https://doi.org/10.1002/adfm.201504142

[22] Y. Yang, L. Chu, S. Yang, H. Zhang, L. Qin, O. Guillaume, D. Eglin, R.G. Richards, T. Tang, Dual-functional 3D-printed composite scaffold for inhibiting bacterial infection and promoting bone regeneration in infected bone defect models, Acta Biomater. 79 (2018) 265–275. https://doi.org/10.1016/j.actbio.2018.08.015

[23] F. Loi, L.A. Córdova, J. Pajarinen, T. hua Lin, Z. Yao, S.B. Goodman, Inflammation, fracture and bone repair, Bone. 86 (2016) 119–130. https://doi.org/10.1016/j.bone.2016.02.020

[24] G. Balasundaram, D.M. Storey, T.J. Webster, Novel nano-rough polymers for cartilage tissue engineering, Int. J. Nanomedicine. 9 (2014) 1845–1853. https://doi.org/10.2147/IJN.S55865

[25] L. Mishnaevsky, E. Levashov, R.Z. Valiev, J. Segurado, I. Sabirov, N. Enikeev, S. Prokoshkin, A. V. Solov'Yov, A. Korotitskiy, E. Gutmanas, I. Gotman, E. Rabkin, S. Psakh'E, L. Dluhoš, M. Seefeldt, A. Smolin, Nanostructured titanium-based materials for medical implants: Modeling and development, Mater. Sci. Eng. R Reports. 81 (2014) 1–19. https://doi.org/10.1016/j.mser.2014.04.002

[26] C.M.A. Rego, A.F. Francisco, C.N. Boeno, M. V. Paloschi, J.A. Lopes, M.D.S. Silva, H.M. Santana, S.N. Serrath, J.E. Rodrigues, C.T.L. Lemos, R.S.S. Dutra, J.N. da Cruz, C.B.R. dos Santos, S. da S. Setúbal, M.R.M. Fontes, A.M. Soares, W.L. Pires, J.P. Zuliani, Inflammasome NLRP3 activation induced by Convulxin, a C-type lectin-like isolated from Crotalus durissus terrificus snake venom, Sci. Rep. 12 (2022) 1–17. https://doi.org/10.1038/s41598-022-08735-7

[27] M.P. Nikolova, M.S. Chavali, Metal oxide nanoparticles as biomedical materials, Biomimetics. 5 (2020) 27. https://doi.org/10.3390/BIOMIMETICS5020027

[28] S.Y. Choi, M.S. Song, P.D. Ryu, A.T.N. Lam, S.W. Joo, S.Y. Lee, Gold nanoparticles promote osteogenic differentiation in human adipose-derived mesenchymal stem cells through the Wnt/β-catenin signaling pathway, Int. J. Nanomedicine. 10 (2015) 4383–4392. https://doi.org/10.2147/IJN.S78775

[29] C. Yi, D. Liu, C.C. Fong, J. Zhang, M. Yang, Gold nanoparticles promote osteogenic differentiation of mesenchymal stem cells through p38 MAPK pathway, ACS Nano. 4 (2010) 6439–6448. https://doi.org/10.1021/nn101373r

[30] H. Samadian, H. Khastar, A. Ehterami, M. Salehi, Bioengineered 3D nanocomposite based on gold nanoparticles and gelatin nanofibers for bone regeneration: in vitro and in vivo study, Sci. Rep. 11 (2021) 1–11. https://doi.org/10.1038/s41598-021-93367-6

[31] Y. Zhang, P. Wang, Y. Wang, J. Li, D. Qiao, R. Chen, W. Yang, F. Yan, Gold nanoparticles promote the bone regeneration of periodontal ligament stem cell sheets through activation of autophagy, Int. J. Nanomedicine. 16 (2021) 61–73. https://doi.org/10.2147/IJN.S282246

[32] Y. Zhang, N. Kong, Y. Zhang, W. Yang, F. Yan, Size-dependent effects of gold nanoparticles on osteogenic differentiation of human periodontal ligament progenitor cells, Theranostics. 7 (2017) 1214–1224. https://doi.org/10.7150/thno.17252

[33] N.L. Rosi, D.A. Giljohann, C.S. Thaxton, A.K.R. Lytton-Jean, M.S. Han, C.A. Mirkin, Oligonucleotide-modified gold nanoparticles for infracellular gene regulation, Science (80-.). 312 (2006) 1027–1030. https://doi.org/10.1126/science.1125559

[34] S. Singh, A. Gupta, I. Qayoom, A.K. Teotia, S. Gupta, P. Padmanabhan, A. Dev, A. Kumar, Biofabrication of gold nanoparticles with bone remodeling potential: an in vitro and in vivo assessment, J. Nanoparticle Res. 22 (2020) 1–15. https://doi.org/10.1007/s11051-020-04883-x

[35] D.N. Heo, W.K. Ko, M.S. Bae, J.B. Lee, D.W. Lee, W. Byun, C.H. Lee, E.C. Kim, B.Y. Jung, I.K. Kwon, Enhanced bone regeneration with a gold nanoparticle-hydrogel complex, J. Mater. Chem. B. 2 (2014) 1584–1593. https://doi.org/10.1039/c3tb21246g

[36] H. Liang, C. Jin, L. Ma, X. Feng, X. Deng, S. Wu, X. Liu, C. Yang, Accelerated Bone Regeneration by Gold-Nanoparticle-Loaded Mesoporous Silica through Stimulating Immunomodulation, ACS Appl. Mater. Interfaces. 11 (2019) 41758–41769. https://doi.org/10.1021/acsami.9b16848

[37] C. Huang, J. Dong, Y. Zhang, S. Chai, X. Wang, S. Kang, D. Yu, P. Wang, Q. Jiang, Gold Nanoparticles-Loaded Polyvinylpyrrolidone/Ethylcellulose Coaxial Electrospun Nanofibers with Enhanced Osteogenic Capability for Bone Tissue Regeneration, Mater. Des. 212 (2021) 110240. https://doi.org/10.1016/j.matdes.2021.110240

[38] C. Huang, Q. Ye, J. Dong, L. Li, M. Wang, Y. Zhang, Y. Zhang, X. Wang, P. Wang, Q. Jiang, Biofabrication of natural Au/bacterial cellulose hydrogel for bone

tissue regeneration via in-situ fermentation, Smart Mater. Med. 4 (2023) 1–14. https://doi.org/10.1016/j.smaim.2022.06.001

[39] B. Murugesan, N. Pandiyan, M. Arumugam, J. Sonamuthu, S. Samayanan, C. Yurong, Y. Juming, S. Mahalingam, Fabrication of palladium nanoparticles anchored polypyrrole functionalized reduced graphene oxide nanocomposite for antibiofilm associated orthopedic tissue engineering, Appl. Surf. Sci. 510 (2020) 145403. https://doi.org/10.1016/j.apsusc.2020.145403

[40] G. Calabrese, S. Petralia, C. Fabbi, S. Forte, D. Franco, S. Guglielmino, E. Esposito, S. Cuzzocrea, F. Traina, S. Conoci, Au, Pd and maghemite nanofunctionalized hydroxyapatite scaffolds for bone regeneration, Regen. Biomater. 7 (2020) 461–469. https://doi.org/10.1093/rb/rbaa033

[41] F. Heidari, F.S. Tabatabaei, M. Razavi, R. Bazargan-Lari, M. Tavangar, G.E. Romanos, D. Vashaee, L. Tayebi, 3D construct of hydroxyapatite/zinc oxide/palladium nanocomposite scaffold for bone tissue engineering, J. Mater. Sci. Mater. Med. 31 (2020) 1–14. https://doi.org/10.1007/s10856-020-06409-2

[42] X. Zhang, G. Cheng, X. Xing, J. Liu, Y. Cheng, T. Ye, Q. Wang, X. Xiao, Z. Li, H. Deng, Near-Infrared Light-Triggered Porous AuPd Alloy Nanoparticles to Produce Mild Localized Heat to Accelerate Bone Regeneration, J. Phys. Chem. Lett. 10 (2019) 4185–4191. https://doi.org/10.1021/acs.jpclett.9b01735

[43] M. Rai, A.P. Ingle, S. Birla, A. Yadav, C.A. Dos Santos, Strategic role of selected noble metal nanoparticles in medicine, Crit. Rev. Microbiol. 42 (2016) 696–719. https://doi.org/10.3109/1040841X.2015.1018131

[44] W.K. Kim, J.C. Kim, H.J. Park, O.J. Sul, M.H. Lee, J.S. Kim, H.S. Choi, Platinum nanoparticles reduce ovariectomy-induced bone loss by decreasing osteoclastogenesis, Exp. Mol. Med. 44 (2012) 432–439. https://doi.org/10.3858/emm.2012.44.7.048

[45] K. Eid, A. Eldesouky, A. Fahmy, A. Shahat, R. AbdElaal, Calcium Phosphate Scaffold Loaded with Platinum Nanoparticles for Bone Allograft, Am. J. Biomed. Sci. 5 (2013) 242–249. https://doi.org/10.5099/aj130400242

[46] J. Radwan-Pragłowska, Ł. Janus, M. Piatkowski, D. Bogdał, D. Matysek, 3D hierarchical, nanostructured chitosan/PLA/HA scaffolds doped with TiO2/Au/Pt NPs with tunable properties for guided bone tissue engineering, Polymers (Basel). 12 (2020) 792. https://doi.org/10.3390/POLYM12040792

[47] N. Fattahi, A. Ramazani, V. Kinzhybalo, Imidazole-Functionalized Fe3O4/Chloro-Silane Core-Shell Nanoparticles: an Efficient Heterogeneous Organocatalyst for Esterification Reaction, Silicon. 11 (2019) 1745–1754. https://doi.org/10.1007/s12633-017-9757-0

[48] N. Fattahi, A. Ramazani, H. Ahankar, P.A. Asiabi, V. Kinzhybalo, Tetramethylguanidine-Functionalized Fe3O4/ Chloro-Silane Core-Shell Nanoparticles:

an Efficient Heterogeneous and Reusable Organocatalyst for Aldol Reaction, Silicon. 11 (2019) 1441–1450. https://doi.org/10.1007/s12633-018-9954-5

[49] D. Zahn, J. Landers, J. Buchwald, M. Diegel, S. Salamon, R. Müller, M. Köhler, G. Ecke, H. Wende, S. Dutz, Ferrimagnetic Large Single Domain Iron Oxide Nanoparticles for Hyperthermia Applications, Nanomaterials. 12 (2022) 343. https://doi.org/10.3390/nano12030343

[50] J. Peng, J. Zhao, Y. Long, Y. Xie, J. Nie, L. Chen, Magnetic Materials in Promoting Bone Regeneration, Front. Mater. 6 (2019) 268. https://doi.org/10.3389/fmats.2019.00268

[51] Y. Li, D. Ye, M. Li, M. Ma, N. Gu, Adaptive Materials Based on Iron Oxide Nanoparticles for Bone Regeneration, ChemPhysChem. 19 (2018) 1965–1979. https://doi.org/10.1002/cphc.201701294

[52] A. Scharf, S. Holmes, M. Thoresen, J. Mumaw, A. Stumpf, J. Peroni, Superparamagnetic iron oxide nanoparticles as a means to track mesenchymal stem cells in a large animal model of tendon injury, Contrast Media Mol. Imaging. 10 (2015) 388–397. https://doi.org/10.1002/cmmi.1642

[53] J. Huang, D. Wang, J. Chen, W. Liu, L. Duan, W. You, W. Zhu, J. Xiong, D. Wang, Osteogenic differentiation of bone marrow mesenchymal stem cells by magnetic nanoparticle composite scaffolds under a pulsed electromagnetic field, Saudi Pharm. J. 25 (2017) 575–579. https://doi.org/10.1016/j.jsps.2017.04.026

[54] P. Jiang, Y. Zhang, C. Zhu, W. Zhang, Z. Mao, C. Gao, Fe3O4/BSA particles induce osteogenic differentiation of mesenchymal stem cells under static magnetic field, Acta Biomater. 46 (2016) 141–150. https://doi.org/10.1016/j.actbio.2016.09.020

[55] H.M. Yun, S.J. Ahn, K.R. Park, M.J. Kim, J.J. Kim, G.Z. Jin, H.W. Kim, E.C. Kim, Magnetic nanocomposite scaffolds combined with static magnetic field in the stimulation of osteoblastic differentiation and bone formation, Biomaterials. 85 (2016) 88–98. https://doi.org/10.1016/j.biomaterials.2016.01.035

[56] A. Bari, N. Bloise, S. Fiorilli, G. Novajra, M. Vallet-Regí, G. Bruni, A. Torres-Pardo, J.M. González-Calbet, L. Visai, C. Vitale-Brovarone, Copper-containing mesoporous bioactive glass nanoparticles as multifunctional agent for bone regeneration, Acta Biomater. 55 (2017) 493–504. https://doi.org/10.1016/j.actbio.2017.04.012

[57] S. D'Mello, S. Elangovan, L. Hong, R.D. Ross, D.R. Sumner, A.K. Salem, Incorporation of copper into chitosan scaffolds promotes bone regeneration in rat calvarial defects, J. Biomed. Mater. Res. - Part B Appl. Biomater. 103 (2015) 1044–1049. https://doi.org/10.1002/jbm.b.33290

[58] Z. Lin, Y. Cao, J. Zou, F. Zhu, Y. Gao, X. Zheng, H. Wang, T. Zhang, T. Wu, Improved osteogenesis and angiogenesis of a novel copper ions doped calcium

Materials Research Forum LLC
https://doi.org/10.21741/9781644902370-7

phosphate cement, Mater. Sci. Eng. C. 114 (2020) 111032. https://doi.org/10.1016/j.msec.2020.111032

[59] Y. Li, Y. Yang, Y. Qing, R. Li, X. Tang, D. Guo, Y. Qin, Enhancing zno-np antibacterial and osteogenesis properties in orthopedic applications: A review, Int. J. Nanomedicine. 15 (2020) 6247–6262. https://doi.org/10.2147/IJN.S262876

[60] S. Ghosh, T.J. Webster, Metallic nanoscaffolds as osteogenic promoters: Advances, challenges and scope, Metals (Basel). 11 (2021) 1356. https://doi.org/10.3390/met11091356

[61] A. Khader, T.L. Arinzeh, Biodegradable zinc oxide composite scaffolds promote osteochondral differentiation of mesenchymal stem cells, Biotechnol. Bioeng. 117 (2020) 194–209. https://doi.org/10.1002/bit.27173

[62] U. Gröber, J. Schmidt, K. Kisters, Magnesium in prevention and therapy, Nutrients. 7 (2015) 8199–8226. https://doi.org/10.3390/nu7095388

[63] K. Zhang, S. Lin, Q. Feng, C. Dong, Y. Yang, G. Li, L. Bian, Nanocomposite hydrogels stabilized by self-assembled multivalent bisphosphonate-magnesium nanoparticles mediate sustained release of magnesium ion and promote in-situ bone regeneration, Acta Biomater. 64 (2017) 389–400. https://doi.org/10.1016/j.actbio.2017.09.039

[64] N. Safari, N. Golafshan, M. Kharaziha, M. Reza Toroghinejad, L. Utomo, J. Malda, M. Castilho, Stable and Antibacterial Magnesium-Graphene Nanocomposite-Based Implants for Bone Repair, ACS Biomater. Sci. Eng. 6 (2020) 6253–6262. https://doi.org/10.1021/acsbiomaterials.0c00613

[65] K.S. Park, B.J. Kim, E. Lih, W. Park, S.H. Lee, Y.K. Joung, D.K. Han, Versatile effects of magnesium hydroxide nanoparticles in PLGA scaffold–mediated chondrogenesis, Acta Biomater. 73 (2018) 204–216. https://doi.org/10.1016/j.actbio.2018.04.022

[66] M. Petretta, A. Gambardella, M. Boi, M. Berni, C. Cavallo, G. Marchiori, M.C. Maltarello, D. Bellucci, M. Fini, N. Baldini, B. Grigolo, V. Cannillo, Composite scaffolds for bone tissue regeneration based on pcl and mg-containing bioactive glasses, Biology (Basel). 10 (2021) 398. https://doi.org/10.3390/biology10050398

[67] N. Raura, A. Garg, A. Arora, M. Roma, Nanoparticle technology and its implications in endodontics: a review, Biomater. Res. 24 (2020) 1–8. https://doi.org/10.1186/s40824-020-00198-z

[68] C.L. Wetteland, N.Y.T. Nguyen, H. Liu, Concentration-dependent behaviors of bone marrow derived mesenchymal stem cells and infectious bacteria toward magnesium oxide nanoparticles, Acta Biomater. 35 (2016) 341–356. https://doi.org/10.1016/j.actbio.2016.02.032

[69] S. Çeşmeli, C. Biray Avci, Application of titanium dioxide (TiO2) nanoparticles in cancer therapies, J. Drug Target. 27 (2019) 762–766. https://doi.org/10.1080/1061186X.2018.1527338

[70] N. Johari, H.R. Madaah Hosseini, A. Samadikuchaksaraei, Optimized composition of nanocomposite scaffolds formed from silk fibroin and nano-TiO2 for bone tissue engineering, Mater. Sci. Eng. C. 79 (2017) 783–792. https://doi.org/10.1016/j.msec.2017.05.105

[71] G. Nechifor, E.E. Totu, A.C. Nechifor, I. Isildak, O. Oprea, C.M. Cristache, Non-resorbable nanocomposite membranes for guided bone regeneration based on polysulfone-quartz fiber grafted with nano-TiO2, Nanomaterials. 9 (2019) 985. https://doi.org/10.3390/nano9070985

[72] E. Bressan, L. Ferroni, C. Gardin, G. Bellin, L. Sbricoli, S. Sivolella, G. Brunello, D. Schwartz-Arad, E. Mijiritsky, M. Penarrocha, D. Penarrocha, C. Taccioli, M. Tatullo, A. Piattelli, B. Zavan, Metal nanoparticles released from dental implant surfaces: Potential contribution to chronic inflammation and peri-implant bone loss, Materials (Basel). 12 (2019) 2036. https://doi.org/10.3390/ma12122036

[73] G.S. Baird, Ionized calcium, Clin. Chim. Acta. 412 (2011) 696–701. https://doi.org/10.1016/j.cca.2011.01.004

[74] J. Sawai, H. Shiga, Kinetic analysis of the antifungal activity of heated scallop-shell powder against Trichophyton and its possible application to the treatment of dermatophytosis, Biocontrol Sci. 11 (2006) 125–128. https://doi.org/10.4265/bio.11.125

[75] C. Silva, F. Bobillier, D. Canales, F.A. Sepúlveda, A. Cament, N. Amigo, L.M. Rivas, M.T. Ulloa, P. Reyes, J.A. Ortiz, T. Gómez, C. Loyo, P.A. Zapata, Mechanical and antimicrobial polyethylene composites with CaO nanoparticles, Polymers (Basel). 12 (2020) 2132. https://doi.org/10.3390/POLYM12092132

[76] E.A. Münchow, D. Pankajakshan, M.T.P. Albuquerque, K. Kamocki, E. Piva, R.L. Gregory, M.C. Bottino, Synthesis and characterization of CaO-loaded electrospun matrices for bone tissue engineering, Clin. Oral Investig. 20 (2016) 1921–1933. https://doi.org/10.1007/s00784-015-1671-5

[77] R. Eivazzadeh-Keihan, E. Bahojb Noruzi, K. Khanmohammadi Chenab, A. Jafari, F. Radinekiyan, S.M. Hashemi, F. Ahmadpour, A. Behboudi, J. Mosafer, A. Mokhtarzadeh, A. Maleki, M.R. Hamblin, Metal-based nanoparticles for bone tissue engineering, J. Tissue Eng. Regen. Med. 14 (2020) 1687–1714. https://doi.org/10.1002/term.3131

[78] B. Yu, S. Fu, Z. Kang, M. Zhu, H. Ding, T. Luo, Y. Zhu, Y. Zhang, Enhanced bone regeneration of 3D printed β-Ca2SiO4 scaffolds by aluminum ions solid solution, Ceram. Int. 46 (2020) 7783–7791. https://doi.org/10.1016/j.ceramint.2019.11.282

[79] E. Toloue, S. Karbasi, H. Salehi, M. Rafienia, Evaluation of mechanical properties and cell viability of poly (3-hydroxybutyrate)-chitosan/Al2O3nanocomposite scaffold for cartilage tissue engineering, J. Med. Signals Sens. 9 (2019) 111–116. https://doi.org/10.4103/jmss.JMSS_56_18

[80] H. Li, P. Xia, S. Pan, Z. Qi, C. Fu, Z. Yu, W. Kong, Y. Chang, K. Wang, D. Wu, X. Yang, The advances of ceria nanoparticles for biomedical applications in orthopaedics, Int. J. Nanomedicine. 15 (2020) 7199–7214. https://doi.org/10.2147/IJN.S270229

[81] F. Wei, C.J. Neal, T.S. Sakthivel, T. Kean, S. Seal, M.J. Coathup, Multi-functional cerium oxide nanoparticles regulate inflammation and enhance osteogenesis, Mater. Sci. Eng. C. 124 (2021) 112041. https://doi.org/10.1016/j.msec.2021.112041

[82] X. Li, M. Qi, X. Sun, M.D. Weir, F.R. Tay, T.W. Oates, B. Dong, Y. Zhou, L. Wang, H.H.K. Xu, Surface treatments on titanium implants via nanostructured ceria for antibacterial and anti-inflammatory capabilities, Acta Biomater. 94 (2019) 627–643. https://doi.org/10.1016/j.actbio.2019.06.023

[83] S.D. Purohit, H. Singh, R. Bhaskar, I. Yadav, C.F. Chou, M.K. Gupta, N.C. Mishra, Gelatin—alginate—cerium oxide nanocomposite scaffold for bone regeneration, Mater. Sci. Eng. C. 116 (2020) 111111. https://doi.org/10.1016/j.msec.2020.111111

[84] A. V. Lukin, G.I. Lukina, A. V. Volkov, A.E. Baranchikov, V.K. Ivanov, A.A. Prokopov, Morphometry Results of Formed Osteodefects When Using Nanocrystalline CeO2 in the Early Stages of Regeneration, Int. J. Dent. 2019 (2019). https://doi.org/10.1155/2019/9416381

[85] G.B. Tomar, J.R. Dave, S.T. Mhaske, S. Mamidwar, P.K. Makar, Applications of Nanomaterials in Bone Tissue Engineering, Nanotechnol. Life Sci. 10 (2020) 209–250. https://doi.org/10.1007/978-3-030-41464-1_10

[86] N.C. da R. Galucio, D. de A. Moysés, J.R.S. Pina, P.S.B. Marinho, P.C. Gomes Júnior, J.N. Cruz, V.V. Vale, A.S. Khayat, A.M. do R. Marinho, Antiproliferative, genotoxic activities and quantification of extracts and cucurbitacin B obtained from Luffa operculata (L.) Cogn, Arab. J. Chem. 15 (2022) 103589. https://doi.org/10.1016/j.arabjc.2021.103589

[87] L. Grausova, L. Bacakova, A. Kromka, S. Potocky, M. Vanecek, M. Nesladek, V. Lisa, Nanodiamond as promising material for bone tissue engineering, J. Nanosci. Nanotechnol. 9 (2009) 3524–3534. https://doi.org/10.1166/jnn.2009.NS26

[88] L. Moore, M. Gatica, H. Kim, E. Osawa, D. Ho, Multi-protein delivery by nanodiamonds promotes bone formation, J. Dent. Res. 92 (2013) 976–981. https://doi.org/10.1177/0022034513504952

[89] S. Choi, S.H. Noh, C.O. Lim, H.J. Kim, H.S. Jo, J.S. Min, K. Park, S.E. Kim, Icariin-functionalized nanodiamonds to enhance osteogenic capacity in vitro, Nanomaterials. 10 (2020) 1–14. https://doi.org/10.3390/nano10102071

[90] S. Prylutska, R. Bilyy, T. Shkandina, D. Rotko, A. Bychko, V. Cherepanov, R. Stoika, V. Rybalchenko, Y. Prylutskyy, N. Tsierkezos, U. Ritter, Comparative study of membranotropic action of single- and multi-walled carbon nanotubes, J. Biosci. Bioeng. 115 (2013) 674–679. https://doi.org/10.1016/j.jbiosc.2012.12.016

[91] B. Zhao, H. Hu, S.K. Mandal, R.C. Haddon, A bone mimic based on the self-assembly of hydroxyapatite on chemically functionalized single-walled carbon nanotubes, Chem. Mater. 17 (2005) 3235–3241. https://doi.org/10.1021/cm0500399

[92] D. Meng, J. Ioannou, A.R. Boccaccini, Bioglass®-based scaffolds with carbon nanotube coating for bone tissue engineering, J. Mater. Sci. Mater. Med. 20 (2009) 2139–2144. https://doi.org/10.1007/s10856-009-3770-9

[93] J. Venkatesan, S.K. Kim, Stimulation of minerals by carbon nanotube grafted glucosamine in mouse mesenchymal stem cells for bone tissue engineering, J. Biomed. Nanotechnol. 8 (2012) 676–685. https://doi.org/10.1166/jbn.2012.1410

[94] A. Fonseca-García, J.D. Mota-Morales, I.A. Quintero-Ortega, Z.Y. García-Carvajal, V. Martínez-Lõpez, E. Ruvalcaba, C. Landa-Solís, L. Solis, C. Ibarra, M.C. Gutiérrez, M. Terrones, I.C. Sanchez, F. Del Monte, M.C. Velasquillo, G. Luna-Bárcenas, Effect of doping in carbon nanotubes on the viability of biomimetic chitosan-carbon nanotubes-hydroxyapatite scaffolds, J. Biomed. Mater. Res. - Part A. 102 (2014) 3341–3351. https://doi.org/10.1002/jbm.a.34893

[95] N. Jamilpour, A. Fereidoon, G. Rouhi, The effects of replacing collagen fibers with carbon nanotubes on the rate of bone remodeling process, J. Biomed. Nanotechnol. 7 (2011) 542–548. https://doi.org/10.1166/jbn.2011.1319

[96] F. Mei, J. Zhong, X. Yang, X. Ouyang, S. Zhang, X. Hu, Q. Ma, J. Lu, S. Ryu, X. Deng, Improved biological characteristics of poly(L-lactic acid) electrospun membrane by incorporation of multiwalled carbon nanotubes/hydroxyapatite nanoparticles, Biomacromolecules. 8 (2007) 3729–3735. https://doi.org/10.1021/bm7006295

[97] H. Zhang, Electrospun poly (lactic-co-glycolic acid)/ multiwalled carbon nanotubes composite scaffolds for guided bone tissue regeneration, J. Bioact. Compat. Polym. 26 (2011) 347–362. https://doi.org/10.1177/0883911511413450

[98] S.W. Crowder, D. Prasai, R. Rath, D.A. Balikov, H. Bae, K.I. Bolotin, H.J. Sung, Three-dimensional graphene foams promote osteogenic differentiation of human mesenchymal stem cells, Nanoscale. 5 (2013) 4171–4176. https://doi.org/10.1039/c3nr00803g

[99] Y. Chen, Y. Qi, Z. Tai, X. Yan, F. Zhu, Q. Xue, Preparation, mechanical properties and biocompatibility of graphene oxide/ultrahigh molecular weight

polyethylene composites, Eur. Polym. J. 48 (2012) 1026–1033. https://doi.org/10.1016/j.eurpolymj.2012.03.011

[100] S. Das, A.S. Wajid, S.K. Bhattacharia, M.D. Wilting, I. V. Rivero, M.J. Green, Electrospinning of polymer nanofibers loaded with noncovalently functionalized graphene, J. Appl. Polym. Sci. 128 (2013) 4040–4046. https://doi.org/10.1002/app.38694

[101] D.A. Heredia, A.M. Durantini, J.E. Durantini, E.N. Durantini, Fullerene C60 derivatives as antimicrobial photodynamic agents, J. Photochem. Photobiol. C Photochem. Rev. 51 (2022) 100471. https://doi.org/10.1016/j.jphotochemrev.2021.100471

[102] L. Bacakova, L. Grausova, J. Vacik, A. Fraczek, S. Blazewicz, A. Kromka, M. Vanecek, V. Svorcik, Improved adhesion and growth of human osteoblast-like MG 63 cells on biomaterials modified with carbon nanoparticles, Diam. Relat. Mater. 16 (2007) 2133–2140. https://doi.org/10.1016/j.diamond.2007.07.015

[103] S.Y. Lim, W. Shen, Z. Gao, Carbon quantum dots and their applications, Chem. Soc. Rev. 44 (2015) 362–381. https://doi.org/10.1039/c4cs00269e

[104] C. Ren, X. Hao, L. Wang, Y. Hu, L. Meng, S. Zheng, F. Ren, W. Bu, H. Wang, D. Li, K. Zhang, H. Sun, Metformin Carbon Dots for Promoting Periodontal Bone Regeneration via Activation of ERK/AMPK Pathway, Adv. Healthc. Mater. 10 (2021) 2100196. https://doi.org/10.1002/adhm.202100196

[105] N. Jin, N. Jin, Z. Wang, L. Liu, L. Meng, D. Li, X. Li, D. Zhou, J. Liu, W. Bu, H. Sun, B. Yang, Osteopromotive carbon dots promote bone regeneration through the PERK-eIF2α-ATF4 pathway, Biomater. Sci. 8 (2020) 2840–2852. https://doi.org/10.1039/d0bm00424c

[106] Q. Yang, H. Yin, T. Xu, D. Zhu, J. Yin, Y. Chen, X. Yu, J. Gao, C. Zhang, Y. Chen, Y. Gao, Engineering 2D Mesoporous Silica@MXene-Integrated 3D-Printing Scaffolds for Combinatory Osteosarcoma Therapy and NO-Augmented Bone Regeneration, Small. 16 (2020) 1906814. https://doi.org/10.1002/smll.201906814

[107] C. Xu, L. Xiao, Y. Cao, Y. He, C. Lei, Y. Xiao, W. Sun, S. Ahadian, X. Zhou, A. Khademhosseini, Q. Ye, Mesoporous silica rods with cone shaped pores modulate inflammation and deliver BMP-2 for bone regeneration, Nano Res. 13 (2020) 2323–2331. https://doi.org/10.1007/s12274-020-2783-z

[108] N. Shadjou, M. Hasanzadeh, Silica-based mesoporous nanobiomaterials as promoter of bone regeneration process, J. Biomed. Mater. Res. - Part A. 103 (2015) 3703–3716. https://doi.org/10.1002/jbm.a.35504

[109] Y. Zhao, Z. Cui, B. Liu, J. Xiang, D. Qiu, Y. Tian, X. Qu, Z. Yang, An Injectable Strong Hydrogel for Bone Reconstruction, Adv. Healthc. Mater. 8 (2019) 1900709. https://doi.org/10.1002/adhm.201900709

[110] B. Gaihre, B. Lecka-Czernik, A.C. Jayasuriya, Injectable nanosilica–chitosan microparticles for bone regeneration applications, J. Biomater. Appl. 32 (2018) 813–825. https://doi.org/10.1177/0885328217741523

[111] Z. Shi, Y. Xu, R. Mulatibieke, Q. Zhong, X. Pan, Y. Chen, Q. Lian, X. Luo, Z. Shi, Q. Zhu, Nano-silicate-reinforced and SDF-1α-loaded gelatin-methacryloyl hydrogel for bone tissue engineering, Int. J. Nanomedicine. 15 (2020) 9337

[112] H. Maleki, M.A. Shahbazi, S. Montes, S.H. Hosseini, M.R. Eskandari, S. Zaunschirm, T. Verwanger, S. Mathur, B. Milow, B. Krammer, N. Hüsing, Mechanically Strong Silica-Silk Fibroin Bioaerogel: A Hybrid Scaffold with Ordered Honeycomb Micromorphology and Multiscale Porosity for Bone Regeneration, ACS Appl. Mater. Interfaces. 11 (2019) 17256–17269. https://doi.org/10.1021/acsami.9b04283

[113] F.S. Alves, J. de A. Rodrigues Do Rego, M.L. Da Costa, L.F. Lobato Da Silva, R.A. Da Costa, J.N. Cruz, D.D.S.B. Brasil, Spectroscopic methods and in silico analyses using density functional theory to characterize and identify piperine alkaloid crystals isolated from pepper (Piper Nigrum L.), J. Biomol. Struct. Dyn. 38 (2020) 2792–2799. https://doi.org/10.1080/07391102.2019.1639547

Nanobiomaterials | Materials Research Forum LLC
Materials Research Foundations 145 (2023) 207-235 | https://doi.org/10.21741/9781644902370-8

Chapter 8

Wound Treatment Using Nanomaterials

J.M. Rajwade[1,*], K. Kawle[2], S. Kulkarni[1], M. Kowshik[3],*

[1]Nanobioscience group, Agharkar Research Institute, G. G. Agharkar Road, Pune 411 004

[2]Department of Biotechnology, Rajarshi Shahu Mahavidyalaya (Autonomous), Latur 413 512

[3]Department of Biological Sciences, BITS Pilani K K Birla Goa Campus, Off N H 17B, Zuarinagar Goa 403726

J. M. Rajwade, jrajwade@aripune.org and M. Kowshik, meenal@goa.bits-pilani.ac.in

Abstract

Skin wounds are categorized as 'acute' or 'chronic' based on the healing process. Wound care is of utmost importance as the break in the skin barrier exposes the internal milieu to various commensal and pathogenic microbes. The usage of nanomaterials is a recent approach to facilitate wound healing by processes that operate singly or in tandem. Metal based nanomaterials show antibacterial, antifungal, and anti-inflammatory activities; and organic-inorganic, organic-organic nanocomposites as 'smart', 'advanced' materials would play a crucial role in wound care. Use of nanomaterials in wound care would revolutionarize treatment, thus lowering the economic burden.

Keywords

Chronic Wounds, Nanoparticles, Wound Healing, Wound Dressings

Contents

 Materials Research Forum LLC
https://doi.org/10.21741/9781644902370-8

1. Introduction

A wound is best defined as a breakdown within the protecting performance of the skin, a loss in the continuity of the epithelial tissue, with or without damage to the underlying host tissue [1]. A wound represents any physical or thermal impact leading to the disintegration of the protective functions of the skin, which may be associated with the loss of connective tissue [2,3]. Wounds may be associated with causes such as diseases, surgical interventions, battlefield injury, accidents etc. [4,5]

2. Types of wounds

Wounds are classified as chronic or acute, depending on the nature of the healing process [6]. Chronic wounds are characterized by endogenous mechanisms primarily due to predisposing conditions, ultimately comprising the loss of both dermal and epidermal tissue integrity [7]. These are often associated with disabilities, affecting the quality of life. Acute wounds usually get healed with time, requiring few interventions, but chronic wounds (resulting from burns, diabetic ulcers, pressure ulcers, and leg ulcers) do not progress through the normal stages of healing [2,8,9].

2.1 Diabetic foot ulcers (DFU)

DFU is characterized by lower extremity vascular disease and neuropathy-related foot discomfort, foot ulcer, foot deformity, and foot gangrene [10]. DFU has a very complex pathogenesis and a wide range of clinical symptoms [11,12]. Lack of local delivery of oxygen and nutrients essential for wound healing via the blood vessels manifests as a DFU. The deepest tissues in the lower limbs are infected, ulcerated, and destroyed by DFUs, with neurologic abnormalities and varying degrees of peripheral vascular dysfunction being the other characteristics [13,14].

2.2 Pressure injury

A localized pressure injury to the skin or subcutaneous soft tissue can appear as intact skin or an open sore and may be accompanied by pain. Usually, it arises at the site of the compression or prominence of the bone caused by medical equipment [15]. Stress and tissue tolerance are the key contributors to injury occurrence. Compressive strength and duration are stress factors; tissue tolerance is typically impacted by the patient's state and the microenvironment [16,17]. Wounds caused by pressure, shear, and friction strength constitute a significant problem for patients and their caregivers [18], which can lead to admission to the clinic .

2.3 Venous leg ulcers (VLU)

Chronic venous leg ulcer (CVLU) is a chronic wound of the lower leg caused by abnormal venous blood flow and shows no tendency to heal after six weeks of appropriate treatment or does not entirely heal after 12 months [19–21]; which are often incurable [22]. The primary contributing factors to VLU creation are venous regurgitation disease, impaired vascular function, weakened venous walls, and poor systolic muscle pump performance [23]. The inflammatory-proteinase activity is persistent, resulting in lower blood supply (chronic venous insufficiency), leading to the development of VLU [24].

2.4 Burns

Due to the lack of an epidermal barrier, bacteria can colonize and rapidly multiply after burn injuries [25]. Despite advances in burn care management, infection remains the primary cause of mortality after severe burns. Surface, partial, and full thickness burns are different types of acute burns [26]. Superficial PTBs involve damage to the papillary dermis and are characterized by intact blisters, moderate edema, a moist surface under the blisters, and a bright pink or red color. Deep PTBs involve damage to the papillary and reticular dermis, characterized as broken blisters, substantial edema, a wet surface, waxy white color [27]. Partial-thickness burns are characterised by blisters, swelling, and redness on the skin. Thus, burns cause skin damage, affecting the epidermis and structures beneath the epidermis, such as blood vessels, hair follicles, and nerves [28]. Under challenging situations, full-thickness burns can even harm the muscles and bones since they impact the entire skin structure [29]. The inflammatory factors in burn wounds are mainly TNF-α, IL-1β, and IL-6 [30].

Materials Research Foundations 145 (2023) 207-235 https://doi.org/10.21741/9781644902370-8

3. Process of wound healing

The wound healing process is a physiological, dynamic, and complex mechanism that aims to correct defects and restore the skin's surface. There is an immediate host response upon an injury [31], and conventionally, the wound healing process is categorized into four phases: hemostasis, inflammation, proliferation, and remodeling [32]. The process of wound healing relies on intrinsic factors such as the nature of the wound, infection status, pressure, and neurological involvement, and systemic factors such as metabolic diseases, nutrition, habits, age, heredity, etc.

3.1 Hemostasis

Hemostasis is the primary injury response, including abatement of hemorrhage and vascular injury. Hemostasis involves three stages: vasoconstriction followed by primary hemostasis; and secondary hemostasis (the formation of cross-linked fibrin due to the action of thrombin). Usually, all the stages are attained through fast, synchronized, and mechanistically interrelated routes [33,34]. The fibrin clot formation prevents blood loss and helps avoid microbial contamination [35].

3.2 Inflammation

The inflammatory phase is concurrent with hemostasis. It involves the recruitment of neutrophils, followed by macrophages and lymphocytes that exert a specific response against microbes. Neutrophils engulf bacteria and decontaminate the wound through proteases and antimicrobial peptides secretion and by producing reactive oxygen intermediates. Macrophages help with the removal of apoptotic neutrophils and other cells and secrete cytokines and multiple growth factors, while (B–and T lymphocytes produce antibodies and secrete cytokines involved in cytolytic activity, respectively [35,36]. Inflammation is associated with increased levels of exudates which cause skin swelling, erythema, pain, and redness [37,38].

3.3 Proliferation

Proliferation starts with granulation of the wound with the connective tissues and ends with epithelialization [39]. Fibroblasts move to the wound site during this phase to produce new blood vessels and ECM components such as fibronectin, hyaluronic acid, collagen, and proteoglycan [40]. In the proliferative phase, the traumatic area of the tissue is diminished. This phase starts during the first 48 h and lasts for up to 14 days [41].

3.4 Maturation

After ECM synthesis and proliferation, the wound repair process goes into the third stage, maturation, also known as remodeling. This usually commences three weeks after injury and can last for years [42]. During maturation, all processes that were activated after injury cease [43]. Put simply, this phase is the balance between synthesis, deposition, and degradation. Several myofibroblasts, endothelial cells, and macrophages help in wound remodeling. Fibroblasts help in cross-linking collagen resulting in collagen realignment

Materials Research Foundations 145 (2023) 207-235
https://doi.org/10.21741/9781644902370-8

into organized grids that upsurge the tensile strength of the tissue, accomplishing about 80% of unwounded/original skin [44].

4. Wound treatment

In a 'real-'life' scenario, minor cuts to life-threatening injuries require wound dressings for treatment. Hundreds of millions of people suffer from wounds each year, and wound care is an economic burden on the healthcare systems worldwide [45].

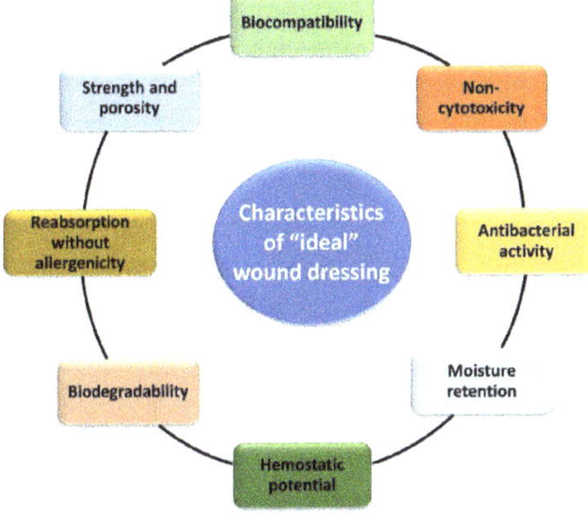

Figure 1: Desirable characteristics of an "'ideal' wound dressing

Wound infections are associated with mortality and morbidity, and hence newer, innovative dressings can help clinicians in effective wound management. The following characteristics are expected for an 'ideal' wound dressing (1) biocompatibility with host tissue (2) non-cytotoxicity (3) antibacterial property to avoid microbial infections (4) moisture retention to promote cell proliferation (5) hemostatic potential (6) biodegradability rate matching with the wound healing (7) good reabsorption without allergic/inflammatory response (8) appropriate physical and mechanical strength (9) appropriate porosity to allow gas exchange [2,46–48] (Fig. 1). Considering rapid advances in technology, attributes such as the presence of sensors to diagnostic functions (determination of analytes, microbes etc.) and real-time monitoring (e.g. imaging) would greatly help the clinicians [47].

Gauze and hydrogel dressings that incorporate antibiotics for curbing infections, and agents that promote angiogenesis, and hemostasis, or are absorbed during healing are available

Materials Research Foundations 145 (2023) 207-235 https://doi.org/10.21741/9781644902370-8

commercially. A review by Simones et al. discusses the recent advances in antibacterial wound dressings for addressing skin and soft tissue infections [49]. With the introduction of 'Nanotechnology,' a plethora of attributes have been imparted to the dressings making them close to "'ideal' wound care materials. Nanomaterials, with their high surface area to volume ratio, could help in reducing the concentration of the drug. Because of their small size, they tend to be compatible with the biological microenvironment and permeate to the appropriate sites, offer a large surface area for conjugation and loading other biomolecules; they could be integrated with other traditional materials in the form of composites, thus improving the characteristics of the latter. The following section presents information on the role of various nanomatrices, such as inorganic, metal, metal oxide, and polymer-based nanoparticles in wound healing.

4.1 Inorganic nanomaterials

4.1.1 Silver nanoparticles (AgNPs)

Silver nanoparticles show multi-level effects on several cellular targets, making them the nanoparticles of choice for developing wound dressings. Silver nanoparticles have garnered particular interest as antibacterial, antifungal, antiviral, anti-inflammatory, and anti-cancer agents [50–52]. Typically, broad-spectrum antibacterial effects are reported with clinically important bacteria which are often implicated in wound infections[51,53]. In gauze-type dressings, silver nanoparticles have been incorporated, and several studies document the release of silver ions (Ag^{+}) which exert bacterial and biofilm inhibitory effects [54–59]. Several natural and synthetic biocompatible polymers have been reported to be compatible with silver nanoparticles, and efforts at the development and evaluation of such wound dressings are reviewed by Kalantari et al. [51]. Collagen, gelatin, silk, keratin, natural rubber latex, chitosan, starch, cellulose, and hyaluronic acid have all been used synergistically with Ag NPs for both *in vitro* and *in vivo* wound-healing applications [60]. Adhesion-resistant gauze incorporating perfluorocarbons and AgNPs is also described [61]. The combination of collagen/chitosan scaffolds with AgNPs (AgNP–collagen/chitosan hybrid scaffolds) was observed to be effective against bacterial infection in the skin [34]. For applications in wound dressings, silver nanoparticles have been prepared using physical and green (chemical, microbial) synthesis methods. Silver-based composites have also been evaluated as wound-healing materials. *Acinetobacter baumannii* and *Pseudomonas aeruginosa* showed high sensitivity against the green synthesized Ag, ZnO, and Ag/ZnO (obtained using *Prosophis fracta* and coffee) [62].

Sponges containing chitosan, hydroxyapatite, and nanosilver have been shown to inhibit MRSA, promoting further research on the development of potential wound dressings for diabetic foot ulcers saturated with antibiotic-resistant bacteria. The combination of AgNPs and chitin nanofibers as wound dressing exhibited strong antimicrobial activity with the same toxicity as the AgNPs alone [63]. Formulations (S-gel®, Silveron®, MegaNano® etc.) containing silver nanoparticles are commercially available. In 2009, Jain et al. reported the bioactivity of silver nanoparticles to be better than commercially available silver sulfadiazine. Tian et al. tested silver nanoparticles and silver sulfadiazine in a thermal injury model, comparing the rate of healing of deep partial-thickness wounds [64]. The authors reported less hypertrophic scarring and nearly normal hair growth on the wound surface treated with silver nanoparticles and confirmed the accelerated wound healing by using silver nanoparticles, compared to silver sulfadiazine. By inducing the differentiation of myofibroblasts from normal fibroblasts, AgNPs promote wound contractility, thus accelerating the healing process. AgNPs stimulate epidermal re-epithelialization through the proliferation and relocation of keratinocytes [65]. Recently, Damle et al. have reviewed the utility of silver nanoparticles for the treatment of orthopedic wounds [66]. A recent study described the one-pot synthesis of magnetic carboxymethyl cellulose-ε-polylysine hybrids (FCE) anchored with silver nanoparticles (Ag NPs). These hybrids were demonstrated to inactivate microbial pathogens and accelerate the bacteria-infected wound healing with the assistance of hydrogen peroxide, which was confirmed by *in vivo* experiments [67].

4.1.2 Gold nanoparticles (AuNPs)

Similar to silver, gold nanoparticles are studied for their wound-healing properties. In a study, gold nanoparticles were observed to penetrate bacterial cells, causing alterations in their membrane and inhibiting the enzyme ATP synthase. This is followed by depletion of ATP and collapse in energy metabolism leading to bacterial cell death. They were demonstrated to inhibit multidrug-resistant pathogens, such as *Staphylococcus aureus* and *Pseudomonas aeruginosa*. Moreover, AuNPs functioned as antioxidants by preventing the production of reactive oxygen species, promoting repair [65,68]. In another study, AuNPs enhanced wound healing by promoting epithelialization, collagen deposition, fast vascularization, and regeneration of the damaged collagen tissue [69]. When AuNPs are cross-linked with collagen, they can be easily integrated with other biomolecules like polysaccharides, growth factors, peptides, and cell adhesion molecules [70].

A study on a rat surgical wound model has shown that chitosan–AuNPs significantly enhanced the hemostasis, formation of epithelial tissue with a high healing rate, and faster closure of wounds in comparison to the standard chitosan and Tegaderm bandages. In another investigation in rats, AuNPs were conjugated with human cryopreserved fibroblasts and applied on the surface of burn wounds. The wounds exhibited an improved recovery rate, decreased inflammation, and enhanced collagen deposition [34]. The addition of AuNPs into cross-linked collagen scaffolds (CS-AuX) enhanced their stability against enzymatic degradation and increased tensile strength. The AuNP-collagen sponge (CS-AuX) group suppressed the inflammation and supported neovascularization [71]. Martinez et al. fabricated a nanocomposite made of Au NPs functionalized with chitosan and calreticulin, which demonstrated wound-healing activity in diabetic mice [72]. As evidenced by in vitro assays, the nanocomposite promoted clonogenicity of keratinocytes, endothelial cells, and fibroblasts and accelerated fibroblast migration.

Materials Research Foundations 145 (2023) 207-235
https://doi.org/10.21741/9781644902370-8

4.1.3 Copper nanoparticles (CuNPs)

Copper is an essential metal and is required in small quantities in many metabolic processes. In fact, under controlled conditions, copper plays a vital role in healing by enhancing the expression of extracellular matrix molecules such as fibrinogen, collagen formation, and integrins, the primary mediators of cell attachment to the extracellular matrix [73]. Copper plays a vital role in wound healing using angiogenesis, where it helps promote VEGF (vascular endothelial growth factor) and upregulates the integrin expression, aiding the wound healing process [74]. CuNPs stimulate angiogenesis by affecting the expression of the hypoxia-inducible factor (HIF-1a) and regulation of the secretion of VEGF, thereby promoting wound healing [34]. According to the findings of Tao et al., a composite hydrogel made of methacrylate-modified gelatin (Gel-MA) and copper chelated with N,N-bis(acryloyl)-cysteine (BACA) exhibited antibacterial efficacy against *Staphylococcus aureus* and *Escherichia coli* [75]. Additionally, it accelerated the chronic wound healing process in the S. aureus-infected mouse, and NIH-3T3 fibroblast proliferation was observed in *in vitro* studies.

4.1.4 Silica nanoparticles (SiNPs)

Among various metalloids, silica has received good attention in wound healing. SiNPs were evaluated for their antibacterial effects and *in vitro* and *in vivo* studies to demonstrate wound healing properties. Silica nanoparticles affect the proliferation and migration of human skin fibroblast cells (CCD-25SK), which is an essential feature of wound repair [74]. Jiang et al. prepared a spaced-oriented scaffold for Si ion release. The scaffolds were coated with silicon-doped amorphous calcium phosphate NPs coating its surface to promote angiogenesis [76]. The Si ions released from the scaffolds promoted wound healing by enhancing angiogenesis, collagen deposition, and re-epithelialization of the diabetic wound. In another study, the *in situ* encapsulation of bFGF (basic fibroblast growth factor) in mesoporous silica nanoparticles (MSNs) was performed by the water-in-oil microemulsion method [77]. NO-releasing silica (SiO_2) NPs demonstrated a very high kill rate against *P. aeruginosa* and *E. coli* (with ≥5 logs of killing for both Gram-negative species), intermediate efficacy against *C. albicans* (3 logs of biofilm killing) and lower efficacy against Gram-positive *S. aureus* and *S. epidermidis* biofilms [78]. Öri et al. developed a dressing using SiO_2 nanoparticles and PVP (polyvinylpyrrolidone), which supported the wound healing process [79]. According to the authors, the SiO_2 increased re-epithelization and decreased scarring.

4.1.5 Zinc oxide nanoparticles (ZnONPs)

Zinc oxides are widely used in skin creams because of their anti-inflammatory and antiseptic conditions. They are widely used in cosmetic and pharmaceutical manufacturing as sunscreens and as wound healing materials [70]. By causing bacterial cell membrane perforations, zinc oxide nanoparticles (ZnONPs) serve as a trustworthy antibacterial agent. Additionally, when added to hydrogel-based wound dressings, the total time in contact is prolonged, encouraging keratinocyte migration and enhancing re-epithelialization [65].

Materials Research Forum LLC
https://doi.org/10.21741/9781644902370-8

The embedded ZnO NPs in chitosan hydrogel, collagen dressing, or cellulose sheets showed antibacterial and tissue regeneration activity, decreasing the risk of infections during wound healing [34]. Compared to chitosan/PVA nanofibrous mats, chitosan/PVA/ZnO nanofibrous membranes were found to possess a greater antioxidant potential and wound healing activities [80]. ZnONP aqueous solutions produced high ROS levels, which demonstrated antibacterial action. Additionally, it was discovered that ZnONPs could kill Mycobacterium by direct contact and ROS formation [81]. It was demonstrated that zinc oxide nanoflowers played a direct role in wound healing by enhancing angiogenesis and cell chemotaxis in both *in vitro* and *in vivo* wound healing and angiogenesis studies [82].

4.1.6 Graphene oxides

Graphene oxide, graphite oxide, graphite, and reduced graphene oxide nanofilms have demonstrated antibacterial activities. Graphene and graphene oxide strongly induce ROS-independent oxidative stress when interacting with bacteria. Graphene oxide sheets enhanced the healing of fungal and bacterial wound infections [83], probably by physical damage to the bacterial cell membrane, destroying the phospholipid bilayer. Thangavel et al., fabricated reduced graphene oxide (rGO)- embedded isabgol nanocomposite scaffolds for diabetic wound treatment [84]. Mitra et al., developed a 3D scaffold by functionalizing nano-GO (NGO) with type-I collagen using an advanced grafting process for tissue engineering purposes [85]. The bio-nanocomposite, viz., Poly (vinyl alcohol)/chitosan/polyethylene glycol-assembled graphene oxide, was developed and extensively characterized for its physicochemical, mechanical characteristics, and biological activities. With good mechanical strength, antibacterial effect, and biodegradability, it was proposed as a suitable nanomaterial for wound healing applications [86]. Khalid et al. described the BC-MWCNT composite film to possess antibacterial activity and demonstrated its *in vivo* wound healing activity in diabetic animals [87]. The composite was shown to increase the expression of VEGF and decrease the pro-inflammatory cytokines IL-1α and TNF-α.

Limited studies report other inorganic materials in the form of nanoparticles with reference to their wound-healing properties. For example, cerium oxide (CeO_2) and yttrium oxide (Y_2O_3) were widely used for the protection against oxidative stress damage for diabetic foot ulcers (DFU). To speed up the healing of diabetic wounds, cerium oxide nanoparticles were conjugated with microRNA-146a (miR-146a). MicroRNAs have been shown to regulate the transcription of specific genes, which in turn controls the production of pro-inflammatory cytokines [88]. At very low concentrations, the non-metallic iodine nanoparticles significantly inhibited bacterial growth, biofilm formation, and wound healing in an *in vivo* study [89]. Ma et al. described hydrogels with MoS2 nanoclusters incorporated in sodium alginate (SA-MS) as having an excellent ability to heal disease-impaired wounds. The released Mo^{4+} ions played a significant role in cell migration and proliferation [90]. Overall the hydrogel was demonstrated to promote angiogenesis and hair follicle regeneration. Inspired by the results of the application of α-gal (Galα1-3Galβ1-(3)4GlcNAc-R) nanoparticles (AGNs) in diabetic wound healing, Samadi et al. reported the usefulness of the topical application of AGNs for treatment of radiation associated injuries [91]. The study was performed in α-1,3-galactosyltransferase knockout (KO) mice, which closely simulates human physiology. Titanium oxide was found to be a promising agent in accelerating the rate of healing in open excision-type wounds. The formulation of TiO_2 nanoparticles containing *Origanum vulgare* was evaluated using the excision wound model, revealed significant wound healing activity [92].

4.2 Polymer-based nanoparticles

Biopolymers (natural or synthetic/man-made) are used in the synthesis of nanoparticles for apparent reasons of biodegradability and biocompatibility. These polymeric nanoparticles are evaluated for applications in wound treatment mostly as hydrogels or as carriers of bioactive components incorporated in dressings.

Among the various natural polymers, alginate, albumin, or chitosan have been widely used in drug formulations. Few others, including collagen, silk fibroin, elastin, gelatin, hyaluronic acid, and bacterial cellulose, have been investigated for wound healing [70,93,94]. Manufactured polymers can be modified during synthesis and can be adapted according to the specific requirements considering their biological applications and hence have the edge over the biopolymers of natural origin [40]. The most widely used synthetic polymers are polylactide, polyethylene, polypropylene, polystyrene, polylactide-polyglycolide copolymers, polycaprolactones, and polyacrylates, with lactide-glycolide copolymers the most extensively explored one. Polymers are explored in various forms, such as sponges, hydrocolloids, films, membranes, and hydrogels, which provide distinct advantages compared to traditionally used gauze dressings. Hydrogel wound dressings are preferred in some instances as they prevent tissue dehydration and avoid any discomfort during change [95]. Recently, Niculescu & Grumezescu have reviewed the usage of biomaterials in wound management[96].

4.2.1 Gelatin

Gelatin, which is derived from collagen, has been utilized predominantly in the production of biodegradable and biocompatible wound dressing materials. Powel and Boyce, showed that the porosity and inter-fiber distance of the gelatin scaffold played a crucial role in skin repair [97]. The topical application of gelatin-based scaffolds to rat wounds resulted in faster wound closure and enhanced overall wound healing [98]. Ye et al., developed gelatin-chitosan-Ag composite, for which, AgNPs and chitosan were combined, cross-linked with tannic acid, and cryodesiccated [99]. This composite exhibited a high density of pores with diameters of 100–250 μm and was found to show wound healing and antibacterial properties while maintaining low cytotoxicity.

4.2.2 Hydrogels

Hydrogels are soft and easily retain copious amount of water. This characteristic can help in preventing tissue dehydration, and hence hydrogels can be used in the preparation of bandages and dressings for burns, acute/chronic wounds, and diabetic foot ulcers [40]. Additionally, hydrogels also display a highly porous structure that allows the accommodation of living cells as well as gases, nutrients, and waste product diffusion [100]. Hydrogels simplify the stages of the healing process, such as granulation hyperplasia, epidermal repair, and removal of excess dead tissue [101]. A fibrin hydrogel nanomaterial with a covalently linked peptide that interacts with the growth factor (neurotrophin-3) was effectively used for treating spinal cord injury [40]. Chitosan hydrogel was reported for its hemostatic properties [102]. Microbial cellulose is another natural nanoscale polymer that is produced as a hydrogel [103,104] useful for wound treatment. As-produced microbial cellulose hydrogel is evaluated as a skin substitute for treating burn wounds [105].

4.2.3 Hyaluronic acid (HA)

HA is a natural polymer that belongs to a group of heteropolysaccharides known as glycosaminoglycans (GAGs), which are found in the human vitreous humor, joints, rooster comb, umbilical cord, skin, and connective tissue [106]. Collagen/ hyaluronic nanofibers encapsulating angiogenic GFs (VEGF, PDGF) and GF-loaded gelatin NPs (EGF and bFGF) were assayed as skin substitutes. The nanofibrous membrane possessed similar mechanical properties to human skin and allowed a faster wound regeneration than the control. The hyaluronan nanofibers had antibacterial activity against Gram-negative bacteria and higher wound repair efficacy compared to controls, proving their efficacy in the wound and chronic ulcers [107]. Silver nanoparticles containing gel formulations were demonstrated to be compatible with hyaluronic acid for improved wound healing [108]. Kenar et al. prepared a blend of HA with COL and poly(L-lactide-co-ε-caprolactone) (PLC), which was then used to produce a nanofibrous membrane able to support cell proliferation and promote the vascularization process [109].

4.2.4 Chitosan

Chitosan, the deacetylated form of chitin, is made up of the aminosugars, D-glucosamine and N-acetyl-D-glucosamine. The polymer is obtained from crustacean shells and fungal cell walls. The amino groups present on the chitin backbone offer many opportunities for functional modifications. Modifications to modulate mechanical stiffness, physiological stability, and biochemical characteristics are reported. Chitosan is bioactive and has inherent characteristics that facilitate wound healing. For example, *in vitro* testing of chitosan nanoparticles showed significant induction of lymphocyte proliferation and nitric oxide (NO) production. NO is involved in wound healing through angiogenesis, migration of epithelial cells, and proliferation of keratinocytes. *In vivo* testing of excision wounds in Sprague–Dawley rats with chitosan nanoparticles showed enhanced NO production, thus supporting the role of chitosan nanoparticles in wound healing [110]. Liu et al. have extensively reviewed the uses of chitosan as wound dressings and drug delivery systems to address specific stages [111]. A new acellular porcine dermal matrix was created in two phases by Chen et al. [112]. To begin with, the acellular network was cross-linked with an oxidized chitosan oligosaccharide that was naturally produced in order to get improved physicochemical features. The intrinsic characteristics of chitosan can favor hemostasis, regulate the activity of inflammatory cells, prevent bacterial contamination, support tissue growth as well as stimulate fibroblast proliferation, collagen deposition, and angiogenesis [113].

In a study, alginate/chitosan-based hydrogel incorporating different concentrations of hesperidin was used to treat skin wounds. Hespiridin (10%) incorporated hydrogels exhibited potent antibacterial activity, were non-cytotoxic, and promoted cell proliferation [114]. Takei et al. reported in their research that they were able to develop chitosan gluconic acid (CG) through physical cross-linking for wound treatment. Combining chitosan (CS) with free fatty acids, such as oleate and linoleate, polymeric micelles CS:OA and CS:LA, respectively, can be obtained via ionic interactions [115]. Both CS:OA and CS:LA were mucoadhesive and had positive effects on cell viability. Because of the amphiphilic nature of micelles, topical delivery of poorly soluble drugs for infection control and wound healing applications were investigated [110]. Natural polymers integrated with the easily applied in situ gel of cefadroxil-loaded chitosan nanoparticles could promote wound healing [116]. Qu et al. designed biocompatible, antibacterial adhesive hydrogel (quaternized chitosan (QCS) and benzaldehyde-terminated Pluronic®F127 (PF127-CHO)), with stretchable and compressive properties, efficient hemostatic performance and accelerated rate of wound healing to be used for joints skin wounds[117]. Shaik et al. incorporated silver nanoparticles as antibacterial agents along with antioxidants in a chitosan-based bilayer wound dressing [118]. This exhibited excellent characteristics in both, *in vitro* and *in vivo* experiments. The reports on chitosan-based asymmetric membranes designed for application in wound healing were reviewed [113]. In such membranes, the outer layer mimics the epidermis of the native skin, and serves as a barrier against external damage, while the inner layer is porous and aids in cell division, migration, and proliferation. Chitosan, due to its biocompatible nature and antibacterial, hemostatic properties, is a material of choice in asymmetric wound dressings. The versatile nature of chitosan and the plethora of applications are recently reviewed [119].

4.2.5 Cellulose

Cellulose is the most abundant natural polymer, the primary source being plants. Because of its good biocompatibility, environmental friendliness, degradable regeneration, and non-toxicity, cellulose has attracted more and more attention in the field of biomaterial applications. Gauze dressings based on cellulose that have been modified with antibiotics have been available commercially for quite some time. A newer process, viz., electrospinning, was used to prepare nanofibers of hydroxypropyl methylcellulose combined with PEO to form nanofibers encapsulating beta-glucan [120]. Nanofibers prepared by the Nanospider technology showed an improved wound healing capacity in diabetic mice. The cellulose obtained from plant biomass is associated with hemicellulose and lignin, thus requiring several steps to obtain it in a pure form. Specific bacterial genera are known to overproduce cellulose extracellularly in the form of mats/pellicles [30,105,121,122]. Bacterial cellulose (BC) has the same chemistry as plant-derived cellulose, but its nanofibrous 3D structure provides better physicomechanical properties than its plant-derived counterparts. Bacterial cellulose (BC), an extracellular polysaccharide, is a versatile biopolymer because of its inherent physio-chemical characteristics, wide range of uses, and outstanding accomplishments in various sectors, particularly in the biomedical field [123]. Bacterial cellulose is produced in a pure form. BC 3D matrices are composed of cellulose fibers of a diameter of 10–100 nm. BC-based scaffolds have been specifically designed to mimic the 3D structures of native tissues to support and provide the microenvironments required for cell adhesion, proliferation, migration, and differentiation [123–125]. When used to treat rat models of diabetic wounds, electrospun chitosan-polyvinyl alcohol nanofibrous blend scaffolds demonstrated faster healing than controls. Low cytotoxicity, long-term antibacterial activity, decreased inflammation, and improved wound healing were all seen in the in vivo investigation using Wistar rats using electrospun nanofiber membranes containing silver nanoparticles [65].

The BC-based composite hydrogels have the flexibility to adapt to the shape of the wound site and reduce the possibility of bacterial infection through the prevention of direct contact between the wound and the external environment [104]. In addition, the BC-based dressings showed better dermal burn treatment than silver sulfadiazine cream. Hydrogels possess high water storage capability inside their 3D polymeric network, permitting them to provide a moist environment to the wound site. The efficiency of BC gel and film in the treatment of chronic venous ulcers (CVU) stimulated reepithelization and significantly reduced the scar area [126].

Cellulose nanofibers (CNF)-based aerogel is a third-generation material, having the advantages such as being light-weight, good flexibility, low cost, and green, environment-friendly material. The Cellulose Nanofibers\Polyvinyl alcohol\Sodium alginate (CPS)-AgO nanocomposite material had a strong bactericidal effect on *E. coli* and *S. aureus* [127]. CNF aerogel microspheres significantly facilitated the growth and proliferation of fibroblasts [128]. CNF incorporated with natural antimicrobial material aerogel (CNF/AM aerogel) could be a potential treatment for chronic diabetic ulcers as it provides sustainable inhibition of microbial growth on wounds. The high porosity of aerogel provides aerobic conditions, preventing the growth of anaerobic bacteria such as *Clostridium perfringens* implicated in gangrene development. Bacterial nanocellulose (BNC) was saturated with antibiotic fusidic acid (FA). BNC/ FA biocomposites exhibited high antibiotic activity against *Staphylococcus aureus* and were demonstrated for use as a wound dressing [129]. Moniri et al. combined bacterial nanocellulose with silver nanoparticles to form BNC-Ag nanocomposite [130]. This was demonstrated to improve wound healing and reduce *Staphylococcus aureus* colonization *in vitro*. The hybrid BNC/AgNPs manifested a strong antibacterial activity against Gram negative and Gram positive as well as unicellular fungus. The results from the *in vitro* study demonstrated that BNC/AgNPs promoted wound healing of chronic ulcers by reducing microbial infection. The BNC/AgNPs were protected from oxidative stressors and played a vital role in improving cell propagation and cell proliferation [131]. The natural polyphenol compound, curcumin, which is a hydrophobic drug with antibacterial activity and wound healing ability, was encapsulated in a pluronic polymer and converted into granules. These were further surface modified with chitosan and BNC. With this, a sustained release of curcumin was achieved, proving prolonged bioactivity of the potential wound dressing [132]. Nanoskin ®, CelMat®, Suprasorb X®, Biofill®, Membracel®, xCell®, Nanoderm, and Nanoderm are a few of the commercialized wound dressing products based on the use of bacterial cellulose [105,133–136].

The involvement of nanomaterials in various phases of wound healing is illustrated (Fig. 2)

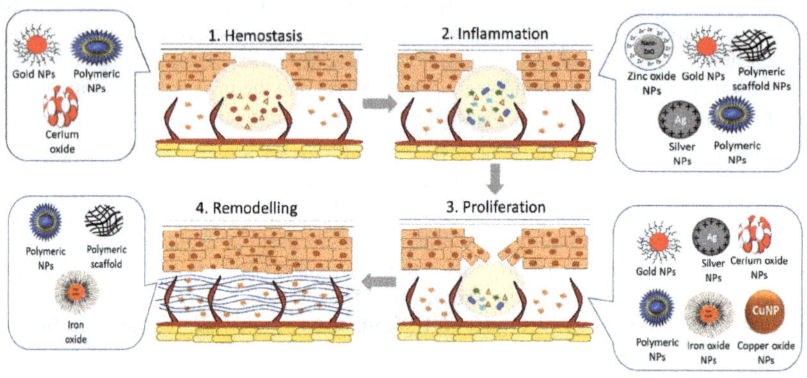

Figure. 2: Possible application of nanomaterials for specific phases in wound healing

New technologies, viz., 3-D printing with 'bio-'ink' have opened up new vistas in creating polymer bandages that could be impregnated with cells and growth factors that can be implanted at the site of injury and expedite the process of healing [137]. Although the incorporation of various nanoparticles in wound dressings is expected to have beneficial effects, aspects such as the cost of nanomaterials synthesis, purification, functionalization, and characterization could be a cause for concern. More research may be required to establish complementing strategies, such as metal nanoparticles combinations with bioactives to treat complex wounds such as those associated with diabetes [63]. In a recent review, Blanco-Fernandez et al., the nanotechnology-based therapies that were developed for the treatment of diabetic wounds were reviewed concerning their efficacy in *in vivo* models of chronic wound healing[138]. The authors state that further studies are critical to providing insights into how nano-therapies can be applied in 'real-world' clinical scenarios. Several products are developed with applications in wound management, but wound

Materials Research Foundations 145 (2023) 207-235
https://doi.org/10.21741/9781644902370-8

presentations are unique and often require 'personalized' materials. Thus, the versatility and effectiveness of existing wound dressings are to be improved, and innovative approaches would be required for the management of non-healing wounds.

5. Toxicity of nanoparticles

There have also been frequent reports of nanomaterials causing oxidative stress, autophagy, and apoptosis in keratinocytes and fibroblasts [4], and concerns of use during pregnancy leading to adverse effects on the new-borns were raised [139]. In addition, it has been reported that nanomaterials cause DNA damage and decrease gene methylation, suggesting the potential to induce cancers [140–142]. Some cases where argyria appeared after treatment of burn wounds with dressings containing nanocrystalline silver were reported. The use of these dressings caused the deposition of silver particles into the mid and deep dermis [143]. Carbon nanotubes and nickel NPs have been reported to cause skin hypersensitivity because of the ions released and surface coverings from NPs [144]. However, there is no convincing indication concerning whether nanoparticles will lead to organ damage and/or tumors in practical use. The toxicity of nanomaterials is highly dependent on the particle size, shape, surface charge, stability, and concentration. Therefore, when developing new nanomaterials for use in wound treatment, the physical and chemical properties should be tailored to reduce toxicity to skin cells [145,146]. Almeida et al., have indicated the necessity for new studies to be undertaken for assessing sensors to be used in 'smart' dressings for monitoring chronic wounds. Careful evaluation of the new nanomaterials developed as wound dressings is of paramount importance. The agencies could decide on a framework of tests and procedures involving animal models that could be best suited for nanomaterials in wound dressings. Currently, the *in vitro* and *in vivo* experiments would require to be conducted under uniform/standardized conditions across all laboratories. All the dressings aimed at treating superficial and deep-seated or chronic wounds would require extensive clinical trials and regulatory approvals before commercialization [147].

Conclusions

This chapter summarizes the importance of nanoparticles in the wound-healing process. Inorganic nanoparticles have shown exceptional, broad-spectrum antimicrobial activities; several of these have been reported to be evaluated in vivo for their ability to inhibit bacteria that cause infection and slow down the wound healing process. Wound dressing materials based on natural biopolymers and synthetic polymeric nanostructures showed enhanced results on cell growth, re-epithelialization, deposition of collagen fibers, tissue regeneration, and ultimately a faster rate of wound closure in chronic wounds. Thus, nanomaterials can be incorporated into wound dressings, conferring antibacterial and wound-healing properties. In the future, 'active' and 'smart' bandages can be prepared for personalized treatment allowing remote monitoring of several parameters by the treating clinician. This would help decrease the economic burden on the healthcare system. Aspects of nanoparticles-related toxicity are discussed so that their use is judicious and DNA damage, skin hypersensitivity, and other harmful effects on skin and other tissues are minimal. The commercialization of the nanomaterials for use in wound repair would warrant regulatory approvals post-clinical trials

References

[1] S. Vyavahare, N. Padole, J. Avari, a Review: Silver Nanoparticles in Wound Healing, Eur. J. Pharm. Med. Res. Www.Ejpmr.Com │. 8 (2015). www.ejpmr.com

[2] E. Rezvani Ghomi, S. Khalili, S. Nouri Khorasani, R. Esmaeely Neisiany, S. Ramakrishna, Wound dressings: Current advances and future directions, J. Appl. Polym. Sci. 136 (2019) 1–12. https://doi.org/10.1002/app.47738

[3] S.N.A. Bakil, H. Kamal, H.Z. Abdullah, M.I. Idris, Sodium alginate-zinc oxide nanocomposite film for antibacterial wound healing applications, Biointerface Res. Appl. Chem. 10 (2020) 6245–6252. https://doi.org/10.33263/BRIAC105.62456252

[4] M. Wang, X. Lai, L. Shao, L. Li, Evaluation of immunoresponses and cytotoxicity from skin exposure to metallic nanoparticles, Int. J. Nanomedicine. 13 (2018) 4445–4459. https://doi.org/10.2147/IJN.S170745

[5] R. Singh, K. Shitiz, A. Singh, Chitin and chitosan: biopolymers for wound management, Int. Wound J. 14 (2017) 1276–1289. https://doi.org/10.1111/iwj.12797

[6] K. Vig, A. Chaudhari, S. Tripathi, S. Dixit, R. Sahu, S. Pillai, V.A. Dennis, S.R. Singh, Advances in skin regeneration using tissue engineering, Int. J. Mol. Sci. 18 (2017) 789. https://doi.org/10.3390/ijms18040789

[7] L.A. Schneider, A. Korber, S. Grabbe, J. Dissemond, Influence of pH on wound-healing: A new perspective for wound-therapy?, Arch. Dermatol. Res. 298 (2007) 413–420. https://doi.org/10.1007/s00403-006-0713-x

[8] S. Dhivya, V.V. Padma, E. Santhini, Wound dressings - A review, Biomed. 5 (2015) 24–28. https://doi.org/10.7603/s40681-015-0022-9

[9] A.R.J.A. de M. Lima, A.S. Siqueira, M.L.S. Möller, R.C. de Souza, J.N. Cruz, A.R.J.A. de M. Lima, R.C. da Silva, D.C.F. Aguiar, J.L. da S.G.V. Junior, E.C. Gonçalves, In silico improvement of the cyanobacterial lectin microvirin and mannose interaction, J. Biomol. Struct. Dyn. (2020). https://doi.org/10.1080/07391102.2020.1821782

[10] R. Fernández-Torres, M. Ruiz-Muñoz, A.J. Pérez-Panero, J.C. García-Romero, M. Gónzalez-Sánchez, Clinician assessment tools for patients with diabetic foot disease: A systematic review, J. Clin. Med. 9 (2020). https://doi.org/10.3390/jcm9051487

[11] J.B. Acosta, D. Garcia Del Barco, D. Cibrian Vera, W. Savigne, P. Lopez-Saura, G. Guillen Nieto, G.S. Schultz, The pro-inflammatory environment in recalcitrant diabetic foot wounds, Int. Wound J. 5 (2008) 530–539. https://doi.org/10.1111/j.1742-481X.2008.00457.x

[12] R. Blakytny, E.B. Jude, Altered molecular mechanisms of diabetic foot ulcers, Int. J. Low. Extrem. Wounds. 8 (2009) 95–104. https://doi.org/10.1177/1534734609337151

[13] M.S. Bader, Diabetic foot infection, Am. Fam. Physician. 78 (2008)

[14] S.C. Mishra, K.C. Chhatbar, A. Kashikar, A. Mehndiratta, Diabetic foot, BMJ. 359 (2017) j5064. https://doi.org/10.1136/bmj.j5064

[15] R. Webb, From decubitus ulcers to pressure injury: What is in a name?, J. Wound Care. 26 (2017) 3. https://doi.org/10.12968/jowc.2017.26.1.3

[16] B. Tirgari, L. Mirshekari, M.A. Forouzi, Pressure Injury Prevention: Knowledge and Attitudes of Iranian Intensive Care Nurses, Adv. Ski. Wound Care. 31 (2018) 1–8. https://doi.org/10.1097/01.ASW.0000530848.50085.ef

[17] C.D. Weller, E.R. Gershenzon, S.M. Evans, V. Team, J.J. McNeil, Pressure injury identification, measurement, coding, and reporting: Key challenges and opportunities, Int. Wound J. 15 (2018) 417–423. https://doi.org/10.1111/iwj.12879

[18] S. Natarajan, D. Williamson, A.J. Stiltz, K.G. Harding, Advances in wound care and healing technology, Am. J. Clin. Dermatol. 1 (2000) 269–275. https://doi.org/10.2165/00128071-200001050-00002

[19] S. Palfreyman, E.A. Nelson, J.A. Michaels, Dressings for venous leg ulcers: Systematic review and meta-analysis, Br. Med. J. 335 (2007) 244–248. https://doi.org/10.1136/bmj.39248.634977.AE

[20] S. Chapman, Venous leg ulcers: An evidence review, Br. J. Community Nurs. 22 (2017) S6–S9. https://doi.org/10.12968/bjcn.2017.22.Sup9.S6

[21] S.A.M. Elgayar, O.A. Hussein, H.A. Mubarak, A.M. Ismaiel, A.M.S. Gomaa, Testing efficacy of the nicotine protection of the substantia nigra pars compacta in a rat Parkinson disease model. Ultrastructure study, Ultrastruct. Pathol. 46 (2022) 37–53. https://doi.org/10.1080/01913123.2021.2015499

[22] J. Boateng, M. Verghese, L.T. Walker, S. Ogutu, Effect of processing on antioxidant contents in selected dry beans (Phaseolus spp. L.), Lwt. 41 (2008) 1541–1547. https://doi.org/10.1016/j.lwt.2007.11.025

[23] F.S. Lozano Sánchez, J. Marinel lo Roura, E. Carrasco Carrasco, J.R. González-Porras, J.R. Escudero Rodríguez, I. Sánchez Nevarez, S. Díaz Sánchez, Venous leg ulcer in the context of chronic venous disease, Phlebology. 29 (2014) 220–226. https://doi.org/10.1177/0268355513480489

[24] J.D. Raffetto, Dermal pathology, cellular biology, and inflammation in chronic venous disease, Thromb. Res. 123 (2009) S66–S71. https://doi.org/10.1016/S0049-3848(09)70147-1

[25] K. Khezri, M.R. Farahpour, S. Mounesi Rad, Efficacy of Mentha pulegium essential oil encapsulated into nanostructured lipid carriers as an in vitro antibacterial

and infected wound healing agent, Colloids Surfaces A Physicochem. Eng. Asp. 589 (2020) 124414. https://doi.org/10.1016/j.colsurfa.2020.124414

[26] D. Stavrou, O. Weissman, A. Tessone, I. Zilinsky, S. Holloway, J. Boyd, J. Haik, Health Related Quality of Life in burn patients - A review of the literature, Burns. 40 (2014) 788–796. https://doi.org/10.1016/j.burns.2013.11.014

[27] S. Hettiaratchy, P. Dziewulski, Pathophysiology and types of burns, Bmj. 328 (2004) 1427. https://doi.org/10.1136/bmj.328.7453.1427

[28] H.J. Klasen, A historical review of the use of silver in the treatment of burns. II. Renewed interest for silver, Burns. 26 (2000) 131–138. https://doi.org/10.1016/S0305-4179(99)00116-3

[29] C. Shi, C. Wang, H. Liu, Q. Li, R. Li, Y. Zhang, Y. Liu, Y. Shao, J. Wang, Selection of Appropriate Wound Dressing for Various Wounds, Front. Bioeng. Biotechnol. 8 (2020) 1–17. https://doi.org/10.3389/fbioe.2020.00182

[30] J. Wang, J. Tavakoli, Y. Tang, Bacterial cellulose production, properties and applications with different culture methods – A review, Carbohydr. Polym. 219 (2019) 63–76. https://doi.org/10.1016/j.carbpol.2019.05.008

[31] S. Saghazadeh, C. Rinoldi, M. Schot, S.S. Kashaf, F. Sharifi, E. Jalilian, K. Nuutila, G. Giatsidis, P. Mostafalu, H. Derakhshandeh, K. Yue, W. Swieszkowski, A. Memic, A. Tamayol, A. Khademhosseini, Drug delivery systems and materials for wound healing applications, Adv. Drug Deliv. Rev. 127 (2018) 138–166. https://doi.org/10.1016/j.addr.2018.04.008

[32] P. Victor, D. Sarada, K.M. Ramkumar, Pharmacological activation of Nrf2 promotes wound healing, Eur. J. Pharmacol. 886 (2020) 173395. https://doi.org/10.1016/j.ejphar.2020.173395

[33] M.T. Matter, S. Probst, S. Läuchli, I.K. Herrmann, Uniting drug and delivery: Metal oxide hybrid nanotherapeutics for skin wound care, Pharmaceutics. 12 (2020) 1–17. https://doi.org/10.3390/pharmaceutics12080780

[34] A. Kushwaha, L. Goswami, B.S. Kim, Nanomaterial-Based Therapy for Wound Healing, Nanomaterials. 12 (2022). https://doi.org/10.3390/nano12040618

[35] F. Thiruvoth, D. Mohapatra, D. Sivakumar, R. Chittoria, V. Nandhagopal, Current concepts in the physiology of adult wound healing, Plast. Aesthetic Res. 2 (2015) 250. https://doi.org/10.4103/2347-9264.158851

[36] L.I.F. Moura, A.M.A. Dias, E. Carvalho, H.C. De Sousa, Recent advances on the development of wound dressings for diabetic foot ulcer treatment - A review, Acta Biomater. 9 (2013) 7093–7114. https://doi.org/10.1016/j.actbio.2013.03.033

[37] H.S. Kim, X. Sun, J.H. Lee, H.W. Kim, X. Fu, K.W. Leong, Advanced drug delivery systems and artificial skin grafts for skin wound healing, Adv. Drug Deliv. Rev. 146 (2019) 209–239. https://doi.org/10.1016/j.addr.2018.12.014

[38] F.S. Alves, J. de A. Rodrigues Do Rego, M.L. Da Costa, L.F. Lobato Da Silva, R.A. Da Costa, J.N. Cruz, D.D.S.B. Brasil, Spectroscopic methods and in silico analyses using density functional theory to characterize and identify piperine alkaloid crystals isolated from pepper (Piper Nigrum L.), J. Biomol. Struct. Dyn. 38 (2020) 2792–2799. https://doi.org/10.1080/07391102.2019.1639547

[39] M. Fronza, G.F. Caetano, M.N. Leite, C.S. Bitencourt, F.W.G. Paula-Silva, T.A.M. Andrade, M.A.C. Frade, I. Merfort, L.H. Faccioli, Hyaluronidase modulates inflammatory response and accelerates the cutaneous wound healing, PLoS One. 9 (2014) 1–12. https://doi.org/10.1371/journal.pone.0112297

[40] N.K. Rajendran, S.S.D. Kumar, N.N. Houreld, H. Abrahamse, A review on nanoparticle based treatment for wound healing, J. Drug Deliv. Sci. Technol. 44 (2018) 421–430. https://doi.org/10.1016/j.jddst.2018.01.009

[41] J. Li, J. Chen, R. Kirsner, Pathophysiology of acute wound healing, Clin. Dermatol. 25 (2007) 9–18. https://doi.org/10.1016/j.clindermatol.2006.09.007

[42] G.L. Brown, L.J. Curtsinger, M. White, R.O. Mitchell, J. Pietsch, R. Nordquist, A. Von Fraunhofer, G.S. Schultz, Acceleration of tensile strength of incisions treated with EGF and TGF-β, Ann. Surg. 208 (1988) 788–794. https://doi.org/10.1097/00000658-198812000-00019

[43] P.I. Morgado, A. Aguiar-Ricardo, I.J. Correia, Asymmetric membranes as ideal wound dressings: An overview on production methods, structure, properties and performance relationship, J. Memb. Sci. 490 (2015) 139–151. https://doi.org/10.1016/j.memsci.2015.04.064

[44] M. Xue, C.J. Jackson, Extracellular Matrix Reorganization During Wound Healing and Its Impact on Abnormal Scarring, Adv. Wound Care. 4 (2015) 119–136. https://doi.org/10.1089/wound.2013.0485

[45] S. Nam, D. Mooney, Polymeric Tissue Adhesives, Chem. Rev. 121 (2021) 11336–11384. https://doi.org/10.1021/acs.chemrev.0c00798

[46] G. Yang, Z. Zhang, K. Liu, X. Ji, P. Fatehi, J. Chen, A cellulose nanofibril-reinforced hydrogel with robust mechanical, self-healing, pH-responsive and antibacterial characteristics for wound dressing applications, J. Nanobiotechnology. 20 (2022) 1–16. https://doi.org/10.1186/s12951-022-01523-5

[47] Q. Zeng, X. Qi, G. Shi, M. Zhang, H. Haick, Wound Dressing: From Nanomaterials to Diagnostic Dressings and Healing Evaluations, ACS Nano. 16 (2022) 1708–1733. https://doi.org/10.1021/acsnano.1c08411

[48] G. Dabiri, E. Damstetter, T. Phillips, Choosing a Wound Dressing Based on Common Wound Characteristics, Adv. Wound Care. 5 (2016) 32–41. https://doi.org/10.1089/wound.2014.0586

Materials Research Foundations 145 (2023) 207-235 https://doi.org/10.21741/9781644902370-8

[49] D. Simões, S.P. Miguel, M.P. Ribeiro, P. Coutinho, A.G. Mendonça, I.J. Correia, Recent advances on antimicrobial wound dressing: A review, Eur. J. Pharm. Biopharm. 127 (2018) 130–141. https://doi.org/10.1016/j.ejpb.2018.02.022

[50] N. Naderi, D. Karponis, A. Mosahebi, A.M. Seifalian, Nanoparticles in wound healing; from hope to promise, from promise to routine, Front. Biosci. - Landmark. 23 (2018) 1038–1059. https://doi.org/10.2741/4632

[51] K. Kalantari, E. Mostafavi, A.M. Afifi, Z. Izadiyan, H. Jahangirian, R. Rafiee-Moghaddam, T.J. Webster, Wound dressings functionalized with silver nanoparticles: Promises and pitfalls, Nanoscale. 12 (2020) 2268–2291. https://doi.org/10.1039/c9nr08234d

[52] X.F. Zhang, W. Shen, S. Gurunathan, Silver nanoparticle-mediated cellular responses in various cell lines: An in vitro model, Int. J. Mol. Sci. 17 (2016) 1–26. https://doi.org/10.3390/ijms17101603

[53] J. Jain, S. Arora, J.M. Rajwade, P. Omray, S. Khandelwal, K.M. Paknikar, Silver nanoparticles in therapeutics: Development of an antimicrobial gel formulation for topical use, Mol. Pharm. 6 (2009) 1388–1401. https://doi.org/10.1021/mp900056g

[54] J. Sheikh, I. Bramhecha, Multi-functionalization of linen fabric using a combination of chitosan, silver nanoparticles and Tamarindus Indica L. seed coat extract, Cellulose. 26 (2019) 8895–8905. https://doi.org/10.1007/s10570-019-02684-7

[55] S. Li, A. Chen, Y. Chen, Y. Yang, Q. Zhang, S. Luo, M. Ye, Y. Zhou, Y. An, W. Huang, T. Xuan, Y. Pan, X. Xuan, H. He, J. Wu, Lotus leaf inspired antiadhesive and antibacterial gauze for enhanced infected dermal wound regeneration, Chem. Eng. J. 402 (2020) 126202. https://doi.org/10.1016/j.cej.2020.126202

[56] K. Brindhadevi, B.H. Elesawy, A. Elfasakhany, I.A. Badruddin, S. Kamangar, Wound dressings coated with silver nanoparticles and essential oil of Labdanum, Appl. Nanosci. (2021). https://doi.org/10.1007/s13204-021-02040-x

[57] C. Rigo, L. Ferroni, I. Tocco, M. Roman, I. Munivrana, C. Gardin, W.R.L. Cairns, V. Vindigni, B. Azzena, C. Barbante, B. Zavan, Active silver nanoparticles for wound healing, Int. J. Mol. Sci. 14 (2013) 4817–4840. https://doi.org/10.3390/ijms14034817

[58] H. Choudhury, M. Pandey, Y.Q. Lim, C.Y. Low, C.T. Lee, T.C.L. Marilyn, H.S. Loh, Y.P. Lim, C.F. Lee, S.K. Bhattamishra, P. Kesharwani, B. Gorain, Silver nanoparticles: Advanced and promising technology in diabetic wound therapy, Mater. Sci. Eng. C. 112 (2020) 110925. https://doi.org/10.1016/j.msec.2020.110925

[59] M. Konop, J. Czuwara, E. Kłodzińska, A.K. Laskowska, D. Sulejczak, T. Damps, U. Zielenkiewicz, I. Brzozowska, A. Sureda, T. Kowalkowski, R.A. Schwartz, L. Rudnicka, Evaluation of keratin biomaterial containing silver nanoparticles as a potential wound dressing in full-thickness skin wound model in diabetic mice, J. Tissue Eng. Regen. Med. 14 (2020) 334–346. https://doi.org/10.1002/term.2998

Materials Research Forum LLC
https://doi.org/10.21741/9781644902370-8

[60] A. Naskar, K.S. Kim, Recent advances in nanomaterial-based wound-healing therapeutics, Pharmaceutics. 12 (2020). https://doi.org/10.3390/pharmaceutics12060499

[61] J. Xiang, R. Zhu, S. Lang, H. Yan, G. Liu, B. Peng, Mussel-inspired immobilization of zwitterionic silver nanoparticles toward antibacterial cotton gauze for promoting wound healing, Chem. Eng. J. 409 (2021) 128291. https://doi.org/10.1016/j.cej.2020.128291

[62] M. Khatami, R.S. Varma, N. Zafarnia, H. Yaghoobi, M. Sarani, V.G. Kumar, Applications of green synthesized Ag, ZnO and Ag/ZnO nanoparticles for making clinical antimicrobial wound-healing bandages, Sustain. Chem. Pharm. 10 (2018) 9–15. https://doi.org/10.1016/j.scp.2018.08.001

[63] A.Y.H. Nor Azlan, H. Katas, M.F. Mh Busra, N.A.M. Salleh, A. Smandri, Metal nanoparticles and biomaterials: The multipronged approach for potential diabetic wound therapy, Nanotechnol. Rev. 10 (2021) 653–670. https://doi.org/10.1515/ntrev-2021-0046

[64] J. Tian, K.K.Y. Wong, C.M. Ho, C.N. Lok, W.Y. Yu, C.M. Che, J.F. Chiu, P.K.H. Tam, Topical delivery of silver nanoparticles promotes wound healing, ChemMedChem. 2 (2007) 129–136. https://doi.org/10.1002/cmdc.200600171

[65] M.M. Mihai, M.B. Dima, B. Dima, A.M. Holban, Nanomaterials for wound healing and infection control, Materials (Basel). 12 (2019) 2176. https://doi.org/10.3390/ma12132176

[66] A. Damle, R. Sundaresan, J.M. Rajwade, P. Srivastava, A. Naik, A concise review on implications of silver nanoparticles in bone tissue engineering, Biomater. Adv. 141 (2022) 213099. https://doi.org/10.1016/j.bioadv.2022.213099

[67] H. Jia, X. Zeng, S. Fan, R. Cai, Z. Wang, Y. Yuan, T. Yue, Silver nanoparticles anchored magnetic self-assembled carboxymethyl cellulose-ε-polylysine hybrids with synergetic antibacterial activity for wound infection therapy, Int. J. Biol. Macromol. 210 (2022) 703–715. https://doi.org/10.1016/j.ijbiomac.2022.04.225

[68] J.G. Leu, S.A. Chen, H.M. Chen, W.M. Wu, C.F. Hung, Y. Der Yao, C.S. Tu, Y.J. Liang, The effects of gold nanoparticles in wound healing with antioxidant epigallocatechin gallate and α-lipoic acid, Nanomedicine Nanotechnology, Biol. Med. 8 (2012) 767–775. https://doi.org/10.1016/j.nano.2011.08.013

[69] P.S. Lau, N. Bidin, S. Islam, W.N.B.W.M. Shukri, N. Zakaria, N. Musa, G. Krishnan, Influence of gold nanoparticles on wound healing treatment in rat model: Photobiomodulation therapy, Lasers Surg. Med. 49 (2017) 380–386. https://doi.org/10.1002/lsm.22614

[70] M.M. Abousamra, A. Mona M, Citation: Abousamra MM. Nanoparticles as Safe and Effective Drug Delivery Systems for Wound Healing, Austin J Nanomed Nanotechnol. 7 (2019) 1–10. www.austinpublishinggroup.com

[71] O. Akturk, K. Kismet, A.C. Yasti, S. Kuru, M.E. Duymus, F. Kaya, M. Caydere, S. Hucumenoglu, D. Keskin, Collagen/gold nanoparticle nanocomposites: A potential skin wound healing biomaterial, J. Biomater. Appl. 31 (2016) 283–301. https://doi.org/10.1177/0885328216644536

[72] S.P.H. Martínez, T.I.R. González, M.A.F. Molina, J.J. Bollain Y Goytia, J.J.M. Sanmiguel, D.G.Z. Triviño, C.R. Padilla, A novel gold calreticulin nanocomposite based on chitosan for wound healing in a diabetic mice model, Nanomaterials. 9 (2019) 75. https://doi.org/10.3390/nano9010075

[73] J. Salvo, C. Sandoval, Role of copper nanoparticles in wound healing for chronic wounds: Literature review, Burn. Trauma. 10 (2022). https://doi.org/10.1093/burnst/tkab047

[74] S.K. Nethi, S. Das, C.R. Patra, S. Mukherjee, Recent advances in inorganic nanomaterials for wound-healing applications, Biomater. Sci. 7 (2019) 2652–2674. https://doi.org/10.1039/c9bm00423h

[75] B. Tao, C. Lin, Y. Deng, Z. Yuan, X. Shen, M. Chen, Y. He, Z. Peng, Y. Hu, K. Cai, Copper-nanoparticle-embedded hydrogel for killing bacteria and promoting wound healing with photothermal therapy, J. Mater. Chem. B. 7 (2019) 2534–2548. https://doi.org/10.1039/C8TB03272F

[76] S. Jiang, B.C. Ma, W. Huang, A. Kaltbeitzel, G. Kizisavas, D. Crespy, K.A.I. Zhang, K. Landfester, Visible light active nanofibrous membrane for antibacterial wound dressing, Nanoscale Horizons. 3 (2018) 439–446. https://doi.org/10.1039/c8nh00021b

[77] R. Jose Varghese, S. Parani, V.R. Remya, R. Maluleke, S. Thomas, O.S. Oluwafemi, Sodium alginate passivated CuInS2/ZnS QDs encapsulated in the mesoporous channels of amine modified SBA 15 with excellent photostability and biocompatibility, Int. J. Biol. Macromol. 161 (2020) 1470–1474. https://doi.org/10.1016/j.ijbiomac.2020.07.240

[78] E.M. Hetrick, J.H. Shin, H.S. Paul, M.H. Schoenfisch, Anti-biofilm efficacy of nitric oxide-releasing silica nanoparticles, Biomaterials. 30 (2009) 2782–2789. https://doi.org/10.1016/j.biomaterials.2009.01.052

[79] F. Öri, R. Dietrich, C. Ganz, M. Dau, D. Wolter, A. Kasten, T. Gerber, B. Frerich, Silicon-dioxide−polyvinylpyrrolidone as a wound dressing for skin defects in a murine model, J. Cranio-Maxillofacial Surg. 45 (2017) 99–107. https://doi.org/10.1016/j.jcms.2016.10.002

[80] R. Ahmed, M. Tariq, I. Ali, R. Asghar, P. Noorunnisa Khanam, R. Augustine, A. Hasan, Novel electrospun chitosan/polyvinyl alcohol/zinc oxide nanofibrous mats with antibacterial and antioxidant properties for diabetic wound healing, Int. J. Biol. Macromol. 120 (2018) 385–393. https://doi.org/10.1016/j.ijbiomac.2018.08.057

[81] R. Pati, R.K. Mehta, S. Mohanty, A. Padhi, M. Sengupta, B. Vaseeharan, C. Goswami, A. Sonawane, Topical application of zinc oxide nanoparticles reduces bacterial skin infection in mice and exhibits antibacterial activity by inducing oxidative stress response and cell membrane disintegration in macrophages, Nanomedicine Nanotechnology, Biol. Med. 10 (2014) 1195–1208. https://doi.org/10.1016/j.nano.2014.02.012

[82] A.K. Barui, S.K. Nethi, C.R. Patra, Investigation of the role of nitric oxide driven angiogenesis by zinc oxide nanoflowers, J. Mater. Chem. B. 5 (2017) 3391–3403. https://doi.org/10.1039/c6tb03323g

[83] M.A. Shalaby, M.M. Anwar, H. Saeed, Nanomaterials for application in wound Healing: current state-of-the-art and future perspectives, Springer Netherlands, 2022. https://doi.org/10.1007/s10965-021-02870-x

[84] P. Thangavel, R. Kannan, B. Ramachandran, G. Moorthy, L. Suguna, V. Muthuvijayan, Development of reduced graphene oxide (rGO)-isabgol nanocomposite dressings for enhanced vascularization and accelerated wound healing in normal and diabetic rats, Elsevier Inc., 2018. https://doi.org/10.1016/j.jcis.2018.01.110

[85] M. Mitra, C. Kulsi, K. Chatterjee, K. Kargupta, S. Ganguly, D. Banerjee, S. Goswami, Reduced graphene oxide-polyaniline composites - Synthesis, characterization and optimization for thermoelectric applications, RSC Adv. 5 (2015) 31039–31048. https://doi.org/10.1039/c5ra01794g

[86] S. Mohammadi, A. Babaei, Poly (vinyl alcohol)/chitosan/polyethylene glycol-assembled graphene oxide bio-nanocomposites as a prosperous candidate for biomedical applications and drug/food packaging industry, Int. J. Biol. Macromol. 201 (2022) 528–538. https://doi.org/10.1016/j.ijbiomac.2022.01.086

[87] A. Khalid, A. Madni, B. Raza, M. ul Islam, A. Hassan, F. Ahmad, H. Ali, T. Khan, F. Wahid, Multiwalled carbon nanotubes functionalized bacterial cellulose as an efficient healing material for diabetic wounds, Int. J. Biol. Macromol. 203 (2022) 256–267. https://doi.org/10.1016/j.ijbiomac.2022.01.146

[88] C. Zgheib, R. Biology, S.A. Hilton, R. Biology, L.C. Dewberry, R. Biology, M.M. Hodges, R. Biology, S. Ghatak, J. Xu, R. Biology, S. Singh, A.M. Processing, A. Centre, HHS Public Access, 228 (2021) 107–115. https://doi.org/10.1016/j.jamcollsurg.2018.09.017.Use

[89] K. Viswanathan, D.B. Babu, G. Jayakumar, G. Dhinakar Raj, Anti-microbial and skin wound dressing application of molecular iodine nanoparticles, Mater. Res. Express. 4 (2017) 104003. https://doi.org/10.1088/2053-1591/aa91e5

[90] B. Ma, W. Dang, Z. Yang, J. Chang, C. Wu, MoS2 Nanoclusters-based biomaterials for disease- impaired wound therapy, Appl. Mater. Today. 20 (2020) 100735. https://doi.org/10.1016/j.apmt.2020.100735

[91] A. Samadi, J. Buro, X. Dong, A. Weinstein, D.O. Lara, K.B. Celie, M.A. Wright, M.A. Gadijko, U. Galili, J.A. Spector, Topical α-Gal Nanoparticles Enhance Wound Healing in Radiated Skin, Skin Pharmacol. Physiol. 35 (2022) 31–40. https://doi.org/10.1159/000518015

[92] R. Sankar, R. Dhivya, K.S. Shivashangari, V. Ravikumar, Wound healing activity of Origanum vulgare engineered titanium dioxide nanoparticles in Wistar Albino rats, J. Mater. Sci. Mater. Med. 25 (2014) 1701–1708. https://doi.org/10.1007/s10856-014-5193-5

[93] L. Houdaille, G. Prévot, H. Ripault, J.Y. Lemonnier, C. Réa, D. Chavanne, J.B. Gauvain, Miliary tuberculosis with crystal deposition disease leading to a diagnosis of tuberculous arthritis [1], Jt. Bone Spine. 69 (2002) 338–340. https://doi.org/10.1016/S1297-319X(02)00405-0

[94] S.P. Ndlovu, K. Ngece, S. Alven, B.A. Aderibigbe, Gelatin-based hybrid scaffolds: Promising wound dressings, Polymers (Basel). 13 (2021). https://doi.org/10.3390/polym13172959

[95] T. de P. de L. Lima, M.F. Passos, Skin wounds, the healing process, and hydrogel-based wound dressings: a short review, J. Biomater. Sci. Polym. Ed. 32 (2021) 1910–1925. https://doi.org/10.1080/09205063.2021.1946461

[96] A.G. Niculescu, A.M. Grumezescu, An Up-to-Date Review of Biomaterials Application in Wound Management, Polymers (Basel). 14 (2022) 1–24. https://doi.org/10.3390/polym14030421

[97] H.M. Powell, S.T. Boyce, Fiber density of electrospun gelatin scaffolds regulates morphogenesis of dermal-epidermal skin substitutes, J. Biomed. Mater. Res. - Part A. 84 (2008) 1078–1086. https://doi.org/10.1002/jbm.a.31498

[98] H. Bilgic, M. Demiriz, M. Ozler, T. Ide, N. Dogan, S. Gumus, A. Kiziltay, T. Endogan, V. Hasirci, N. Hasirci, Gelatin based scaffolds and effect of EGF dose on wound healing, J. Biomater. Tissue Eng. 3 (2013) 205–211. https://doi.org/10.1166/jbt.2013.1077

[99] H. Ye, J. Cheng, K. Yu, In situ reduction of silver nanoparticles by gelatin to obtain porous silver nanoparticle/chitosan composites with enhanced antimicrobial and wound-healing activity, Elsevier B.V, 2019. https://doi.org/10.1016/j.ijbiomac.2018.10.056

[100] T.D. Nguyen, T.T. Nguyen, K.L. Ly, A.H. Tran, T.T.N. Nguyen, M.T. Vo, H.M. Ho, N.T.N. Dang, V.T. Vo, D.H. Nguyen, T.T.H. Nguyen, T.H. Nguyen, In vivo study of the antibacterial chitosan/polyvinyl alcohol loaded with silver nanoparticle hydrogel for wound healing applications, Int. J. Polym. Sci. 2019 (2019). https://doi.org/10.1155/2019/7382717

[101] D. Xiang, X. Wu, W. Cao, B. Xue, M. Qin, Y. Cao, W. Wang, Hydrogels With Tunable Mechanical Properties Based on Photocleavable Proteins, Front. Chem. 8 (2020) 1–9. https://doi.org/10.3389/fchem.2020.00007

[102] G. Patil, R. Pawar, S. Jadhav, V. Ghormade, A chitosan based multimodal "soft" hydrogel for rapid hemostasis of non-compressible hemorrhages and its mode of action, Carbohydr. Polym. Technol. Appl. 4 (2022) 100237. https://doi.org/10.1016/j.carpta.2022.100237

[103] J.V. Kumbhar, J.M. Rajwade, K.M. Paknikar, Fruit peels support higher yield and superior quality bacterial cellulose production, Appl. Microbiol. Biotechnol. 99 (2015) 6677–6691. https://doi.org/10.1007/s00253-015-6644-8

[104] H. Moradpoor, H. Mohammadi, M. Safaei, H.R. Mozaffari, R. Sharifi, P. Gorji, A.B. Sulong, N. Muhamad, M. Ebadi, Recent Advances on Bacterial Cellulose-Based Wound Management: Promises and Challenges, Int. J. Polym. Sci. 2022 (2022). https://doi.org/10.1155/2022/1214734

[105] J.M. Rajwade, K.M. Paknikar, J. V. Kumbhar, Applications of bacterial cellulose and its composites in biomedicine, Appl. Microbiol. Biotechnol. 99 (2015) 2491–2511. https://doi.org/10.1007/s00253-015-6426-3

[106] A. Fallacara, E. Baldini, S. Manfredini, S. Vertuani, Hyaluronic acid in the third millennium, Polymers (Basel). 10 (2018). https://doi.org/10.3390/polym10070701

[107] A.M. Abdel-Mohsen, J. Jancar, R.M. Abdel-Rahman, L. Vojtek, P. Hyršl, M. Dušková, H. Nejezchlebová, A novel in situ silver/hyaluronan bio-nanocomposite fabrics for wound and chronic ulcer dressing: In vitro and in vivo evaluations, Int. J. Pharm. 520 (2017) 241–253. https://doi.org/10.1016/j.ijpharm.2017.02.003

[108] K.M. Paknikar, Therapeutic applications of silver nanoparticles, in: Appl. Nanomater., 2013

[109] H. Kenar, C.Y. Ozdogan, C. Dumlu, E. Doger, G.T. Kose, V. Hasirci, Microfibrous scaffolds from poly(L-lactide-co-ε-caprolactone) blended with xeno-free collagen/hyaluronic acid for improvement of vascularization in tissue engineering applications, Mater. Sci. Eng. C. 97 (2019) 31–44. https://doi.org/10.1016/j.msec.2018.12.011

[110] M. Parani, G. Lokhande, A. Singh, A.K. Gaharwar, Engineered Nanomaterials for Infection Control and Healing Acute and Chronic Wounds, ACS Appl. Mater. Interfaces. 8 (2016) 10049–10069. https://doi.org/10.1021/acsami.6b00291

[111] H. Liu, C. Wang, C. Li, Y. Qin, Z. Wang, F. Yang, Z. Li, J. Wang, A functional chitosan-based hydrogel as a wound dressing and drug delivery system in the treatment of wound healing, RSC Adv. 8 (2018) 7533–7549. https://doi.org/10.1039/c7ra13510f

[112] Y. Chen, N. Dan, W. Dan, X. Liu, L. Cong, A novel antibacterial acellular porcine dermal matrix cross-linked with oxidized chitosan oligosaccharide and modified by in situ synthesis of silver nanoparticles for wound healing applications, Mater. Sci. Eng. C. 94 (2019) 1020–1036. https://doi.org/10.1016/j.msec.2018.10.036

[113] S.P. Miguel, A.F. Moreira, I.J. Correia, Chitosan based-asymmetric membranes for wound healing: A review, Int. J. Biol. Macromol. 127 (2019) 460–475. https://doi.org/10.1016/j.ijbiomac.2019.01.072

[114] Z. Bagher, A. Ehterami, M.H. Safdel, H. Khastar, H. Semiari, A. Asefnejad, S.M. Davachi, M. Mirzaii, M. Salehi, Wound healing with alginate/chitosan hydrogel containing hesperidin in rat model, J. Drug Deliv. Sci. Technol. 55 (2020) 101379. https://doi.org/10.1016/j.jddst.2019.101379

[115] T. Takei, H. Nakahara, H. Ijima, K. Kawakami, Synthesis of a chitosan derivative soluble at neutral pH and gellable by freeze-thawing, and its application in wound care, Acta Biomater. 8 (2012) 686–693. https://doi.org/10.1016/j.actbio.2011.10.005

[116] M. Basha, M.M. AbouSamra, G.A. Awad, S.S. Mansy, A potential antibacterial wound dressing of cefadroxil chitosan nanoparticles in situ gel: Fabrication, in vitro optimization and in vivo evaluation, Int. J. Pharm. 544 (2018) 129–140. https://doi.org/10.1016/j.ijpharm.2018.04.021

[117] J. Qu, X. Zhao, Y. Liang, T. Zhang, P.X. Ma, B. Guo, Antibacterial adhesive injectable hydrogels with rapid self-healing, extensibility and compressibility as wound dressing for joints skin wound healing, Biomaterials. 183 (2018) 185–199. https://doi.org/10.1016/j.biomaterials.2018.08.044

[118] M.M. Shaik, A. Dapkekar, J.M. Rajwade, S.H. Jadhav, M. Kowshik, Antioxidant-antibacterial containing bi-layer scaffolds as potential candidates for management of oxidative stress and infections in wound healing, J. Mater. Sci. Mater. Med. 30 (2019). https://doi.org/10.1007/s10856-018-6212-8

[119] Y. Gao, Y. Wu, Recent advances of chitosan-based nanoparticles for biomedical and biotechnological applications, Int. J. Biol. Macromol. 203 (2022) 379–388. https://doi.org/10.1016/j.ijbiomac.2022.01.162

[120] J. Grip, R.E. Engstad, I. Skjæveland, N. Škalko-Basnet, J. Isaksson, P. Basnet, A.M. Holsæter, Beta-glucan-loaded nanofiber dressing improves wound healing in diabetic mice, Eur. J. Pharm. Sci. 121 (2018) 269–280. https://doi.org/10.1016/j.ejps.2018.05.031

[121] F. Esa, S.M. Tasirin, N.A. Rahman, Overview of Bacterial Cellulose Production and Application, Agric. Agric. Sci. Procedia. 2 (2014) 113–119. https://doi.org/10.1016/j.aaspro.2014.11.017

[122] M.B. Noremylia, M.Z. Hassan, Z. Ismail, Recent advancement in isolation, processing, characterization and applications of emerging nanocellulose: A review, Int.

 Materials Research Forum LLC
https://doi.org/10.21741/9781644902370-8

J. Biol. Macromol. 206 (2022) 954–976. https://doi.org/10.1016/j.ijbiomac.2022.03.064

[123] S. Khan, M. Ul-Islam, M.W. Ullah, Y. Zhu, K.B. Narayanan, S.S. Han, J.K. Park, Fabrication strategies and biomedical applications of three-dimensional bacterial cellulose-based scaffolds: A review, Int. J. Biol. Macromol. 209 (2022) 9–30. https://doi.org/10.1016/j.ijbiomac.2022.03.191

[124] J.D.P. de Amorim, C.J.G. da Silva Junior, A.D.M. de Medeiros, H.A. do Nascimento, M. Sarubbo, T.P.M. de Medeiros, A.F. de S. Costa, L.A. Sarubbo, Bacterial Cellulose as a Versatile Biomaterial for Wound Dressing Application, Molecules. 27 (2022). https://doi.org/10.3390/molecules27175580

[125] F. Jabbari, V. Babaeipour, S. Bakhtiari, Bacterial cellulose-based composites for nerve tissue engineering, Int. J. Biol. Macromol. 217 (2022) 120–130. https://doi.org/10.1016/j.ijbiomac.2022.07.037

[126] M.M.P. Silva, M.I.F. de Aguiar, A.B. Rodrigues, M.D.C. Miranda, M.Â.M. Araújo, I.L.T.P. Rolim, A.M.A. e Souza, The use of nanoparticles in wound treatment: A systematic review, Rev. Da Esc. Enferm. 51 (2017) 1–9. https://doi.org/10.1590/S1980-220X2016043503272

[127] B. Wang, P. Ji, Y. Ma, J. Song, Z. You, S. Chen, Bacterial cellulose nanofiber reinforced poly(glycerol-sebacate) biomimetic matrix for 3D cell culture, Cellulose. 28 (2021) 8483–8492. https://doi.org/10.1007/s10570-021-04053-9

[128] J. Cai, J. Chen, Q. Zhang, M. Lei, J. He, A. Xiao, C. Ma, S. Li, H. Xiong, Well-aligned cellulose nanofiber-reinforced polyvinyl alcohol composite film: Mechanical and optical properties, Carbohydr. Polym. 140 (2016) 238–245. https://doi.org/10.1016/j.carbpol.2015.12.039

[129] E. Liyaskina, V. Revin, E. Paramonova, M. Nazarkina, N. Pestov, N. Revina, S. Kolesnikova, Nanomaterials from bacterial cellulose for antimicrobial wound dressing, J. Phys. Conf. Ser. 784 (2017) 012034. https://doi.org/10.1088/1742-6596/784/1/012034

[130] M. Moniri, A.B. Moghaddam, S. Azizi, R.A. Rahim, S.W. Zuhainis, M. Navaderi, R. Mohamad, In vitro molecular study of wound healing using biosynthesized bacteria nanocellulose/ silver nanocomposite assisted by bioinformatics databases, Int. J. Nanomedicine. 13 (2018) 5097–5112. https://doi.org/10.2147/IJN.S164573

[131] T.I. Shaheen, I. Capron, Formulation of re-dispersible dry o/w emulsions using cellulose nanocrystals decorated with metal/metal oxide nanoparticles, RSC Adv. 11 (2021) 32143–32151. https://doi.org/10.1039/d1ra06054f

[132] A. Rojewska, A. Karewicz, K. Boczkaja, K. Wolski, M. Kępczyński, S. Zapotoczny, M. Nowakowska, Modified bionanocellulose for bioactive wound-healing dressing, Eur. Polym. J. 96 (2017) 200–209. https://doi.org/10.1016/j.eurpolymj.2017.09.010

[133] R. Portela, C.R. Leal, P.L. Almeida, R.G. Sobral, Bacterial cellulose: a versatile biopolymer for wound dressing applications, Microb. Biotechnol. 12 (2019) 586–610. https://doi.org/10.1111/1751-7915.13392

[134] W.K. Czaja, D.J. Young, M. Kawecki, R.M. Brown, The future prospects of microbial cellulose in biomedical applications, Biomacromolecules. 8 (2007) 1–12. https://doi.org/10.1021/bm060620d

[135] T. Wild, M. Bruckner, M. Payrich, C. Schwarz, T. Eberlein, A. Andriessen, Eradication of methicillin-resistant Staphylococcus aureus in pressure ulcers comparing a polyhexanide-containing cellulose dressing with polyhexanide swabs in a prospective randomized study, Adv. Ski. Wound Care. 25 (2012) 17–22. https://doi.org/10.1097/01.ASW.0000410686.14363.ea

[136] G.F. Picheth, C.L. Pirich, M.R. Sierakowski, M.A. Woehl, C.N. Sakakibara, C.F. de Souza, A.A. Martin, R. da Silva, R.A. de Freitas, Bacterial cellulose in biomedical applications: A review, Int. J. Biol. Macromol. 104 (2017) 97–106. https://doi.org/10.1016/j.ijbiomac.2017.05.171

[137] W. Wang, H. Sheng, D. Cao, F. Zhang, W. Zhang, F. Yan, D. Ding, N. Cheng, S-nitrosoglutathione functionalized polydopamine nanoparticles incorporated into chitosan/gelatin hydrogel films with NIR-controlled photothermal/NO-releasing therapy for enhanced wound healing, Int. J. Biol. Macromol. 200 (2022) 77–86. https://doi.org/10.1016/j.ijbiomac.2021.12.125

[138] B. Blanco-Fernandez, O. Castaño, M.Á. Mateos-Timoneda, E. Engel, S. Pérez-Amodio, Nanotechnology Approaches in Chronic Wound Healing, Adv. Wound Care. 10 (2021) 234–256. https://doi.org/10.1089/wound.2019.1094

[139] N. Hadrup, A.K. Sharma, K. Loeschner, Toxicity of silver ions, metallic silver, and silver nanoparticle materials after in vivo dermal and mucosal surface exposure: A review, Regul. Toxicol. Pharmacol. 98 (2018) 257–267. https://doi.org/10.1016/j.yrtph.2018.08.007

[140] A. Ali, M. Suhail, S. Mathew, M.A. Shah, S.M. Harakeh, S. Ahmad, Z. Kazmi, M.A.R. Alhamdan, A. Chaudhary, G.A. Damanhouri, I. Qadri, Nanomaterial induced immune responses and cytotoxicity, J. Nanosci. Nanotechnol. 16 (2016) 40–57. https://doi.org/10.1166/jnn.2016.10885

[141] K. Sooklert, S. Nilyai, R. Rojanathanes, D. Jindatip, N. Sae-Liang, N. Kitkumthorn, A. Mutirangura, A. Sereemaspun, N-acetylcysteine reverses the decrease of DNA methylation status caused by engineered gold, silicon, and chitosan nanoparticles, Int. J. Nanomedicine. 14 (2019) 4573–4587. https://doi.org/10.2147/IJN.S204372

[142] N.C. da R. Galucio, D. de A. Moysés, J.R.S. Pina, P.S.B. Marinho, P.C. Gomes Júnior, J.N. Cruz, V.V. Vale, A.S. Khayat, A.M. do R. Marinho, Antiproliferative, genotoxic activities and quantification of extracts and cucurbitacin B obtained from

Luffa operculata (L.) Cogn, Arab. J. Chem. 15 (2022) 103589. https://doi.org/10.1016/j.arabjc.2021.103589

[143] M. Walker, C.A. Cochrane, P.G. Bowler, D. Parsons, P. Bradshaw, Silver deposition and tissue staining associated with wound dressings containing silver, Ostomy Wound Manag. 52 (2006) 42–50

[144] B.C. Palmer, S.J. Phelan-Dickenson, L.A. Delouise, Multi-walled carbon nanotube oxidation dependent keratinocyte cytotoxicity and skin inflammation, Part. Fibre Toxicol. 16 (2019) 1–16. https://doi.org/10.1186/s12989-018-0285-x

[145] S. Hashempour, S. Ghanbarzadeh, H.I. Maibach, M. Ghorbani, H. Hamishehkar, Skin toxicity of topically applied nanoparticles, Ther. Deliv. 10 (2019) 383–396. https://doi.org/10.4155/tde-2018-0060

[146] C.M.A. Rego, A.F. Francisco, C.N. Boeno, M. V. Paloschi, J.A. Lopes, M.D.S. Silva, H.M. Santana, S.N. Serrath, J.E. Rodrigues, C.T.L. Lemos, R.S.S. Dutra, J.N. da Cruz, C.B.R. dos Santos, S. da S. Setúbal, M.R.M. Fontes, A.M. Soares, W.L. Pires, J.P. Zuliani, Inflammasome NLRP3 activation induced by Convulxin, a C-type lectin-like isolated from Crotalus durissus terrificus snake venom, Sci. Rep. 12 (2022) 1–17. https://doi.org/10.1038/s41598-022-08735-7

[147] I. Barros Almeida, L. Garcez Barretto Teixeira, F. Oliveira De Carvalho, É. Ramos Silva, P. Santos Nunes, M.R. Viana Dos Santos, A. Antunes De Souza Araújo, Smart dressings for wound healing: A review, Adv. Ski. Wound Care. 34 (2021) 1–8. https://doi.org/10.1097/01.ASW.0000725188.95109.68

Materials Research Foundations 145 (2023) 236-249 | https://doi.org/10.21741/9781644902370-9

Chapter 9

In silico Methods for Evaluating the Mode of Interaction of Nanoparticles with Molecular Target

Suraj N. Mali[1,*], Jorddy Neves Cruz[2,*], Akshay R. Yadav[3]

[1]Department of Pharmaceutical Sciences and Technology, Institute of Chemical Technology, Mumbai, India

[2]Institute of Biological Sciences, Federal University of Pará, Belém, Pará., Brazil

[3]Department of Pharmaceutical Chemistry, Rajarambapu College of Pharmacy, Kasegaon, Dist. Sangli, Maharashtra, India

jorddy.cruz@icb.ufpa.br (J.N.C) and mali.suraj1695@gmail.com (S.N.M)

Abstract

An essential tool for structure-based drug design is molecular docking. As mentioned at the outset, docking aims to anticipate the most advantageous ligand-target spatial arrangement and quantify the associated complex free energy. However, precise scoring systems are still difficult to come across. This chapter overviews current molecular docking techniques and their uses in nanostructures.

Keywords

Nanoparticle, Docking, *in silico*, Drug Design, Methodologies

Contents

Materials Research Foundations 145 (2023) 236-249 https://doi.org/10.21741/9781644902370-9

1. Introduction

The last ten years have seen a surge in interest in nanotechnology, which has significantly advanced several fields of science and technology [1]. In medicine, nanotechnology has many uses, such as creating antimicrobials [2], bioimaging, and treating cancer [2]. Currently, various FDA-approved nanoparticles (NPs) are available for clinical application [3]. When a nanostructure enables a particular material's distinctive phenomena with a size between 1 and 100 nm, the term "nanotechnology" is used in the scientific community [4] to say that the NPs may be dangerous to multicellular organisms because they have unique properties that come from their high surface-to-volume ratios. Additionally, engineered nanoparticles are already commonplace and utilized daily [5]. These products include but are not limited to, nutritional supplements, soaps, lotions, toothpaste, and shampoos [6,7].

Additionally, NPs may be found in the air, water from industrial discharges, wastewater treatment effluents, and soil from fertilizers [8]. These artificial nanoparticles have the potential to interact with biological macromolecules and harm living cells [9]. To learn more about the interactions and dynamics of NPs in biological systems, computational analyses such as molecular docking [10,11], density functional theory calculations [12], Monte Carlo, kinetic mean-field model, and coarse-grained molecular dynamics simulations [13]. The physiological alterations brought on by the interactions of NPs with macromolecules might result in various disease phenotypes [14]. The lengthy and detailed process of looking for a lead molecule might be discouraging since so many options exist. Fortunately, computational technologies have saved the day and unquestionably played a crucial part in streamlining the drug discovery process. Of all the methods, molecular docking has been essential in computer-aided drug design. It has rapidly risen to occupy an important place in the current landscape of structure-based drug design [15,16]. The principles, sampling techniques, scoring functions, and several molecular docking software options have all been outlined in this chapter. We demonstrate how docking, traditional structure-based design methods, and X-ray crystallography interact throughout the drug development process. We also focus on a few of the constraints that docking research encounter. The future of docking studies has also been examined, and a few instances of how it has aided in the development of new medications have been provided [17,18].

2. Molecular docking

Let alone the interaction between a protein and another protein or between a protein and a small molecule; two molecules can interact in various ways. By predicting the intermolecular structure formed between a protein and a small molecule or a protein and protein, molecular docking enables us to identify the binding strategies that inhibit the protein [19]. One needs a high-resolution X-ray, N.M.R., or homology-modeled structure of the biomolecule with a known or predicted binding site to accurately conduct docking studies [19]. 148,827 are now accessible in the database (PDB). Docking algorithms score how well a ligand fits into a binding site by considering and optimizing factors like steric, hydrophobic, and electrostatic complementarity and calculating the binding free energy [20].

The sampling method and the scoring function are the two fundamental elements that set the various docking software apart from one another and are covered in depth here.

2.1 Sampling algorithm used in the docking

There are several ways for two molecules to bind together, as was previously said. Even with advancements in parallel processing and higher clock speeds, creating every possible combination would be costly and take a long time. As a result, methods were required to separate valuable conformations from pointless ones [21,22]. Different algorithms have been created in this area and may be categorized according to how many degrees of freedom they disregard [23]. The degree of freedom was reduced to only six by the most straightforward approach, which regarded the molecules as two rigid entities (three translational and three rotational [24]. DOCK is a well-known instance of a program that makes use of this method. This algorithm was created to identify compounds with a great deal of form similarity with binding sites or pockets. It shows alleged binding sites on the protein's surface [25,26]. This picture comprises several overlapping spheres with different radii that only contact the macromolecule's molecular surface twice. The ligand molecule may be considered a collection of spheres that roughly occupy the ligand's space. Once the sphere models of the protein surface and the ligand are done, the matching rule is used [27,28]. The pairing rule is based on the principle that a ligand sphere can be paired with a protein sphere if the internal distances of all the spheres in the ligand set match all the internal distances within the protein set, allowing some user-specified tolerance [29]. Thus, it will enable the program to identify geometrically similar clusters of spheres on the protein site and the ligand. Many other programs were developed later, using matching algorithms (M.A.), including LibDock, L.I.D.A.E.U.S., PhDOCK, Ph4DOCK, Q-fit, S.A.N.D.O.C.K., etc. All these programs based on M.A. have the advantage of speed. Still, they have several limitations, such as the prior need for detailed receptor geometry and a lack of molecular flexibility, which does not accurately define many aspects of ligand-protein interactions [30].

The second algorithm is incremental construction (I.C.), wherein the ligand is fragmented from rotatable bonds into various segments. One of the segments is anchored to the receptor surface [31]. The anchor is generally considered to be the fragment that shows maximum interactions with the receptor surface, has the minimum number of alternate conformations and is fairly rigid, such as the ring system. Once the base/anchor has been established, the next step is adding each fragment. Ideally, those fragments are added first, which have a greater chance of showing interactions like hydrogen bonding since they are directional and responsible for the ligand's specificity. In addition, hydrogen bonds lead to a more accurate prediction of geometry [32].

The Monte Carlo algorithm is yet another helpful one (MC). The method is rapid and reliable because when a particular fragment is inserted, the postures with the lowest energies are considered for the next iteration [33]. Programs like DOCK 4.0, FlexX, Hammerhead, SLIDE and eHiTS, S.K.E.L.G.E.N., ProPose, PatchDock, MacDock, FLOG, etc., have all utilized I.C. in some way. One significant drawback of this software is that it can only be used with medium-sized ligands since large-sized ligands would yield too many fragments, which would be problematic [34]. This method involves translation or rotation of the whole ligand and bond rotation to modify the ligand progressively. A specific conformation may also be obtained by simultaneously changing many parameters [35]. The confirmation is then assessed at the binding site using a molecular mechanics energy calculation. Depending on Boltzmann's probability constant, it is either accepted or rejected for the next iteration. The change in energy for a parameter T, which is physically understood as temperature, determines whether the conformation is accepted or rejected (simulated annealing) [36]. This approach stands out from the others due to the approval or rejection criteria. While the other algorithm favors energy reduction, the MC approach also allows for increases. Increases are more familiar with higher levels of T [37]. Small energy barriers may be overcome, and the configuration can advance beyond local minima if one begins at a high value of T. This makes it especially helpful when a global minimum is sought amid several local minima [38]. The Tabu search, which keeps track of the search area of the binding site that has previously been visited and guarantees that the binding site is thoroughly investigated, is an intriguing offshoot of the MC technique [39]. Programs like DockVision 1.0.3, F.D.S., GlamDock, I.C.M., M.C.D.O.C.K., P.R.O.D.O.C.K., Q.X.P., ROSETTALIGAND, RiboDock, Yucca, AutoDock, etc. employ the MC method. The uncertainty of convergence, which is one of the main issues with the MC technique, may be reduced by carrying out many independent runs [40].

A genetic algorithm (G.A.), which is essentially used to determine the global minima, is pretty similar to the MC approach. These have a significant evolutionary theory bias. A population of ligands with a corresponding fitness that is specified by the scoring function is maintained by GA [41]. Each ligand represents a possible hit. Through mutation or crossing, the G.A. modifies the population of ligands. The more suitable ligands from the previous phase are accessed, and a new population is produced by choosing them. The modification process involves changing the population's members [42].

While the crossover operator facilitates communication between two (rarely more) population members, the mutation operator generates new ligands from a single ligand by arbitrarily altering a representational fragment [43]. Programs like Autodock 4.0, DARWIN, D.I.V.A.L.I., FITTED, FLIPDock, GAMBLER, GAsDock, GOLD 3.1, and PSI-DOCK have included G.A. Like the MC method; the G.A. method has a limitation where the uncertainty of convergence is a big problem [44]. The hierarchical technique is an additional strategy (Franklin 2007). The low-energy conformations of the ligand are pre-calculated and aligned in this method. Similar conformations are positioned next to one another within the hierarchy by merging the populations of the pre-generated ligand conformations [45]. The docking algorithm will then employ this hierarchical data

Materials Research Foundations 145 (2023) 236-249 https://doi.org/10.21741/9781644902370-9

structure when the ligand is rotated or translated to minimize the results. Let's use a straightforward example to illustrate. Suppose an atom close to the rigid center of the ligand is discovered to collide with the protein in a specific rotation or translation. In that case, this method can rule out all conformations below the conformation under scrutiny in the hierarchy because the descendants must also contain the same clash. The GLIDE program employs a hierarchical approach [46,47].

2.2 Software used for NPs molecular docking

In general, many phases are included in the molecular docking simulation process to accurately simulate the suggested interaction between the NPs and the macromolecules. To recreate the NPs' accurate form and size, the structure must first be meticulously constructed. The chemical structure of the NPs could be found online in places like the Cambridge Cluster Database [48], or it could be made from scratch using software like Material Studio [49] and Chem Draw [50]. Then, using the proper method and degree of theory, the geometry of those NPs must be optimized by energy minimization. Recent scholarship has made extensive use of software modules for N.P. optimization. Gaussian, Forcite, and C.A.S.T.E.P. are a few examples [51]. In 1970, Carnegie Mellon University created the general-purpose computational chemistry software program known as Gaussian. It can carry out ab initio, DFT, semi-empirical computations, and molecular mechanics calculations utilizing specific force fields [52]. With the classic molecular mechanic tool Forcite, periodic systems and crystalline structures may be optimized while maintaining crystal symmetry [53]. To simulate the solids, interfaces, and surfaces of numerous material classes, including ceramics, semiconductors, and metals, C.A.S.T.E.P. uses an ab initio quantum mechanical approach. Most studies use NPs less than 100 nm in size and come in various forms, including sheets, lattices, and spheres [4]. Additionally, several investigations have focused on tiny single units with sizes in the region of an angstrom, which are typical in real life. It is important to carefully replicate the main shape and size of the NPs in biological fluids while preparing the NPs structure. For example, certain NPs, like gold, have a tendency to cluster fast in solution [54]. To predict the binding mode, the docking program should be given a thoroughly produced, parametrized copy of the actual structure of the NPs under physiological settings. As previously noted, NPs with precise size and form might be created *in silico* using experimental data from N.M.R. and Dynamic Light Scattering (D.L.S.) studies or retrieved from internet resources like the Cambridge Cluster Database [55]. Many databases are already offering experimental data for the common NPs, including InterNano, Nano-EHS, NanoHUB, and NANO by Springer Nature [56]. For instance, researchers from Heidelberg University hypothesized that gold nanoparticles have a mean diameter of 12 nm. In contrast, the proteins are only 3 nm (explained below), and ubiquitin-protein interprets them as flat surfaces. A single three-atom molecule of titanium dioxide (TiO2) was employed for the docking simulation in recent research on the interaction between human serum albumin (H.S.A.) and titanium dioxide nanoparticles (TiO2 NPs) [57]. However, TiO2 NPs often have a rutile or anatase crystal structure, in which the titanium is coordinated by 6 oxygen atoms in an octahedral configuration. The simplification in this scenario might invalidate the docking findings since it misrepresents the true nature and size of TiO2 NPs. The biological macromolecule has to be prepared similarly. The RSCB. Protein Data Bank is typically used to get the proper crystal structure (PDB) [58]. Without such software, a homology model must be created using tools like SwissModel, Modeler (B.I.O.V.I.A.), and Prime (Schrödinger), among many others. Notably, homology modeling is a further degree of approximation that ought to be avoided when a fine crystal structure is available [59]. The ligand and macromolecule are prepared, and the program of choice is used to complete the docking procedure. It is important to note that no unique software programs exist to handle the docking simulation of NPs. The AutoDock software with the Lamarckian genetic algorithm has been employed in most current research [60]. Additionally, NPs docking has lately made use of various online docking servers. These include Patchdock, which utilizes an algorithm based on shape complementarity, and HEX 6.3, which employs a method based on the Fourier transform. We have also observed that most of the published experiments employed a stiff docking technique with fixed protein residues. The visualization of the interaction between NPs and macromolecules utilizing several platforms, including Discovery Studio, Pymol, iGEMDOCK [61], and U.C.S.F. Chimera, is the last phase [62]). The desolvation of the NPs upon protein binding is a last and significant issue. The docking method's scoring function should consider the desolvation event before contact with biological molecules. Because some proteins and NPs interact favorably in energy, this interaction should lower the system's enthalpy. However, doing so requires releasing the hydration water linked to the NPs, which increases the system's entropy [63]. As was already noted, the scoring system employed in the docking simulation will determine how to account for this entropy change. The majority of widely used docking software can take the desolvation impact into account and include it in the binding score shown to the user [64]. In the reported NPs docking, for example, the scoring function of AutoDock software, which is frequently used, is based on the AMBER force field and includes a desolvation term as well as a term dependent on the number of sp3 bonds to account for unfavourable entropy of ligand binding due to restricted rotations [65–67].

2.3 Molecular docking as a promising predictive model for silver nanoparticle mediated inhibition of cytochrome P450 enzymes

Many xenobiotics undergo oxidative metabolism thanks to cytochrome P450 (C.Y.P.) enzymes. We investigated the interactions of silver nanoparticles (AgNPs) and silver ions (Ag^+) with six C.Y.P. isoforms, namely CYP1A2, CYP2C9, CYP2C19, CYP2D6, CYP2E1, and CYP3A4, using molecular docking and quantum mechanical (Q.M.) simulations. The docking findings showed that critical amino acids of CYP2C9, CYP2C19, and CYP2D6 interacted with the Ag3 cluster, not Ag^+, at a distance of around 3Å. Also, the Q.M. analysis showed that the amino acid residues of these C.Y.P. enzymes interacted strongly with the Ag3 cluster, which helped explain how the C.Y.P. enzyme activity might be stopped [68]. Interestingly, these outcomes support earlier in vitro research showing that AgNPs decreased CYP2C and CYP2D activity in rat liver microsomes. The Ag_3 cluster is proposed as the smallest unit of AgNPs for *in silico* modeling. In conclusion, we showed that molecular docking, together with Q.M. analysis, is an effective method for anticipating AgNP-mediated C.Y.P. inhibition. These techniques might be applied to different nanomaterials and help gain a greater knowledge of response processes [54].

2.4 Limitations

The main drawback of molecular docking is the lack of trust in scoring functions' capacity to provide precise binding energies. This is due to the difficulty in precisely predicting certain intermolecular interaction variables, such as the solvation effect and entropy change. Additionally, several intermolecular interactions that have been shown to be significant are seldom taken into account by scoring methods. For instance, guanidine-arginine interactions and halogen bonding are shown to affect the affinity of the protein-ligand binding but are not taken into account. For two reasons, accurately handling the water molecules in the binding pocket during the docking process is still an open issue that requires much attention shortly. First, since smaller atoms scatter light inefficiently, hydrogen coordinate information cannot be found in x-ray crystal structures. Inaccuracies in the identification of water molecules that may be serving as a bridge molecule between the ligand and the receptor result from not knowing the precise location of hydrogen. Second, it is impossible to correctly forecast how ligands will influence water molecules and how strong that influence will be.

Furthermore, it is hard to estimate, given the state of our knowledge, how many of the water molecules in the binding pocket would be replaced by potential ligands and how ligand binding would affect the hydrogen bonding network. The stiff receptor is one of the most difficult problems in docking. Depending on the ligand they bind to, proteins may take on a variety of distinct shapes. As a consequence, docking using a stiff receptor will only represent one receptor conformation, which often results in false negatives when the ligand is subsequently shown to be active. This happens because proteins can stay in constant motion between different structural states with similar energies.

3. Future status and opportunities for improvement

For a long time, molecular docking has been a valuable tool for probing the interaction of biologically active molecules with their targets. Other structure-based techniques, such as M.D. simulation and free energy perturbation of protein complexes, could also provide insights into the mode of binding and the mechanism of action of the therapeutic agents of interest. Recently, significant progress has been achieved in the structure-based *in silico* design of NPs and the prediction of their nanotoxicity. However, there are still some shortcomings and an excellent potential for improvement of these techniques for the study of NPs cytotoxicity. We can summarize our recommendation for future studies in the following points: (1) The solubility and redox potential of the NPs should be carefully addressed, as many toxic effects of metal oxide NPs are mediated by their ability to release metal cations and increase the levels of R.O.S. (2) Prior to molecular docking investigations, the exact shape and dimensions of NPs should be reproduced. The actual size of NPs in biological fluids could greatly vary. For example, it is necessary to know if

Materials Research Forum LLC
https://doi.org/10.21741/9781644902370-9

the NPs are docked within a protein pocket or if the whole protein molecule adheres to the NPs surface. (3) The effect of protein corona should be considered. Usually, this corona has a dynamic behavior, and its composition varies depending on the cellular environment. (4) Energy expressions specialized for studying the interaction of protein-metal interfaces, such as INTERFACE FF should be utilized instead of standard harmonic force fields. (5) There are still very few M.D. simulation studies of the protein-NPs complexes despite the availability of several software and online tools. This technique could help in understanding the time-dependent behavior of NPs-macromolecule binding. (6) A combination of structure-based and ligand-based techniques, such as Q.S.A.R. modeling, should be used for studying the interaction of NPs to abundant proteins with large and/or flexible binding sites such as H.S.A., Ubiquitin, and cytochromes. Docking into these large elusive protein pockets could be challenging and might not be sufficient to understand the mechanism of toxicity.

References

[1] J. Adams, M. Wright, H. Wagner, J. Valiente, D. Britt, A. Anderson, Cu from dissolution of CuO nanoparticles signals changes in root morphology, Plant Physiol. Biochem. 110 (2017) 108–117. https://doi.org/10.1016/j.plaphy.2016.08.005

[2] A. Ambrosone, C. Tortiglione, Methodological approaches for nanotoxicology using cnidarian models, Toxicol. Mech. Methods. 23 (2013) 207–216. https://doi.org/10.3109/15376516.2012.747117

[3] A.C. Anselmo, S. Mitragotri, Nanoparticles in the clinic: An update, Bioeng. Transl. Med. 4 (2019). https://doi.org/10.1002/btm2.10143

[4] S. Ates, E. Zor, I. Akin, H. Bingol, S. Alpaydin, E.G. Akgemci, Discriminative sensing of DOPA enantiomers by cyclodextrin anchored graphene nanohybrids, Anal. Chim. Acta. 970 (2017) 30–37. https://doi.org/10.1016/j.aca.2017.03.052

[5] M. Sotelo-Boyás, Z. Correa-Pacheco, S. Bautista-Baños, Y. Gómez y Gómez, Release study and inhibitory activity of thyme essential oil-loaded chitosan nanoparticles and nanocapsules against foodborne bacteria, Int. J. Biol. Macromol. 103 (2017) 409–414. https://doi.org/10.1016/j.ijbiomac.2017.05.063

[6] E. Baranowska-Wójcik, D. Szwajgier, P. Oleszczuk, A. Winiarska-Mieczan, Effects of Titanium Dioxide Nanoparticles Exposure on Human Health—a Review, Biol. Trace Elem. Res. 193 (2020) 118–129. https://doi.org/10.1007/s12011-019-01706-6

[7] F. Barbero, L. Russo, M. Vitali, J. Piella, I. Salvo, M.L. Borrajo, M. Busquets-Fité, R. Grandori, N.G. Bastús, E. Casals, V. Puntes, Formation of the Protein Corona: The Interface between Nanoparticles and the Immune System, Semin. Immunol. 34 (2017) 52–60. https://doi.org/10.1016/j.smim.2017.10.001

[8] H. Bayraktar, C.C. You, V.M. Rotello, M.J. Knapp, Facial control of nanoparticle binding to cytochrome c, J. Am. Chem. Soc. 129 (2007) 2732–2733. https://doi.org/10.1021/ja067497i

[9] S.F. Bellah, H. Akbar, S.M.S. Billah, D.M. Sedzro, The Role of CCL18 protein in Breast Cancer Development and Progression, Cell Biol. Res. Ther. 07 (2018). https://doi.org/10.4172/2324-9293.1000138

[10] L. Bertilsson, M.L. Dahl, P. Dalén, A. Al-Shurbaji, Molecular genetics of CYP2D6: Clinical relevance with focus on psychotropic drugs, Br. J. Clin. Pharmacol. 53 (2002) 111–122. https://doi.org/10.1046/j.0306-5251.2001.01548.x

[11] F. Bertoli, D. Garry, M.P. Monopoli, A. Salvati, K.A. Dawson, The Intracellular Destiny of the Protein Corona: A Study on its Cellular Internalization and Evolution, ACS Nano. 10 (2016) 10471–10479. https://doi.org/10.1021/acsnano.6b06411

[12] G. Brancolini, D.B. Kokh, L. Calzolai, R.C. Wade, S. Corni, Docking of ubiquitin to gold nanoparticles, ACS Nano. 6 (2012) 9863–9878. https://doi.org/10.1021/nn303444b

[13] C. Buzea, I.I. Pacheco, K. Robbie, Nanomaterials and nanoparticles: sources and toxicity., Biointerphases. 2 (2007) MR17-71. https://doi.org/10.1116/1.2815690

[14] J. Cao, Y. Pan, Y. Jiang, R. Qi, B. Yuan, Z. Jia, J. Jiang, Q. Wang, Computer-aided nanotoxicology: risk assessment of metal oxide nanoparticlesvianano-QSAR, Green Chem. 22 (2020) 3512–3521. https://doi.org/10.1039/d0gc00933d

[15] F. Carnal, A. Clavier, S. Stoll, Polypeptide-nanoparticle interactions and corona formation investigated by monte carlo simulations, Polymers (Basel). 8 (2016). https://doi.org/10.3390/polym8060203

[16] F.S. Alves, J. de A. Rodrigues Do Rego, M.L. Da Costa, L.F. Lobato Da Silva, R.A. Da Costa, J.N. Cruz, D.D.S.B. Brasil, Spectroscopic methods and in silico analyses using density functional theory to characterize and identify piperine alkaloid crystals isolated from pepper (Piper Nigrum L.), J. Biomol. Struct. Dyn. 38 (2020) 2792–2799. https://doi.org/10.1080/07391102.2019.1639547

[17] T. Cedervall, I. Lynch, S. Lindman, T. Berggård, E. Thulin, H. Nilsson, K.A. Dawson, S. Linse, Understanding the nanoparticle-protein corona using methods to quntify exchange rates and affinities of proteins for nanoparticles, Proc. Natl. Acad. Sci. U. S. A. 104 (2007) 2050–2055. https://doi.org/10.1073/pnas.0608582104

[18] J.N. Cruz, S.N. Mali, Antimalarial Hemozoin Inhibitors (β-Hematin Formation Inhibition): Latest Updates, Comb. Chem. High Throughput Screen. 25 (2022) 1987–1990. https://doi.org/10.2174/1386207325666220117145351

[19] C.I. Chang, W.J. Lee, T.F. Young, S.P. Ju, C.W. Chang, H.L. Chen, J.G. Chang, Adsorption mechanism of water molecules surrounding Au nanoparticles of different sizes, J. Chem. Phys. 128 (2008). https://doi.org/10.1063/1.2897931

[20] Y.C. Chen, Beware of docking!, Trends Pharmacol. Sci. 36 (2015) 78–95. https://doi.org/10.1016/j.tips.2014.12.001

[21] R.S. Kalash, V.K. Lakshmanan, C.S. Cho, I.K. Park, Theranostics, Biomater. Nanoarchitectonics. (2016) 197–215. https://doi.org/10.1016/B978-0-323-37127-8.00012-1

[22] A.R.J.A. de M. Lima, A.S. Siqueira, M.L.S. Möller, R.C. de Souza, J.N. Cruz, A.R.J.A. de M. Lima, R.C. da Silva, D.C.F. Aguiar, J.L. da S.G.V. Junior, E.C. Gonçalves, In silico improvement of the cyanobacterial lectin microvirin and mannose interaction, J. Biomol. Struct. Dyn. (2020). https://doi.org/10.1080/07391102.2020.1821782

[23] N. Chowdhury, A. Bagchi, Molecular insight into the activity of LasR protein from Pseudomonas aeruginosa in the regulation of virulence gene expression by this organism, Gene. 580 (2016) 80–87. https://doi.org/10.1016/j.gene.2015.12.067

[24] A.J. Clark, P. Tiwary, K. Borrelli, S. Feng, E.B. Miller, R. Abel, R.A. Friesner, B.J. Berne, Prediction of Protein-Ligand Binding Poses via a Combination of Induced Fit Docking and Metadynamics Simulations, J. Chem. Theory Comput. 12 (2016) 2990–2998. https://doi.org/10.1021/acs.jctc.6b00201

[25] R. Concu, V. V. Kleandrova, A. Speck-Planche, M.N.D.S. Cordeiro, Probing the toxicity of nanoparticles: a unified in silico machine learning model based on perturbation theory, Nanotoxicology. 11 (2017) 891–906. https://doi.org/10.1080/17435390.2017.1379567

[26] V.M. Almeida, Ê.R. Dias, B.C. Souza, J.N. Cruz, C.B.R. Santos, F.H.A. Leite, R.F. Queiroz, A. Branco, Methoxylated flavonols from Vellozia dasypus Seub ethyl acetate active myeloperoxidase extract: in vitro and in silico assays, J. Biomol. Struct. Dyn. 40 (2022) 7574–7583. https://doi.org/10.1080/07391102.2021.1900916

[27] M.M. D'Elios, F. Vallese, N. Capitani, M. Benagiano, M.L. Bernardini, M. Rossi, G.P. Rossi, M. Ferrari, C.T. Baldari, G. Zanotti, M. De Bernard, G. Codolo, The Helicobacter cinaedi antigen CAIP participates in atherosclerotic inflammation by promoting the differentiation of macrophages in foam cells, Sci. Rep. 7 (2017). https://doi.org/10.1038/srep40515

[28] C.M.A. Rego, A.F. Francisco, C.N. Boeno, M. V. Paloschi, J.A. Lopes, M.D.S. Silva, H.M. Santana, S.N. Serrath, J.E. Rodrigues, C.T.L. Lemos, R.S.S. Dutra, J.N. da Cruz, C.B.R. dos Santos, S. da S. Setúbal, M.R.M. Fontes, A.M. Soares, W.L. Pires, J.P. Zuliani, Inflammasome NLRP3 activation induced by Convulxin, a C-type lectin-like isolated from Crotalus durissus terrificus snake venom, Sci. Rep. 12 (2022) 1–17. https://doi.org/10.1038/s41598-022-08735-7

[29] K.M. Darwish, I. Salama, S. Mostafa, M.S. Gomaa, E.S. Khafagy, M.A. Helal, Synthesis, biological evaluation, and molecular docking investigation of benzhydrol-

and indole-based dual PPAR-γ/FFAR1 agonists, Bioorganic Med. Chem. Lett. 28 (2018) 1595–1602. https://doi.org/10.1016/j.bmcl.2018.03.051

[30] T.R. De Kievit, R. Gillis, S. Marx, C. Brown, B.H. Iglewski, Quorum-Sensing Genes in Pseudomonas aeruginosa Biofilms: Their Role and Expression Patterns, Appl. Environ. Microbiol. 67 (2001) 1865–1873. https://doi.org/10.1128/AEM.67.4.1865-1873.2001

[31] L.B. de O. Freitas, L. de M. Corgosinho, J.A.Q.A. Faria, V.M. dos Santos, J.M. Resende, A.S. Leal, D.A. Gomes, E.M.B. de Sousa, Multifunctional mesoporous silica nanoparticles for cancer-targeted, controlled drug delivery and imaging, Microporous Mesoporous Mater. 242 (2017) 271–283. https://doi.org/10.1016/j.micromeso.2017.01.036

[32] A.R. DeBofsky, R.H. Klingler, F.X. Mora-Zamorano, M. Walz, B. Shepherd, J.K. Larson, D. Anderson, L. Yang, F. Goetz, N. Basu, J. Head, P. Tonellato, B.M. Armstrong, C. Murphy, M.J. Carvan, Female reproductive impacts of dietary methylmercury in yellow perch (Perca flavescens) and zebrafish (Danio rerio), Chemosphere. 195 (2018) 301–311. https://doi.org/10.1016/j.chemosphere.2017.12.029

[33] R.J. Deeth, N. Fey, B. Williams-Hubbard, DommiMOE: An implementation of ligand field molecular mechanics in the molecular operating environment, J. Comput. Chem. 26 (2005) 123–130. https://doi.org/10.1002/jcc.20137

[34] C.A. Dinarello, Overview of the IL-1 family in innate inflammation and acquired immunity., Immunol. Rev. 281 (2018) 8–27. https://doi.org/10.1111/imr.12621

[35] C.P. Profaci, R.N. Munji, R.S. Pulido, R. Daneman, The blood-brain barrier in health and disease: Important unanswered questions., J. Exp. Med. 217 (2020). https://doi.org/10.1084/jem.20190062

[36] A. Elkashif, M.N. Seleem, Investigation of auranofin and gold-containing analogues antibacterial activity against multidrug-resistant Neisseria gonorrhoeae, Sci. Rep. 10 (2020). https://doi.org/10.1038/s41598-020-62696-3

[37] H. Mirzaei, S. Emami, Recent advances of cytotoxic chalconoids targeting tubulin polymerization: Synthesis and biological activity, Eur. J. Med. Chem. 121 (2016) 610–639. https://doi.org/10.1016/j.ejmech.2016.05.067

[38] V.E. Fako, D.Y. Furgeson, Zebrafish as a correlative and predictive model for assessing biomaterial nanotoxicity, Adv. Drug Deliv. Rev. 61 (2009) 478–486. https://doi.org/10.1016/j.addr.2009.03.008

[39] Z. Fei Yin, L. Wu, H. Gui Yang, Y. Hua Su, Recent progress in biomedical applications of titanium dioxide, Phys. Chem. Chem. Phys. 15 (2013) 4844–4858. https://doi.org/10.1039/c3cp43938k

[40] K.R. Feingold, Introduction to Lipids and Lipoproteins., in: K.R. Feingold, B. Anawalt, A. Boyce, G. Chrousos, W.W. de Herder, K. Dhatariya, K. Dungan, J.M. Hershman, J. Hofland, S. Kalra, G. Kaltsas, C. Koch, P. Kopp, M. Korbonits, C.S. Kovacs, W. Kuohung, B. Laferrère, M. Levy, E.A. McGee, R. McLachlan, J.E. Morley, M. New, J. Purnell, R. Sahay, F. Singer, M.A. Sperling, C.A. Stratakis, D.L. Trence, D.P. Wilson (Eds.), South Dartmouth (MA), 2000

[41] C.C. Fleischer, C.K. Payne, Nanoparticle-cell interactions: Molecular structure of the protein corona and cellular outcomes, Acc. Chem. Res. 47 (2014) 2651–2659. https://doi.org/10.1021/ar500190q

[42] F. Föger, W. Noonpakdee, B. Loretz, S. Joojuntr, W. Salvenmoser, M. Thaler, A. Bernkop-Schnürch, Inhibition of malarial topoisomerase II in Plasmodium falciparum by antisense nanoparticles., Int. J. Pharm. 319 (2006) 139–146. https://doi.org/10.1016/j.ijpharm.2006.03.034

[43] A. Strini, G. Roviello, L. Ricciotti, C. Ferone, F. Messina, L. Schiavi, D. Corsaro, R. Cioffi, TiO2-based photocatalytic geopolymers for nitric oxide degradation, Materials (Basel). 9 (2016). https://doi.org/10.3390/ma9070513

[44] S. Fraga, A. Brandão, M.E. Soares, T. Morais, J.A. Duarte, L. Pereira, L. Soares, C. Neves, E. Pereira, M. de L. Bastos, H. Carmo, Short- and long-term distribution and toxicity of gold nanoparticles in the rat after a single-dose intravenous administration, Nanomedicine Nanotechnology, Biol. Med. 10 (2014) 1757–1766. https://doi.org/10.1016/j.nano.2014.06.005

[45] E.L. Fuchs, E.D. Brutinel, A.K. Jones, N.B. Fulcher, M.L. Urbanowski, T.L. Yahr, M.C. Wolfgang, The Pseudomonas aeruginosa Vfr regulator controls global virulence factor expression through cyclic AMP-dependent and -independent mechanisms, J. Bacteriol. 192 (2010) 3553–3564. https://doi.org/10.1128/JB.00363-10

[46] M. Dhayalan, L. Anitha Jegadeeshwari, N. Nagendra Gandhi, Biological activity sources from traditionally usedtribe and herbal plants material, Asian J. Pharm. Clin. Res. 8 (2015) 11–23

[47] S. Gandhi, I. Roy, Synthesis and characterization of manganese ferrite nanoparticles, and its interaction with bovine serum albumin: A spectroscopic and molecular docking approach, J. Mol. Liq. 296 (2019). https://doi.org/10.1016/j.molliq.2019.111871

[48] M. Geetha, A.K. Singh, R. Asokamani, A.K. Gogia, Ti based biomaterials, the ultimate choice for orthopaedic implants - A review, Prog. Mater. Sci. 54 (2009) 397–425. https://doi.org/10.1016/j.pmatsci.2008.06.004

[49] R. Ghadari, A study on the interactions of amino acids with nitrogen doped graphene; Docking, MD simulation, and QM/MM studies, Phys. Chem. Chem. Phys. 18 (2016) 4352–4361. https://doi.org/10.1039/c5cp06734k

[50] Z.N. Gheshlaghi, G.H. Riazi, S. Ahmadian, M. Ghafari, R. Mahinpour, Toxicity and interaction of titanium dioxide nanoparticles with microtubule protein, Acta Biochim. Biophys. Sin. (Shanghai). 40 (2008) 777–782. https://doi.org/10.1111/j.1745-7270.2008.00458.x

[51] E. GIACOBINI, Histochemical Demonstration of AChE Activity in Isolated Nerve Cells, Acta Physiol. Scand. 36 (1956) 276–290. https://doi.org/10.1111/j.1748-1716.1956.tb01325.x

[52] K. Giannousi, G. Geromichalos, D. Kakolyri, S. Mourdikoudis, C. Dendrinou-Samara, Interaction of ZnO Nanostructures with Proteins: In Vitro Fibrillation/Antifibrillation Studies and in Silico Molecular Docking Simulations, ACS Chem. Neurosci. 11 (2020) 436–444. https://doi.org/10.1021/acschemneuro.9b00642

[53] M. González-Durruthy, A.K. Giri, I. Moreira, R. Concu, A. Melo, J.M. Ruso, M.N.D.S. Cordeiro, Computational modeling on mitochondrial channel nanotoxicity, Nano Today. 34 (2020) 100913. https://doi.org/https://doi.org/10.1016/j.nantod.2020.100913

[54] L. Gonzalez-Moragas, A. Roig, A. Laromaine, C. elegans as a tool for in vivo nanoparticle assessment, Adv. Colloid Interface Sci. 219 (2015) 10–26. https://doi.org/10.1016/j.cis.2015.02.001

[55] M.M. Mihai, M.B. Dima, B. Dima, A.M. Holban, Nanomaterials for wound healing and infection control, Materials (Basel). 12 (2019) 2176. https://doi.org/10.3390/ma12132176

[56] A. Kushwaha, L. Goswami, B.S. Kim, Nanomaterial-Based Therapy for Wound Healing, Nanomaterials. 12 (2022). https://doi.org/10.3390/nano12040618

[57] S. Smulders, K. Luyts, G. Brabants, K. Van Landuyt, C. Kirschhock, E. Smolders, L. Golanski, J. Vanoirbeek, P.H.M. Hoet, Toxicity of nanoparticles embedded in paints compared with pristine nanoparticles in mice, Toxicol. Sci. 141 (2014) 132–140. https://doi.org/10.1093/toxsci/kfu112

[58] A. Gupta, A.T. Müller, B.J.H. Huisman, J.A. Fuchs, P. Schneider, G. Schneider, Generative Recurrent Networks for De Novo Drug Design, Mol. Inform. 37 (2018) 1700111. https://doi.org/10.1002/minf.201700111

[59] T.I. Adelusi, A.-Q.K. Oyedele, I.D. Boyenle, A.T. Ogunlana, R.O. Adeyemi, C.D. Ukachi, M.O. Idris, O.T. Olaoba, I.O. Adedotun, O.E. Kolawole, Y. Xiaoxing, M. Abdul-Hammed, Molecular modeling in drug discovery, Informatics Med. Unlocked. 29 (2022) 100880. https://doi.org/https://doi.org/10.1016/j.imu.2022.100880

[60] S. Dallakyan, A.J. Olson, Small-molecule library screening by docking with PyRx, Methods Mol. Biol. 1263 (2015) 243–250. https://doi.org/10.1007/978-1-4939-2269-7_19

[61] D. Nath, F. Singh, R. Das, X-ray diffraction analysis by Williamson-Hall, Halder-Wagner and size-strain plot methods of CdSe nanoparticles- a comparative study, Mater. Chem. Phys. 239 (2020) 122021. https://doi.org/10.1016/j.matchemphys.2019.122021

[62] E.F. Pettersen, T.D. Goddard, C.C. Huang, G.S. Couch, D.M. Greenblatt, E.C. Meng, T.E. Ferrin, UCSF Chimera - A visualization system for exploratory research and analysis, J. Comput. Chem. 25 (2004) 1605–1612. https://doi.org/10.1002/jcc.20084

[63] H. Li, P. Xia, S. Pan, Z. Qi, C. Fu, Z. Yu, W. Kong, Y. Chang, K. Wang, D. Wu, X. Yang, The advances of ceria nanoparticles for biomedical applications in orthopaedics, Int. J. Nanomedicine. 15 (2020) 7199–7214. https://doi.org/10.2147/IJN.S270229

[64] H.J. Wiggers, J.R. Rocha, J. Cheleski, C.A. Montanari, Integration of ligand- and target-based virtual screening for the discovery of cruzain inhibitors, Mol. Inform. 30 (2011) 565–578. https://doi.org/10.1002/minf.201000146

[65] N.C. da R. Galucio, D. de A. Moysés, J.R.S. Pina, P.S.B. Marinho, P.C. Gomes Júnior, J.N. Cruz, V.V. Vale, A.S. Khayat, A.M. do R. Marinho, Antiproliferative, genotoxic activities and quantification of extracts and cucurbitacin B obtained from Luffa operculata (L.) Cogn, Arab. J. Chem. 15 (2022) 103589. https://doi.org/10.1016/j.arabjc.2021.103589

[66] J. Wang, R.M. Wolf, J.W. Caldwell, P.A. Kollman, D.A. Case, Development and testing of a general Amber force field, J. Comput. Chem. 25 (2004) 1157–1174. https://doi.org/10.1002/jcc.20035

[67] A.K. Malde, L. Zuo, M. Breeze, M. Stroet, D. Poger, P.C. Nair, C. Oostenbrink, A.E. Mark, An Automated force field Topology Builder (ATB) and repository: Version 1.0, J. Chem. Theory Comput. 7 (2011) 4026–4037. https://doi.org/10.1021/ct200196m

[68] M. González-Durruthy, S. Manske Nunes, J. Ventura-Lima, M.A. Gelesky, H. González-Díaz, J.M. Monserrat, R. Concu, M.N.D.S. Cordeiro, MitoTarget Modeling Using ANN-Classification Models Based on Fractal SEM Nano-Descriptors: Carbon Nanotubes as Mitochondrial F0F1-ATPase Inhibitors, J. Chem. Inf. Model. 59 (2019) 86–97. https://doi.org/10.1021/acs.jcim.8b00631

Materials Research Foundations 145 (2023) 250-280 https://doi.org/10.21741/9781644902370-10

Chapter 10

Applications of Nanomaterials in Cancer Diagnosis

Syeda Mariam Fatima[1], Usman Ali Ashfaq[1], Muhammad Qasim[1], Sehar Aslam[1], Samman Munir[1], Muhammad Hassan Sarfraz[2], Muhammad Bilal[3], Mohsin Khurshid[2,*]

[1]Department of Bioinformatics and Biotechnology, Government College University, Faisalabad, Pakistan

[2] Institute of Microbiology, Government College University, Faisalabad, Pakistan

[3]School of Life Sciences and Food Engineering, Huaiyin Institute of Technology, Huaian, China

mohsinkhurshid@gcuf.edu.pk

Abstract

The early detection of cancer is extremely important in the fight against cancer. However, the detection of various cancers in their early stages has been impeded by the inherent capacity of conventional diagnostic methods. Nanotechnology has provided new materials and contrast agents for an earlier and more accurate diagnosis of cancers as well as the prognosis of cancers during cancer treatment. Therefore, it is believed that nanotechnology could provide a highly sensitive, specific, and multiplexed capacity to measure extracellular cancer biomarkers and can be able to detect cancer cells. In this chapter, the recent developments in the application of nanotechnology for the diagnosis of cancers have been summarized. Further, the key issues which hinder the clinical application of nanotechnology-based diagnostic approaches have been discussed.

Keywords

Biomarkers, Cancer, Ultrasound, Imaging, Diagnosis, Prognosis

Contents

Materials Research Foundations 145 (2023) 250-280 https://doi.org/10.21741/9781644902370-10

1. Introduction

The incidence and mortality rate of cancer has been increasing throughout the world. According to GLOBOCAN 2018, there will be 18.1 million new cases of cancer while the death toll will reach 9.6 million. It is estimated that many new cases of cancer will be 18.1 million while the number of deaths related to cancer will reach 9.6 million according to GLOBOCAN 2018. Moreover, it is suggested that thirty million people will die from cancer every year by 2030. So, the first step against the battle of cancer involves early detection that ensures successful cancer treatment because the cancer mortality rate is significantly reduced. For example, patients with breast cancer exhibiting distant metastasis have a five-year survival rate of 27% while the patients at the local stage exhibit a relative five-year survival rate of 90% [1].

Good patient prognosis is linked with cancer detection at early stages as it allows immediate meditation to avert the progression of cancer at early stages. Although this is not possible in every circumstance as cancer would only be detectable when it reached a certain size. In such conditions, imaging or biopsy may be performed to confirm the diagnosis by examining whether any cell is precancerous or cancerous. However, the detection of cancer becomes difficult due to heterogeneity within the cancer lesion that retards the evaluation of the sample at the molecular or cellular level [2-4]. So, the sample must be homogenized after evaluating the samples at the molecular level, as the normal cell may give false results.

The objective of cancer detection is to detect cancer at the earliest stage when small numbers of cells were transformed into the cancerous cell as well as to monitor the rate of progression of cancer and analyze the treatment response [5]. Nowadays, cytology, histopathology, and imaging techniques are used for the early detection of cancer. Currently available cancer detection techniques like ultrasound, X-ray analysis, magnetic resonance imaging, and endoscopy can be used for detection only if a noteworthy change in the tissues is present. So, these techniques can detect cancers at the stage when there is a possibility of metastasis by many cells. Moreover, currently available techniques are not able to discriminate benign lesions from malignant lesions [6]. Furthermore, histopathology and cytology couldn't be able to detect cancer at an earlier stage effectively. Hence, there is a major challenge to develop such technologies that could aid in cancer detection at earlier stages [7].

The clinical use of nanotechnology to detect cancers is not yet in practice because it is still subjected to several medical tests and screening procedures. Nanoparticles (NPs) are used to inactive cancer biomarkers (CB) such as circulating tumor cells (CTCs), exosomes, cancer-associated proteins, and circulating tumor DNA (ctDNA) [8]. Nanoparticles can be explored for cancer detection owing to their properties like a large surface area to volume ratio, which provides a binding site for various biological entities such as antibodies, aptamers, peptides, etc. These biological moieties, attached to the NPS, can then be used for identifying cancer cells. Therefore, introducing various ligands to cancer cells enabled to achieve the multivalent effects that can improve the sensitivity and selectivity of an assay [9]. Nanotechnology-dependent diagnostics are considered promising tools for cost-effective, real-time, and convenient cancer detection and diagnosis [10]. This chapter highlighted the recent progress of nanotechnology and discussed the latest applications of nanotechnology in cancer diagnosis.

2. Characteristics of nanoparticles and nanomaterials

Several nanomaterials such as metal or metal oxide-based, carbon-based, polymer-based as well as liposomes, and quantum dots (QDs) have been considered efficient and reliable in the diagnosis of cancer. These nanomaterials possess certain characteristics such as

Materials Research Forum LLC
https://doi.org/10.21741/9781644902370-10

definite size and shape (cubes, rods, stars, and clusters) along with surface area characteristics which allow them to bind with biological moieties for specific cell targeting [5]. Along with cell detection, the NPs can also be complexed with the therapeutic agent as a potential treatment option at the designated site. Moreover, nanoparticles could also act as signal generators as they can intensify the signals [11]. These aspects of nanoparticles lead to upgraded imaging technologies. For example, surface plasmon-enhanced light scattering (SP-LS) for in-vitro imaging could be enabled by metallic nanoparticles (MNPs) such as aluminum, gold, and silver. These MNPs depending on the size and surface attributes boost up fluorescent yield in imaging and therefore they are called metal enhanced fluorescence based imaging [12, 13]. Moreover, MNPs like gold and silver play the role of amplifiers to generate the Raman spectrum and thus can be used in surface-enhanced Raman spectroscopy (SERS) to enable deep tissue penetration [14]. For example, nanoshells and carbon nanotubes can be used in deep tissue imaging techniques, involving near-infrared radiation (NIR), where the ultrasound is produced, collected, and converted to an electric signal [15, 16]. Another example of a deep tissue imaging technique is the detection of luminescence of gold nanostructures by using in-vivo or in-vitro two-photon luminescence (TPL) imaging techniques for cancer diagnosis [11]. Hence, these imaging techniques were considered efficient to use for biomedical imaging due to the remarkable characteristics of nanomaterials.

QDs are fluorescent semiconductor nanocrystals that are photostable, have broad absorption spectra and high quantum yields as well as resistant to photobleaching. Hence these properties of QDs enabled them to improve fluorescent imaging in-vitro and in-vivo [17, 18].

Nanoparticles may be conjugated or encapsulated with other imaging molecules so that's why they are called multifunctional. Moreover, multiplexed, or multimodal imaging is referred to the use of more than one imaging technique at the same time. This technique involves the use of two or more imaging modalities for precise information about the metabolic and molecular behavior, localization, and metastasis of cancer. So, the benefit of using multi-purpose nanoparticles is that they can be employed both for diagnosis and therapy purposes, by attaching the imaging and therapeutic agent to the nanoparticles at the same time [5].

3. Nanotechnology for cancer diagnosis

Nanotechnology is a promising methodology that creates new detection platforms using nanofabrication and nanomaterials that improved the early diagnosis of cancer by enhancing imaging techniques and screening (Table 1). Moreover, the discomfort linked with the endoscopic techniques is relieved by the process of miniaturization. There are two methodologies for the diagnosis of cancer that have been discussed below.

Materials Research Foundations 145 (2023) 250-280 | https://doi.org/10.21741/9781644902370-10

3.1 Nanotechnology-based biosensing

The sensitivity of different transducers that are used in biosensing such as field effect transistor (FET), quartz crystal microbalance (QCM), and surface plasmon resolution (SPR) microscopy could be enhanced by nanotechnology. Due to the increased surface area-to-volume ratio, the nanomaterials used for the transducers can help in the higher loading of the molecular recognition elements (MREs) [19, 20]. Biosensing platforms employ various nanoparticles or nanomaterials that use different bottom-up or top-down approaches for fabrication. However, some of these biosensing platforms used nanoscale components while the rest represents some common nano-based biosensing platforms.

3.1.1 Lab-on-a-chip and microarrays

Lab-on-a-chip uses micro and nanofluidic components by integrating nano and micro-electronics, actuators, mechanical elements, sensors, reaction chambers, and microchannels on a silicon substrate [21]. Moreover, it also employs nanowires and nanoelectrodes that are present in microfluidic channels and are used for biosensing in compact devices. Most of the structures are created by top-down approaches such as chemical etching or photolithography techniques. The purpose of using such devices is based on the notion of miniaturizing various lab procedures to perform them on a chip at a small scale. There are various benefits of this miniaturization such as high reproducibility, lower dead volumes, and increased analysis speed, low consumption of reagents and samples that conduct various analyses simultaneously by enabling high throughput screening in a short time. Unique conductive, optical, thermal, or magnetic properties, as well as multifunctionality of several nanomaterials such as MNPs, QDs, smart polymers, or carbon nanotubes (CNTs), allowed them to incorporate into microfluidics for biosensing [22-24]. Microarrays are one of the commonly used devices for the detection of biological molecules like DNA, RNA, proteins, etc. There are several consequences of using microarrays such as it could potentially improve the isolation of specific cancer cells, drug screening, proteomic and genomic profiling, and initiation of in-vitro cancer models [25-27].

3.1.2 Sphere-based platforms

Various spherical platforms involve the use of different nanoparticles for imaging or biosensing such as the use of gold nanoparticles (AuNPs) as biosensors because of their ability to change color because of aggregation. Moreover, clusters are formed when DNA-functionalized AuNPs bind to their target DNA, and as a result changes in SERS, color, and light scattering take place [28]. Furthermore, silver (Ag) or AuNPs are either confined to a surface or in solution form or could be able to modulate the optical properties, specifically of the localized SPR, that depend on the biomolecular interactions happening on their surfaces when these NPs combined with the stimuli-responsive polymer [29, 30]. Similarly, this platform also uses anionic green fluorescent protein (GFP) that binds with cationic AuNPs that are responsible for suppressing the fluorescence of GFP. While the addition of cancer cell lysates to NPs causes replacement of GFP and ultimately fluorescence detection. Moreover, interactions of functionalizing AuNPs with proteins result in achieving protein signatures that help in developing hydrogen or hydrophobic bonding [31]. In addition, another spherical-based biosensing platform uses liposomes for the detection of biomolecular interactions by supporting the formation of lipid bilayers on sphere-generated platforms. Additionally, they are quite potent to use for biomarker screening [32].

3.1.3 Magnetic-based assays

Various magnetic screening assays efficiently use functionalized magnetic NPs as sensors such as in the diagnostic magnetic resonance technique; several biomarkers could be detected by functionalized magnetic NPs by detecting variations in the effects of magnetic resonance [33, 34]. Similarly, magneto resistance (MR) based biosensors are another way to effectively use magnetic NPs that employ MR devices. These devices are a fusion of magnetic and nanomagnetic structures that use magnetic NPs conjugated with an analyte or an antigen as well as MRE-containing MR devices to detect minor changes in the magnetic field [35].

3.1.4 Other platforms

Along with the multifunctional devices mentioned above, there are some other nano-based platforms like cantilevers, nanoelectrodes, and self-assembled monolayers (SAMs), that can be used for biosensing applications. Micro and nano-cantilevers are made up of CNTs or silicon and they are used for detecting selective bio interactions and can be employed for lab-free detection assays [36, 37]. The nanoelectrodes, consisting of metal oxides or CNTs, can be used as sensors for in-vitro detection of cancer cells by detecting various biomolecules [37-39]. On the other hand, SAMs are long chains of alkane thiols that are adsorbed to the surface of gold by forming a covalent bond between the sulfur of the thiol group and gold to form ordered monolayer film that is used for printing and patterning on substrates.

3.2 Nanotechnology for screening of biomarkers

CB is a biological molecule that is found in various tissues, blood, or body fluids such as urine and saliva and could act as an indicator of cancer in the body [40]. There are different biomarkers for cancer detection which include nucleic acids (DNA or circulating tumor miRNA), carbohydrates, and proteins (either cell surface protein or secreted proteins) [41-43]. Cancer can be detected earlier and the efficiency of therapy could be monitored by measuring certain levels of cancer biomarkers. However, several obstacles limited the use of biomarkers such as diversity in the abundance of biomarkers, the small amount available in body fluids, and their timing within the patients make it difficult to carry out further studies [44]. As nanotechnology has high sensitivity and selectivity so it can conduct measurements of multiple targets at once. Nanomaterials or nanoparticles are used to improve biosensors for specific targeting [45]. Furthermore, the use of nanoparticles in biosensors provides sensitivity as they have a high surface-to-volume ratio that helps them

Materials Research Foundations 145 (2023) 250-280
https://doi.org/10.21741/9781644902370-10

to achieve the demands of specialized molecular diagnostic techniques. Polymer dots (PDs), AuNPs, and QDs are the most common probes of nanoparticles for cancer diagnostics [46].

3.2.1 Protein detection

The various proteins combined with aptamers, antibodies, or antibody fragments help in the detection of cancer such as CA-125 (ovarian cancer), CEA (colorectal cancer), PSA (prostate cancer), and AFP (liver cancer). A measurable quantifiable signal could be generated as a result of this interaction. QD-based biosensors have been used for the detection of cancer biomarkers in recent years as they possess unique properties such as high resistance to degradation and photobleaching, high efficient Stokes shifts, outstanding molar extinction factor, and quantum yield [47-49]. The protein biomarkers can be detected by a sandwiched-based assay where a primary capture antibody detects the biomarker, followed by a secondary antibody that binds with the capture antibody. Various methods such as fluorescence and staining help to visualize secondary antibodies [50]. Two QD-conjugated antibodies can be used for diagnosing two different biomarkers (carcinoembryonic antigen and neuron-specific enolase) having a limit of detection (LOD) for each reaching 1.0 ng/mL [51]. Similarly, an immunoassay of a QD-based sandwich of zinc oxide (ZnO) is developed for nanowire substrates of ZnO that provide multiple binding sites and a large surface area for detection. The carcinoembryonic antigen can be used to anticipate the phenomenon of tumor recurrence even after surgery in late-stage cancer patients as well as to monitor treatment against cancer and is therefore considered an important cancer biomarker. On the other hand, NSE is an enzyme that coverts 2-phosphoglycerate into phosphoenolpyruvate which describes the link between small cell lung carcinoma, carcinoids, and islet cell tumors. They could be detected at concentrations of more than 15ng/mL after secretion.

Similarly, another example of QDs based immune-sensor is antibody-coated ZnO QDs against carbohydrate antigen (CA) 19-9. Due to the high isoelectric point of ZnO, the immobilization process favored by electrostatic absorption depends upon the high isoelectric point of ZnO and the immune reaction of antigens and antibodies of CA 19-9 that are responsible for the formation of sandwich-like structure. This immunological recognition of CA 19-9 was converted into quantifiable signals that are detected by inherent photoluminescence (PL) as well as square wave stripping voltammetry (SWV). The dynamic range of optical spectral detection and the electrochemical assay was 1–180 U/ml and 0.1– 180 U/ml while their LOD reached 0.25 U/ml and 0.04 U/ml respectively [52].

Peptides are used frequently to target in vivo cancerous tissues. The Arginylglycylaspartic acid (RGD) is a common peptide motif that has been recognized on the cell surfaces involved in angiogenesis and cancer metastasis by a receptor integrin $\alpha v\beta 3$ that is used to target cancerous tissues for diagnosis [53]. A study reported the co-administration of an enzyme-mediated gold nanoparticle system with iRGD (tumor penetration peptide) and obtained higher penetration in 4 T1 mammary tumors [54].

Materials Research Forum LLC
https://doi.org/10.21741/9781644902370-10

Aptamers are RNA sequences or single-stranded DNA (ssDNA) that have strong binding specificity and high affinity for their specified e.g., in the case of prostate cancer, detection can be done by nanoparticle conjugated with 10 RNA aptamer that can bind with prostate-specific membrane antigen (PSMA) due to the incorporation of Cy5. Cy5-PLA/aptamer NPs could only bind to canine prostate adenocarcinoma cells and LNCaP cells that are positive for PSMA. Furthermore, Cy5-PLA nanoparticles have been utilized in balb/c mice and possess lower background fluorescence but excellent signals in different organs [55].

3.2.2 ctDNA detection

Circulating tumor-derived DNA (ctDNA) express tumor-derived fragments in the blood stream that are almost 100-200 base pairs long [56]. CTCs or primary tumors release ctDNA which allows cancer detection via cancer-associated genetic abnormalities that could detect cancer before any sign or symptom. Cancer-specific genetic anomalies could also be detected by specific nucleic acid hybridization probes having complementary sequences [57, 58]. Moreover, for the detection of a single exon in the BRCA1 gene in breast cancer a fluorescent probe of DNA silver nanocrystal (NC) was developed whose LOD was increased to 6.4×10^{-11}M in an optimized environment. NC fluorescence is used to detect deletion mutations in BRCA1, and the fluorescent probe of DNA silver NC enhances AgNC fluorescence which distinguished the BRCA1 deletions. The emission peaks of fluorescence intensity of DNA-AgNCs reached 440nm while LOD reached 6.4×10^{-11} M. Higher fluorescence was achieved for the deletion types while low fluorescence was obtained for the normal type in the sensing system [59].

3.2.3 MicroRNA detection

Prostate cancer can be diagnosed by the detection of miR-141 using a two-step biosensing platform system. The biosensing platform used CdSe/ZnS QDs in the first step that is modified with a FRET quencher that possesses covalent binding with functionalized nucleic acids having telomerase primer sequence along with recognition sequence for miR-141. When the probe hybridized miR-141 a duplex was created that cleaved the duplex-specific nuclease (DSN) and as the consequence, QDs activated and exposed the telomerase primer sequence. In the next step, elongation of the primer unit occurred with the help of telomerase/dNTPs. Moreover, the incorporation of hemin and luminol/H2O2 created chemiluminescence. So, it aids in differentiating prostate cancer of patients from healthy people by detecting miR-141 in the serum samples [60].

3.2.4 DNA methylation detection

The change in global methylation levels was reported by methylscape which is a common cancer diagnostic tool. Based on the gold affinity of DNA and DNA solvation, the authors distinguished normal genomes from cancerous ones. In this novel methodology, authors discriminate cancer genomes from normal genomes based on DNA solvation and DNA gold affinity so, in this way quite sensitive, efficient, and simple colorimetric or electrochemical one-step assays are developed for the detection of cancer [61].

3.2.5 Extracellular vesicle detection

The molecular state of the tumor cells can be accessed from various biological molecules (DNA, protein, mRNA, etc.) that are packed inside extracellular vehicles (EVs). Recent studies show that scientists developed novel magnetic nanopore sequencing technology that could isolate particular EVs from plasma [62]. Moreover, EV miRNA biomarkers are identified by RNA-sequencing and machine-learning algorithms. A mouse model having pancreatic ductal adenocarcinoma (PDAC) is used to test this technique and it leads to the identification of eleven EV miRNAs biomarkers. In recent years, a sensor-platform is used that comprises 5 aptamers panels and 13nm AuNPs to profile proteins on the surface of exosomes with high specificity and affinity in a few minutes. Further, the complex of AuNPs with aptamers averts the accumulation of NPs in a high salt solution, and exosomes helped in breaking the non-specific and weak binding between aptamers and AuNP. However, the specific and strong binding among surface proteins of exosomes and aptamers disturbed aptamers from the surface of AuNP and thus aiding the accumulation of AuNPs. The color of AuNPs changed from red to blue due to aggregation and it indicates the binding of aptamers with exosomal surface proteins. The aggregation of AuNPs (A650/A520) demonstrated the intensity as an indicator of relative abundance expressed by exosomal surface target proteins [63].

3.2.6 Clinical applications of nanotechnology in biosensing

The aspect of biosensing has also been applied in the field of cancer diagnostics with the purpose to develop the element of sensitivity in the detection. The idea was generated from the fact that trained dogs can detect disease based on the difference in the breath odor. So, the role of potential volatile key molecules was examined that could be detected by dogs by using their sniffing sense [64, 65]. This work influenced the scientists to develop specific MREs that are utilized in biosensing technologies of bioelectronic nose for the detection of volatile molecules in breath, excretions, body fluids, and secretions of the patients [66, 67].

Artificial nanoscale noses can be used for ovarian, gastric, and lung cancer by detecting volatile biomarkers of cancer through a biosensing approach by employing nanowires, porous materials, gold nanoparticles, and different transducers. A clinical trial for a nano-based nose sensor made up of CNTs and AuNPs is under study for detecting volatile cancer biomarkers in the breath of patients suffering from gastric cancer [68, 69] while another similar study carried out trial for glandular prostate cancer. The work in the field of nanotechnology for ameliorating the quantification and detection of miRNA is on-going but more progress is required that ensure reproducibility with the help of sample processing standardization and agreement on internal standards of miRNA [70]. Currently, some challenges have been faced in terms of optimizing standardization and sample processing. For example, lab-on-a-chip technology has been facing many challenges such as a lack of interest by researchers to expedite clinician adoption, economic misunderstanding, standardization, and reproducibility [71].

3.3 Nanotechnology for cancer imaging

Nanomaterials have applications in targeted molecular imaging as well as either enabling new modalities or enhancing the existing ones for cancer imaging. Functionalized nanomaterials combined with MREs can be employed for cancer detection by using ligand-receptor interactions (Table 2). However, many other target cancers may be gathered at the cancer site due to extravasation. In addition, these conditions would cause leakage of nanoparticles; having the size of 20-400nm, through the endothelial gaps that may serve as the site for leucocyte extravasation into the cancer tissues and retained there. As compared to the formulations of control solutions this leads to a better accumulation at the site of cancer tissue [72, 73]. As the consequence, the nanoparticles segregated into cancer resulting in a large number of imaging agents and ultimately the increasing contrast between cancer and the surrounding tissue. Meanwhile, other factors are also reviewed that may be patient-or tumor microenvironment (TME) specific and could have an impact on extravasation [74]. Furthermore, malfunctioning of nanoparticles enabled the use of functionalizing therapeutic agents along with imaging agents to attain an adequate theragnostic level that leads to the monitoring of therapy response.

3.3.1 Targeted molecular imaging

Techniques of targeted molecular imaging involve targeted imaging of gene expression as well as certain cancer-associated cell markers. Appropriate biomarkers are pre-requisite for molecular imaging that are recognized and targeted by designed probes or MREs, made up of targeting ligands linked with an imaging tag or reporter. The tag or reporter used in any of the imaging methodologies could be a nanomaterial. Various molecular targets in cancers include VEGF receptor (VEGFR), folate receptor alpha, matrix metalloproteases, epidermal growth factor receptor (EGFR), prostate-specific antigen (PSA), human epidermal growth factor receptor-2 (HER2), transferrin, and integrin receptors [75-78]. Targeted molecular imaging is used to target cancer biomarkers and it is noticed that target tumors contain even less than 5% of NPs. Targeted molecular imaging use of nanomaterials has various advantages such as combining imaging and targeting in a platform through nanocarrier multifunctionality, using novel imaging modalities, and having the characteristic of multifunctionalizing the nanocarrier with a therapeutic agent.

Gene expression of cancer-specified antigens can be examined via molecular imaging where the promoter of the cancer reporter gene is paired with an imaging probe which, upon activation of the promoter (PSA, carcinoembryonic antigen, and alpha-fetoprotein), provides detection of the gene [79]. Molecular imaging holds the potential for both diagnostic and treatment purposes where the imaging can ensure the tumor microfoci debulking.

3.3.2 Types of imaging enabled by nanomaterials

Nanomaterials exhibit unique characteristics in enabling biomedical imaging methodologies such as TPL, SPR, SERS, and PAI that have not been used before. These molecular imaging techniques employ NPs to target certain markers for cancer imaging

Materials Research Forum LLC
https://doi.org/10.21741/9781644902370-10

through ligand-receptor interaction. For example, SPR microscopy can be used for the in vitro analysis of cancerous cells by targeting EGFR with AuNPs [80, 81]. The expression of particular cancer markers is studied by the evaluation of biopsies from patients. PAI employs different AuNPs, gold nanorods coated with silica, and gold nanoshells for deep tissue and in-vivo imaging [11, 82]. SERS imaging utilizes AuNPs and is used for both in-vivo and in-vitro analysis where it has been successfully used in solid cancers and deep tissue imaging. Furthermore, TPL used gold-based nanostructures for the cancerous tissue [83].

3.3.3 Imaging techniques enhanced by nanomaterials

Various techniques for imaging can be enhanced by nanomaterials where different nanostructures can be employed in MRI, sonography, NIR, and fluorescence imaging to provide in-depth analysis. The different imaging modalities and their nanoparticulate contrast agents are summarized in Table 3.

3.3.3.1 Microbubbles in sonography

The polymer and lipid-based microbubbles contain intravenously directed contrast agents (a gas normally) that passed through microvasculature and are detected by contrast-enhanced ultrasound (CE-US) and thus improved sonography [84]. Different cell markers can be targeted by the microbubbles for localizing them at the inflammation site for imaging [85]. Moreover, the functionalization of microbubbles with PEG boosts their half-life in circulation and averts their accumulation. Additionally, CE-US has been used to evaluate microvascular changes in ovarian cancer and to determine the success of anti-angiogenic therapy. Hence, this technique is considered appropriate for monitoring the anti-angiogenic therapy response in cancers or detecting neovascularities at the metastatic sites of patients [86, 87].

3.3.3.2 Nanomaterials as enhancing contrast agents for MRI

Superparamagnetic iron oxide nanoparticles (SPIONS) and gadolinium (Gd) based NPs are used to enhance the sensitivity of MRI by nanotechnology [75]. Various NPs including micelles, dendrimers, liposomes, and polymers have utilized Gd which increased its imaging time by increasing half-life, enabled theranostics, reduced Gd toxicity, and enhanced MRI sensitivity [88]. SPIONS are modified by combining them with polymers such as dendrimers or PEG, liposomes, and micelles while unmodified magnetic iron oxide nanoparticles (MIONS) use FeO in MRI [89, 90]. Half-life in circulation and solubility of FeO is improved by PEGylation which helps in escaping uptake and recognition by the reticuloendothelial system [91]. In addition to several cancer-specific markers such as transferrin receptor, folate receptor alpha, and integrins are used in combination with FeO for molecular imaging [75].

3.3.3.3 Quantum dots for fluorescent imaging

 Materials Research Forum LLC
https://doi.org/10.21741/9781644902370-10

Quantum dots (QDs) can be used for both in-vitro and in-vivo analytical techniques own to their photostability. QDs can produce fluorescent emissions upon excitation with single excitation light which aids in detection. Furthermore, multiple molecular events can also be detected in the live cells by using multiple colored QDs which show different colored emissions for each detection [92, 93]. They are also used in-vivo to target cancer with the help of ligand-receptor interactions or extravasations. Various NPs such as CNTs, micelles, polymers, and liposomes are used in combination with QDs for theranostic purposes or imaging [94]. However, the major issue regarding the use of QDs is the toxicity that remains even after functionalization with different biomolecules.

3.3.3.4 Nanomaterials for NIR deep tissue imaging

Natural NIR optical absorbance is beneficial for in-vivo imaging of cancerous tissues using various nanomaterials such as carbon and gold-based nanomaterials, QDs, and upconverting nanoparticles (UCNPs). Simultaneously, the absorbance of NIR by these nanomaterials results in heat generation as a consequence of which the death of cells in the surrounding tissue takes place. Due to these characteristics of nanomaterials, they are efficiently used in photo ablation therapy combined with imaging techniques in such a manner that they have a theranostic value [95, 96].

3.3.3.5 Nano-based multimodal imaging

Multimodal imaging produces various signals and therefore can be employed for multiple imaging techniques. For example, single photon emission computer tomography (SPECT), positron emission tomography (PET) scan and MRI could be detected by combined using of radioactive, magnetic, and optical reporters. The objective of this technique is to enhance localization and enable the timely detection of cancers [97]. Moreover, multiple analyses can also be carried out simultaneously and the progression of cellular events can be observed in real-time. This technique employs simultaneous MRI and CT imaging using iron oxide and AuNPs in hepatoma-bearing mice and simultaneous SERS-based imaging and NIR fluorescence imaging using Au-QD hybrids or gold nanorods [98, 99]. However, recent studies show that graphene iron oxide NPs are used for conducting in vivo triple mode fluorescent imaging, MRI, and PAI [100].

4. Limitations

Biocompatibility and toxicity are the two major factors that are needed to be fulfilled by the nanomaterial to be employed for the detection techniques. These two issues are always the foremost concern for the use of nanomaterials. Generalizations should be avoided, and the individual materials are studied separately. Moreover, the biodistribution, safety, and effectiveness of nanomaterials are needed to be evaluated further by in-vivo and in-vitro studies. Nanomaterials or some imaging materials could be retained in patients' bodies and this involves major concerns in the clearance of nanomaterials as they result in toxicity [91, 101]. Additionally, the biomarkers like polyethylene glycol (PEG) or peptides or

 Materials Research Forum LLC
https://doi.org/10.21741/9781644902370-10

oligonucleotides, or proteins, to be used for targeting the target site of the cell, can be immunogenic and thus can elicit an immune response.

5. Clinical applications

Small-size tumors could be detected earlier by the use of nanomaterials in targeted molecular imaging. On the other hand, nanomaterials are also responsible for developing novel imaging modes or enhancing the existing ones. Imaging technique during surgeries allows real-time monitoring of successful residual removal phenomenon. Hence, the clinical applications of cancer imaging are vast. In addition, SPECT and MRI nuclear imaging are approved by FDA using colloidal NPs and SPIONs respectively. These colloidal nanoparticles involve albumin and sulfur colloids that lead to a high labeling yield. However, researchers are facing a severe challenge in the clinical use of nanomaterials in imaging due to the rising concerns of nanomaterials regarding safety and toxicity [102, 103].

Various studies have reported the potential of nanotechnology for a cancer diagnosis; however, ultimate rigorous clinical trials are required. At present, multiple cancer diagnostic approaches based on nanotechnology are under experimentation such as silica hybrid NPs (C dots) combined with 124I-labeled cyclo-[Arg-Gly-Asp-Tyr] (cRGDY) peptides have been developed for PET imaging of the patients suffering from malignant brain tumors or metastatic melanoma. These cRGDY peptides could selectively bind with integrins to probe the tumor cells. Further, to visualize lymph nodes with cancer during surgery fluorescent cRGDY C-dots have been developed for lymph node mapping. Hence it is obvious that as the research proceeds, more and more cancer diagnostic approaches that are nanotechnology-based will move toward clinics [104].

Conclusion and future outlooks

Cancer diagnostics are improved by enhancing imaging techniques and biosensing of cancer markers through nanotechnology. The enhancement of biosensing of cancer biomarkers involves high sensitivity, cheap and rapid screening, less requirement of power, miniaturization, and high throughput analysis. The improved methods will expedite the identification of cancer markers in biosensing for the development of cancer vaccines. The examination of proteomes requires great importance by introducing different functional groups on NPs for screening. Different functional groups are introduced on NPs to examine the proteome for screening. This feature is usually absent in current methods but it will help in the staging of cancers and distinguishing the normal tissues from cancerous tissues [5].

Noninvasive imaging methods are enhanced due to novel multimodal or single imaging methodologies. The use of new photostable materials or novel imaging techniques in nanotechnology improved in vitro imaging as they enabled real-time imaging for a long period of time. Recently, multimodal imaging is a matter of interest as it improves the early detection of cancer by functionalizing various NPs to attain theranostic value which paves

Materials Research Foundations 145 (2023) 250-280 https://doi.org/10.21741/9781644902370-10

the way toward personalized cancer therapy. Through nanotechnology, new modalities for cancer imaging have been created for the early detection of cancer as well as personalized cancer therapy and resection of cancer become possible using simultaneous imaging along with real-time monitoring of the patients to observe therapy response in them. In addition, nanotechnology is responsible for the detection of residual cancer or micrometastasis as well as improving nodal staging in cancer [5].

Various challenges in the clinical use of the above techniques include uncontrollable processes for scale-up production, differences between findings of clinical studies in humans versus preclinical studies in animals, changes in batch-to-batch reproducibility, the inadequacy of standardization, lack of interest in research in clinical adoption and concerns related with the toxicity of some NPs. However, this field remains focused to find different ways and means for fruitfully utilizing nanomaterials for cancer diagnosis and therapy [105, 106]. It is a dire need to focus on specific cancers and develop such ways to employ nanotechnology in the diagnostics and therapeutics of these cancers. Thus, it is suggested that clinical applications will be successful through the collaboration between physical and biomedical scientists to get novel nano-inventions and further get approval for their clinical use by studying the efficacy, clearance, biodistribution, and safety of NPs before using them for imaging applications.

Table 1: Nanoparticles for diagnosis of various cancer types

Type of Cancer	Nanoparticle	Specificity of Ligand	Type of Affinity Probe	References
Breast Cancer	Gold nanoparticles	Her2	Aptamer	[107]
	Magnetic nanoparticles	EpCAM	Antibody	[108]
	Nanofibers			[109]
	Nanoparticle-coated silicon bead	EpCAM/CD146		[110]
	Nanorod arrays	EpCAM	DNA aptamer	[111]
	Polymer dots		Antibody	[112]
	Upconversion NPs	Her2	Antibody	[113]
Colon Cancer	Magnetic nanoparticles	EpCAM		[114]
Colorectal Cancer	Nanoparticle-coated silicon bead	EpCAM/CD146		[110]
Leukemia	Gold nanoparticles	Cd2/cd3		[115]
	Quantum dots	PTK7	Aptamer	[116]
Liver Cancer	Carbon nanotubes	EpCAM	Antibody	[117]
	Magnetic nanoparticles			[118]
Lung Cancer				[114]

Table 2: Ongoing clinical trials for cancer diagnostic imaging

Phases	NCT Number	Type of cancer	Imaging modalities
Early Phase 1	NCT03280277	Stage IIIA/IIIB/IIIC rectal cancer AJCC-7	Contrast enhanced MRI using Ferumoxytol
	NCT02857218	Stage IIB/IIIA/IIIB/IIIC esophageal cancer AJCC-7/	MRI using Ferumoxytol
Phase 1	NCT04167969	Prostate cancer	MRI/PET using 64Cu-NOTA-PSMAi-PEG-Cy5.5-C' dots or (64Cu)-labeled PSMA-targeting particle
	NCT04167722	prostate cancer	Robotic radical prostatectomy
	NCT03967652	Cancer diagnosis	Nanomaterial-based sensors
Phase 1/2	NCT04300673	Lymph node metastases/ Prostate cancer	Radio guided surgery using Indium-labeled prostate-specific membrane antigen (PSMA)
	NCT03134846	Head and neck squamous cell carcinoma	Cetuximab-IRDye800CW
	NCT02106598	Neck and head melanoma	Fluorescent cRGDY-PEG-Cy5.5-C dots
Phase 3	NCT04261777	Prostate cancer/metastasis/prostatecto my	Ferumoxtran-10
	NCT02751606	Breast/rectal neoplasms	MRI using Ferumoxtran-10
	NCT04239105	Circulating tumor cells/ Breast neoplasms	Microfluidic and Raman spectrum
Not Applicable	NCT04825002	Prostate cancer	Urinary multimarker sensor
	NCT04661176		Sentinel™ PCC4 Assay
	NCT04482803	Biopsy of axillary lymph nodes	Carbon NPs suspension
	NCT03817307	Neck and head squamous cell carcinoma	USPIO-enhanced MRI
	NCT01411904	Leukemia	MagProbe™

Table 3: Overview of different imaging modalities and their nanoparticulate contrast agents

Imaging techniques	**Nanoparticulate contrast agents**	**Advantages**	**Disadvantages**
Computed Tomography (CT)	Bismuth NPs [119]	Rapid, Low costs, Unlimited penetrating depth, Higher resolution (~ 20-200μm), Good soft tissue contrast following injection of contrast agent	Lower sensitivity to contrast agents, Radiation exposure
	Barium NPs [120]		
	Gold NPs [121]		
	Liposomes and iodine-based micelles [122]		
Ultrasound (US)	Nanobubbles [123]	Rapid, Low costs, Real-time imaging, Higher temporal, and spatial resolution (~50-100μm),	User dependent, inappropriate for complete body imaging
	Air-releasing polymers [124]		
	Gas-filled microbubbles [84, 125, 126]		
Magnetic resonance imaging (MRI)	Hyperpolarized probes [127]	Very high penetration, High spatial resolution (~10-500μm), Excellent soft tissue contrast, Multiple options for structural, functional, and metabolic characterization	High cost, Time consuming, Low sensitivity to contrast agents
	Gadolinium-containing probes [128]		
	Superparamagnetic iron oxide NPs (SPION) [129]		
	paraCEST agents [130]		
	Paramagnetic polymers & liposomes [131, 132]		
Optical Imaging (OI)	Fluorescent NP probes [133]	Low cost, broader probe range, Higher sensitivity for contrast agents	Sensitive to artifacts, Low penetrating (<10cm), High background signal
	Near-infrared fluorochrome labeled NPs [134]		
	Quantum dots [135]		

Photoacoustic imaging (PAI)	Fluorescent/dye-loaded NPs [16]	Low-cost, real-time imaging, highly sensitive	Low penetration depth (up to ~5-6cm), relatively low specificity to contrast agents
	Nanorods and NPs [136]		
	Carbon nanotubes [137]		
Positron emission tomography (PET)	Polymeric NPs [138]	High sensitivity, quantitative, deeper penetration	High cost, radiation exposure, no anatomical information, low spatial resolution (1-2mm)
	Radiolabeled gold nanoshells as contrast agents [139]		
Single photon emission computed tomography (SPECT)	Nano and microcolloids [140]	High sensitivity, very high penetration depth, long-circulating radionuclides	High cost, radioactive probes, low spatial resolution (1-2mm), no anatomical information
	Technetium-labeled gold NPs [141]		
	Indium-labeled liposomes [142]		

References

[1] A. Rezaianzadeh, M. Jalali, A. Maghsoudi, A.M. Mokhtari, S.H. Azgomi, S.L. Dehghani, The overall 5-year survival rate of breast cancer among Iranian women: A systematic review and meta-Analysis of published studies, Breast Dis. 37 (2017) 63–68. https://doi.org/10.3233/BD-160244

[2] S. Bhatia, J. V. Frangioni, R.M. Hoffman, A.J. Iafrate, K. Polyak, The challenges posed by cancer heterogeneity, Nat. Biotechnol. 30 (2012) 604–610. https://doi.org/10.1038/nbt.2294

[3] D. Dornan, J. Settleman, Dissecting cancer heterogeneity, Nat. Biotechnol. 29 (2011) 1095–1096. https://doi.org/10.1038/nbt.2063

[4] N.C. da R. Galucio, D. de A. Moysés, J.R.S. Pina, P.S.B. Marinho, P.C. Gomes Júnior, J.N. Cruz, V.V. Vale, A.S. Khayat, A.M. do R. Marinho, Antiproliferative, genotoxic activities and quantification of extracts and cucurbitacin B obtained from Luffa operculata (L.) Cogn, Arab. J. Chem. 15 (2022) 103589. https://doi.org/10.1016/j.arabjc.2021.103589

[5] R. Zeineldin, Nanotechnology for cancer screening and diagnosis: from innovations to clinical applications, Biomater. Cancer Ther. (2020) 261–289. https://doi.org/10.1016/b978-0-08-102983-1.00010-7

[6] Y.E. Choi, J.W. Kwak, J.W. Park, Nanotechnology for early cancer detection, Sensors. 10 (2010) 428–455. https://doi.org/10.3390/s100100428

[7] A.B. Chinen, C.M. Guan, J.R. Ferrer, S.N. Barnaby, T.J. Merkel, C.A. Mirkin, Nanoparticle Probes for the Detection of Cancer Biomarkers, Cells, and Tissues by Fluorescence, Chem. Rev. 115 (2015) 10530–10574. https://doi.org/10.1021/acs.chemrev.5b00321

[8] S. Jia, R. Zhang, Z. Li, J. Li, Clinical and biological significance of circulating tumor cells, circulating tumor DNA, and exosomes as biomarkers in colorectal cancer, Oncotarget. 8 (2017) 55632–55645. https://doi.org/10.18632/oncotarget.17184

[9] X.J. Chen, X.Q. Zhang, Q. Liu, J. Zhang, G. Zhou, Nanotechnology: A promising method for oral cancer detection and diagnosis, J. Nanobiotechnology. 16 (2018) 52. https://doi.org/10.1186/s12951-018-0378-6

[10] T.L. Doane, C. Burda, The unique role of nanoparticles in nanomedicine: Imaging, drug delivery and therapy, Chem. Soc. Rev. 41 (2012) 2885–2911. https://doi.org/10.1039/c2cs15260f

[11] J.R. Lakowicz, Radiative decay engineering 5: Metal-enhanced fluorescence and plasmon emission, Anal. Biochem. 337 (2005) 171–194. https://doi.org/10.1016/j.ab.2004.11.026

[12] M. Schmelzeisen, Y. Zhao, M. Klapper, K. Müllen, M. Kreiter, Fluorescence enhancement from individual plasmonic gap resonances, ACS Nano. 4 (2010) 3309–3317. https://doi.org/10.1021/nn901655v

[13] A.S. Thakor, J. Jokerst, C. Zavaleta, T.F. Massoud, S.S. Gambhir, Gold nanoparticles: A revival in precious metal administration to patients, Nano Lett. 11 (2011) 4029–4036. https://doi.org/10.1021/nl202559p

[14] D. Pan, M. Pramanik, S.A. Wickline, L. V. Wang, G.M. Lanza, Recent advances in colloidal gold nanobeacons for molecular photoacoustic imaging, Contrast Media Mol. Imaging. 6 (2011) 378–388. https://doi.org/10.1002/cmmi.449

[15] W.J. Akers, C. Kim, M. Berezin, K. Guo, R. Fuhrhop, G.M. Lanza, G.M. Fischer, E. Daltrozzo, A. Zumbusch, X. Cai, L. V. Wang, S. Achilefu, Noninvasive photoacoustic and fluorescence sentinel lymph node identification using dye-loaded perfluorocarbon nanoparticles, ACS Nano. 5 (2011) 173–182. https://doi.org/10.1021/nn102274q

[16] M. Bruchez, M. Moronne, P. Gin, S. Weiss, A.P. Alivisatos, Semiconductor nanocrystals as fluorescent biological labels, Science (80-.). 281 (1998) 2013–2016. https://doi.org/10.1126/science.281.5385.2013

[17] K.E. Sapsford, T. Pons, I.L. Medintz, H. Mattoussi, Biosensing with luminescent semiconductor quantum dots, Sensors. 6 (2006) 925–953. https://doi.org/10.3390/s6080925

[18] R. Antiochia, P. Bollella, G. Favero, F. Mazzei, Nanotechnology-Based Surface Plasmon Resonance Affinity Biosensors for in Vitro Diagnostics, Int. J. Anal. Chem. 2016 (2016) 2981931. https://doi.org/10.1155/2016/2981931

[19] P. Abdul Rasheed, N. Sandhyarani, Quartz crystal microbalance genosensor for sequence specific detection of attomolar DNA targets, Anal. Chim. Acta. 905 (2016) 134–139. https://doi.org/10.1016/j.aca.2015.11.033

[20] M. Medina-Sánchez, S. Miserere, A. Merkoçi, Nanomaterials and lab-on-a-chip technologies, Lab Chip. 12 (2012) 1932–1943. https://doi.org/10.1039/c2lc40063d

[21] C.H. Vannoy, A.J. Tavares, M. Omair Noor, U. Uddayasankar, U.J. Krull, Biosensing with quantum dots: A microfluidic approach, Sensors. 11 (2011) 9732–9763. https://doi.org/10.3390/s111009732

[22] A.A. Ghazani, J.A. Lee, J. Klostranec, Q. Xiang, R.S. Dacosta, B.C. Wilson, M.S. Tsao, W.C.W. Chan, High throughput quantification of protein expression of cancer antigens in tissue microarray using quantum dot nanocrystals, Nano Lett. 6 (2006) 2881–2886. https://doi.org/10.1021/nl062111n

[23] S.W. Dutse, N.A. Yusof, Microfluidics-based lab-on-chip systems in DNA-based biosensing: An overview, Sensors. 11 (2011) 5754–5768. https://doi.org/10.3390/s110605754

[24] L. Wang, P.C.H. Li, Microfluidic DNA microarray analysis: A review, Anal. Chim. Acta. 687 (2011) 12–27. https://doi.org/10.1016/j.aca.2010.11.056

[25] S.J. Maerkl, Next generation microfluidic platforms for high-throughput protein biochemistry, Curr. Opin. Biotechnol. 22 (2011) 59–65. https://doi.org/10.1016/j.copbio.2010.08.010

[26] C.S. Thaxton, D.G. Georganopoulou, C.A. Mirkin, Gold nanoparticle probes for the detection of nucleic acid targets, Clin. Chim. Acta. 363 (2006) 120–126. https://doi.org/10.1016/j.cccn.2005.05.042

[27] M. Tagliazucchi, M.G. Blaber, G.C. Schatz, E.A. Weiss, I. Szleifer, Optical properties of responsive hybrid Au@polymer nanoparticles, ACS Nano. 6 (2012) 8397–8406. https://doi.org/10.1021/nn303221y

[28] S. Rana, A.K. Singla, A. Bajaj, S.G. Elci, O.R. Miranda, R. Mout, B. Yan, F.R. Jirik, V.M. Rotello, Array-based sensing of metastatic cells and tissues using nanoparticle-fluorescent protein conjugates, ACS Nano. 6 (2012) 8233–8240. https://doi.org/10.1021/nn302917e

[29] S. Chemburu, K. Fenton, G.P. Lopez, R. Zeineldin, Biomimetic silica microspheres in biosensing, Molecules. 15 (2010) 1932–1957. https://doi.org/10.3390/molecules15031932

[30] D. Issadore, C. Min, M. Liong, J. Chung, R. Weissleder, H. Lee, Miniature magnetic resonance system for point-of-care diagnostics, Lab Chip. 11 (2011) 2282–2287. https://doi.org/10.1039/c1lc20177h

[31] J.B. Haun, T.J. Yoon, H. Lee, R. Weissleder, Molecular detection of biomarkers and cells using magnetic nanoparticles and diagnostic magnetic resonance, Methods Mol. Biol. 726 (2011) 33–49. https://doi.org/10.1007/978-1-61779-052-2_3

[32] X. Sun, D. Ho, L.M. Lacroix, J.Q. Xiao, S. Sun, Magnetic nanoparticles for magnetoresistance-based biodetection, IEEE Trans. Nanobioscience. 11 (2012) 46–53. https://doi.org/10.1109/TNB.2011.2176509

[33] B.N. Johnson, R. Mutharasan, Biosensing using dynamic-mode cantilever sensors: A review, Biosens. Bioelectron. 32 (2012) 1–18. https://doi.org/10.1016/j.bios.2011.10.054

[34] F. Huber, H.P. Lang, J. Zhang, D. Rimoldi, C. Gerber, Nanosensors for cancer detection, Swiss Med. Wkly. 145 (2015). https://doi.org/10.4414/smw.2015.14092

[35] N. Triroj, P. Jaroenapibal, H. Shi, J.I. Yeh, R. Beresford, Microfluidic chip-based nanoelectrode array as miniaturized biochemical sensing platform for prostate-specific antigen detection, Biosens. Bioelectron. 26 (2011) 2927–2933. https://doi.org/10.1016/j.bios.2010.11.039

[36] S. Viswanathan, C. Rani, S. Ribeiro, C. Delerue-Matos, Molecular imprinted nanoelectrodes for ultra sensitive detection of ovarian cancer marker, Biosens. Bioelectron. 33 (2012) 179–183. https://doi.org/10.1016/j.bios.2011.12.049

[37] J.R. Chevillet, I. Lee, H.A. Briggs, Y. He, K. Wang, Issues and prospects of microRNA-based biomarkers in blood and other body fluids, Molecules. 19 (2014) 6080–6105. https://doi.org/10.3390/molecules19056080

[38] H. Ma, J. Liu, M.M. Ali, M.A.I. Mahmood, L. Labanieh, M. Lu, S.M. Iqbal, Q. Zhang, W. Zhao, Y. Wan, Nucleic acid aptamers in cancer research, diagnosis and therapy, Chem. Soc. Rev. 44 (2015) 1240–1256. https://doi.org/10.1039/c4cs00357h

[39] C.A.K. Borrebaeck, Precision diagnostics: Moving towards protein biomarker signatures of clinical utility in cancer, Nat. Rev. Cancer. 17 (2017) 199–204. https://doi.org/10.1038/nrc.2016.153

[40] L.C. Hull, D. Farrell, P. Grodzinski, Highlights of recent developments and trends in cancer nanotechnology research-view from NCI alliance for nanotechnology in cancer, Biotechnol. Adv. 32 (2014) 666–678. https://doi.org/10.1016/j.biotechadv.2013.08.003

[41] M. Sharifi, M.R. Avadi, F. Attar, F. Dashtestani, H. Ghorchian, S.M. Rezayat, A.A. Saboury, M. Falahati, Cancer diagnosis using nanomaterials based electrochemical nanobiosensors, Biosens. Bioelectron. 126 (2019) 773–784. https://doi.org/10.1016/j.bios.2018.11.026

[42] H. Zhang, J. Lv, Z. Jia, Efficient fluorescence resonance energy transfer between quantum dots and gold nanoparticles based on porous silicon photonic crystal for DNA detection, Sensors (Switzerland). 17 (2017) 1–12. https://doi.org/10.3390/s17051078

[43] I.L. Medintz, H.T. Uyeda, E.R. Goldman, H. Mattoussi, Quantum dot bioconjugates for imaging, labelling and sensing, Nat. Mater. 4 (2005) 435–446. https://doi.org/10.1038/nmat1390

[44] R. Freeman, I. Willner, Optical molecular sensing with semiconductor quantum dots (QDs), Chem. Soc. Rev. 41 (2012) 4067–4085. https://doi.org/10.1039/c2cs15357b

[45] V. Anagnostou, K.N. Smith, P.M. Forde, N. Niknafs, R. Bhattacharya, J. White, T. Zhang, V. Adleff, J. Phallen, N. Wali, C. Hruban, V.B. Guthrie, K. Rodgers, J. Naidoo, H. Kang, W. Sharfman, C. Georgiades, F. Verde, P. Illei, Q.K. Li, E. Gabrielson, M. V. Brock, C.A. Zahnow, S.B. Baylin, R.B. Scharpf, J.R. Brahmer, R. Karchin, D.M. Pardoll, V.E. Velculescu, Evolution of neoantigen landscape during immune checkpoint blockade in non-small cell lung cancer, Cancer Discov. 7 (2017) 264–276. https://doi.org/10.1158/2159-8290.CD-16-0828

[46] H. Li, Z. Cao, Y. Zhang, C. Lau, J. Lu, Simultaneous detection of two lung cancer biomarkers using dual-color fluorescence quantum dots, Analyst. 136 (2011) 1399–1405. https://doi.org/10.1039/c0an00704h

[47] B. Gu, C. Xu, C. Yang, S. Liu, M. Wang, ZnO quantum dot labeled immunosensor for carbohydrate antigen 19-9, Biosens. Bioelectron. 26 (2011) 2720–2723. https://doi.org/10.1016/j.bios.2010.09.031

[48] C. Puig-Saus, L.A. Rojas, E. Laborda, A. Figueras, R. Alba, C. Fillat, R. Alemany, IRGD tumor-penetrating peptide-modified oncolytic adenovirus shows enhanced tumor transduction, intratumoral dissemination and antitumor efficacy, Gene Ther. 21 (2014) 767–774. https://doi.org/10.1038/gt.2014.52

[49] R. Tong, V.J. Coyle, L. Tang, A.M. Barger, T.M. Fan, J. Cheng, Polylactide nanoparticles containing stably incorporated cyanine dyes for in vitro and in vivo imaging applications, Microsc. Res. Tech. 73 (2010) 901–909. https://doi.org/10.1002/jemt.20824

[50] M.C. Schwaederlé, S.P. Patel, H. Husain, M. Ikeda, R.B. Lanman, K.C. Banks, A.A. Talasaz, L. Bazhenova, R. Kurzrock, Utility of genomic assessment of blood-derived circulating tumor DNA (ctDNA) in patients with advanced lung adenocarcinoma, Clin. Cancer Res. 23 (2017) 5101–5111. https://doi.org/10.1158/1078-0432.CCR-16-2497

[51] S.J. Tan, T. Yeo, S.A. Sukhatme, S.L. Kong, W.T. Lim, C.T. Lim, Personalized treatment through detection and monitoring of genetic aberrations in single circulating tumor cells, Adv. Exp. Med. Biol. 994 (2017) 255–273. https://doi.org/10.1007/978-3-319-55947-6_14

[52] R. Mehra, S.A. Tomlins, J. Yu, X. Cao, L. Wang, A. Menon, M.A. Rubin, K.J. Pienta, R.B. Shah, A.M. Chinnaiyan, Characterization of TMPRSS2-ETS gene aberrations in androgen-independent metastatic prostate cancer, Cancer Res. 68 (2008) 3584–3590. https://doi.org/10.1158/0008-5472.CAN-07-6154

[53] A.F.J. Jou, C.H. Lu, Y.C. Ou, S.S. Wang, S.L. Hsu, I. Willner, J.A.A. Ho, Diagnosing the miR-141 prostate cancer biomarker using nucleic acid-functionalized CdSe/ZnS QDs and telomerase, Chem. Sci. 6 (2015) 659–665. https://doi.org/10.1039/c4sc02104e

[54] J. Ko, N. Bhagwat, T. Black, S.S. Yee, Y.J. Na, S. Fisher, J. Kim, E.L. Carpenter, B.Z. Stanger, D. Issadore, MiRNA profiling of magnetic nanopore-isolated extracellular vesicles for the diagnosis of pancreatic cancer, Cancer Res. 78 (2018) 3688–3697. https://doi.org/10.1158/0008-5472.CAN-17-3703

[55] Y. Jiang, M. Shi, Y. Liu, S. Wan, C. Cui, L. Zhang, W. Tan, Aptamer/AuNP Biosensor for Colorimetric Profiling of Exosomal Proteins, Angew. Chemie - Int. Ed. 56 (2017) 11916–11920. https://doi.org/10.1002/anie.201703807

[56] T.L. Edwards, C. Giezen, C.M. Browne, Influences of indication response requirement and target prevalence on dogs' performance in a scent-detection task, Appl. Anim. Behav. Sci. 253 (2022) 105657. https://doi.org/10.1016/j.applanim.2022.105657

[57] T. Jezierski, M. Walczak, T. Ligor, J. Rudnicka, B. Buszewski, Study of the art: Canine olfaction used for cancer detection on the basis of breath odour. Perspectives and limitations, J. Breath Res. 9 (2015) 27001. https://doi.org/10.1088/1752-7155/9/2/027001

[58] A. Krilaviciute, J.A. Heiss, M. Leja, J. Kupcinskas, H. Haick, H. Brenner, Detection of cancer through exhaled breath: A systematic review, Oncotarget. 6 (2015) 38643–38657. https://doi.org/10.18632/oncotarget.5938

[59] K. Ranjan, R. Singh, Dog Nose to E-Nose in Disease Diagnosis, J. Adv. Biol. 11 (2018) 2294–2306. https://doi.org/10.24297/jab.v11i0.7959

[60] Diagnosis and Classification of 17 Diseases from 1404 Subjects via Pattern Analysis of Exhaled Molecules-Supporting Information, (2009) 198–201. https://doi.org/10.1021/acsnano.6b04930.s001

[61] R. Fiammengo, Can nanotechnology improve cancer diagnosis through miRNA detection?, Biomark. Med. 11 (2017) 69–86. https://doi.org/10.2217/bmm-2016-0195

[62] M.P. McRae, G. Simmons, J.T. McDevitt, Challenges and opportunities for translating medical microdevices: Insights from the programmable bio-nano-chip, Bioanalysis. 8 (2016) 905–919. https://doi.org/10.4155/bio-2015-0023

[63] H. Maeda, G.Y. Bharate, J. Daruwalla, Polymeric drugs for efficient tumor-targeted drug delivery based on EPR-effect, Eur. J. Pharm. Biopharm. 71 (2009) 409–419. https://doi.org/10.1016/j.ejpb.2008.11.010

[64] J. Shi, P.W. Kantoff, R. Wooster, O.C. Farokhzad, Cancer nanomedicine: Progress, challenges and opportunities, Nat. Rev. Cancer. 17 (2017) 20–37. https://doi.org/10.1038/nrc.2016.108

[65] J.E. Rosen, L. Chan, D. Bin Shieh, F.X. Gu, Iron oxide nanoparticles for targeted cancer imaging and diagnostics, Nanomedicine Nanotechnology, Biol. Med. 8 (2012) 275–290. https://doi.org/10.1016/j.nano.2011.08.017

[66] L. Fass, Imaging and cancer: A review, Mol. Oncol. 2 (2008) 115–152. https://doi.org/10.1016/j.molonc.2008.04.001

[67] L.M.A. Crane, M. Van Oosten, R.G. Pleijhuis, A. Motekallemi, S.C. Dowdy, W.A. Cliby, A.G.J. Van Der Zee, G.M. Van Dam, Intraoperative imaging in ovarian cancer: Fact or fiction?, Mol. Imaging. 10 (2011) 248–257. https://doi.org/10.2310/7290.2011.00004

[68] H.E.C. Bhang, M.G. Pomper, Cancer imaging: Gene transcription-based imaging and therapeutic systems, Int. J. Biochem. Cell Biol. 44 (2012) 684–689. https://doi.org/10.1016/j.biocel.2012.02.001

[69] I.H. El-Sayed, X. Huang, M.A. El-Sayed, Surface plasmon resonance scattering and absorption of anti-EGFR antibody conjugated gold nanoparticles in cancer diagnostics: Applications in oral cancer, Nano Lett. 5 (2005) 829–834. https://doi.org/10.1021/nl050074e

[70] J.S. Aaron, J. Oh, T.A. Larson, S. Kumar, T.E. Milner, K. V. Sokolov, Increased optical contrast in imaging of epidermal growth factor receptor using magnetically actuated hybrid gold/iron oxide nanoparticles, Opt. Express. 14 (2006) 12930. https://doi.org/10.1364/oe.14.012930

[71] J. V. Jokerst, M. Thangaraj, P.J. Kempen, R. Sinclair, S.S. Gambhir, Photoacoustic imaging of mesenchymal stem cells in living mice via silica-coated gold nanorods, ACS Nano. 6 (2012) 5920–5930. https://doi.org/10.1021/nn302042y

[72] M.B. Dowling, L. Li, J. Park, G. Kumi, A. Nan, H. Ghandehari, J.T. Fourkas, P. Deshong, Multiphoton-absorption-induced-luminescence (MAIL) imaging of tumor-targeted gold nanoparticles, Bioconjug. Chem. 21 (2010) 1968–1977. https://doi.org/10.1021/bc100115m

[73] F. Kiessling, J. Bzyl, S. Fokong, M. Siepmann, G. Schmitz, M. Palmowski, Targeted Ultrasound Imaging of Cancer: An Emerging Technology on its Way to Clinics, Curr. Pharm. Des. 18 (2012) 2184–2199. https://doi.org/10.2174/138161212800099900

[74] J.M. Warram, A.G. Sorace, R. Saini, H.R. Umphrey, K.R. Zinn, K. Hoyt, A triple-targeted ultrasound contrast agent provides improved localization to tumor vasculature, J. Ultrasound Med. 30 (2011) 921–931. https://doi.org/10.7863/jum.2011.30.7.921

[75] A.G. Sorace, R. Saini, M. Mahoney, K. Hoyt, Molecular ultrasound imaging using a targeted contrast agent for assessing early tumor response to antiangiogenic therapy, J. Ultrasound Med. 31 (2012) 1543–1550. https://doi.org/10.7863/jum.2012.31.10.1543

[76] F. Cavalieri, M. Zhou, M. Ashokkumar, The Design of Multifunctional Microbubbles for Ultrasound Image-Guided Cancer Therapy, Curr. Top. Med. Chem. 10 (2010) 1198–1210. https://doi.org/10.2174/156802610791384180

[77] A. de M. Lima, A.S. Siqueira, M.L.S. Möller, R.C. de Souza, J.N. Cruz, A.R.J. Lima, R.C. da Silva, D.C.F. Aguiar, J.L. da S.G.V. Junior, E.C. Gonçalves, In silico improvement of the cyanobacterial lectin microvirin and mannose interaction, J. Biomol. Struct. Dyn. (2020). https://doi.org/10.1080/07391102.2020.1821782

[78] D. Kozlowska, P. Foran, P. MacMahon, M.J. Shelly, S. Eustace, R. O'Kennedy, Molecular and magnetic resonance imaging: The value of immunoliposomes, Adv. Drug Deliv. Rev. 61 (2009) 1402–1411. https://doi.org/10.1016/j.addr.2009.09.003

[79] X.H. Peng, X. Qian, H. Mao, A.Y. Wang, Z.G. Chen, S. Nie, D.M. Shin, Targeted magnetic iron oxide nanoparticles for tumor imaging and therapy, Int. J. Nanomedicine. 3 (2008) 311–321. https://doi.org/10.2147/ijn.s2824

[80] X. Michalet, R. Colyer, J. Antelman, O. Siegmund, A. Tremsin, J. Vallerga, S. Weiss, Single-Quantum Dot Imaging with a Photon Counting Camera, Curr. Pharm. Biotechnol. 10 (2009) 543–557. https://doi.org/10.2174/138920109788922100

[81] M. Longmire, P.L. Choyke, H. Kobayashi, Clearance properties of nano-sized particles and molecules as imaging agents: Considerations and caveats, Nanomedicine. 3 (2008) 703–717. https://doi.org/10.2217/17435889.3.5.703

[82] G. Iyer, X. Michalet, Y.P. Chang, S. Weiss, Tracking Single Proteins in Live Cells Using Single-Chain Antibody Fragment-Fluorescent Quantum Dot Affinity Pair, Methods Enzymol. 475 (2010) 61–79. https://doi.org/10.1016/S0076-6879(10)75003-5

[83] Y. Wang, L. Chen, Quantum dots, lighting up the research and development of nanomedicine, Nanomedicine Nanotechnology, Biol. Med. 7 (2011) 385–402. https://doi.org/10.1016/j.nano.2010.12.006

[84] J.T. Robinson, G. Hong, Y. Liang, B. Zhang, O.K. Yaghi, H. Dai, In vivo fluorescence imaging in the second near-infrared window with long circulating carbon nanotubes capable of ultrahigh tumor uptake, J. Am. Chem. Soc. 134 (2012) 10664–10669. https://doi.org/10.1021/ja303737a

[85] V.J. Pansare, S. Hejazi, W.J. Faenza, R.K. Prud'Homme, Review of long-wavelength optical and NIR imaging materials: Contrast agents, fluorophores, and multifunctional nano carriers, Chem. Mater. 24 (2012) 812–827. https://doi.org/10.1021/cm2028367

[86] H. Kobayashi, M.R. Longmire, M. Ogawa, P.L. Choyke, S. Kawamoto, Multiplexed imaging in cancer diagnosis: Applications and future advances, Lancet Oncol. 11 (2010) 589–595. https://doi.org/10.1016/S1470-2045(10)70009-7

[87] J. Qian, L. Jiang, F. Cai, D. Wang, S. He, Fluorescence-surface enhanced Raman scattering co-functionalized gold nanorods as near-infrared probes for purely optical in

Materials Research Forum LLC
https://doi.org/10.21741/9781644902370-10

vivo imaging, Biomaterials. 32 (2011) 1601–1610. https://doi.org/10.1016/j.biomaterials.2010.10.058

[88] Z. Wang, H. Wu, C. Wang, S. Xu, Y. Cui, Gold aggregates- and quantum dots-embedded nanospheres: Switchable dual-mode image probes for living cells, J. Mater. Chem. 21 (2011) 4307–4313. https://doi.org/10.1039/c0jm03884a

[89] K. Yang, L. Hu, X. Ma, S. Ye, L. Cheng, X. Shi, C. Li, Y. Li, Z. Liu, Multimodal imaging guided photothermal therapy using functionalized graphene nanosheets anchored with magnetic nanoparticles, Adv. Mater. 24 (2012) 1868–1872. https://doi.org/10.1002/adma.201104964

[90] N.M. Rofsky, A.D. Sherry, R.E. Lenkinski, Nephrogenic systemic fibrosis: A chemical perspective, Radiology. 247 (2008) 608–612. https://doi.org/10.1148/radiol.2473071975

[91] A.S. Thakor, J. V. Jokerst, P. Ghanouni, J.L. Campbell, E. Mittra, S.S. Gambhir, Clinically Approved Nanoparticle Imaging Agents, J. Nucl. Med. 57 (2016) 1833–1837. https://doi.org/10.2967/jnumed.116.181362

[92] C.M.A. Rego, A.F. Francisco, C.N. Boeno, M. V. Paloschi, J.A. Lopes, M.D.S. Silva, H.M. Santana, S.N. Serrath, J.E. Rodrigues, C.T.L. Lemos, R.S.S. Dutra, J.N. da Cruz, C.B.R. dos Santos, S. da S. Setúbal, M.R.M. Fontes, A.M. Soares, W.L. Pires, J.P. Zuliani, Inflammasome NLRP3 activation induced by Convulxin, a C-type lectin-like isolated from Crotalus durissus terrificus snake venom, Sci. Rep. 12 (2022) 4706. https://doi.org/10.1038/s41598-022-08735-7

[93] Y. Zhang, M. Li, X. Gao, Y. Chen, T. Liu, Nanotechnology in cancer diagnosis: Progress, challenges and opportunities, J. Hematol. Oncol. 12 (2019) 137. https://doi.org/10.1186/s13045-019-0833-3

[94] J. Youkhanna, J. Syoufjy, M. Rhorer, O. Oladeinde, R. Zeineldin, Toward nanotechnology-based solutions for a particular disease: Ovarian cancer as an example, Nanotechnol. Rev. 2 (2013) 473–484. https://doi.org/10.1515/ntrev-2013-0008

[95] R. Zeineldin, J. Syoufjy, Cancer nanotechnology: Opportunities for prevention, diagnosis, and therapy, Methods Mol. Biol. 1530 (2017) 3–12. https://doi.org/10.1007/978-1-4939-6646-2_1

[96] W. Hong, S. Lee, H.J. Chang, E.S. Lee, Y. Cho, Multifunctional magnetic nanowires: A novel breakthrough for ultrasensitive detection and isolation of rare cancer cells from non-metastatic early breast cancer patients using small volumes of blood, Biomaterials. 106 (2016) 78–86. https://doi.org/10.1016/j.biomaterials.2016.08.020

[97] X. Wu, T. Xiao, Z. Luo, R. He, Y. Cao, Z. Guo, W. Zhang, Y. Chen, A micro-/nano-chip and quantum dots-based 3D cytosensor for quantitative analysis of circulating

tumor cells, J. Nanobiotechnology. 16 (2018) 65. https://doi.org/10.1186/s12951-018-0390-x

[98] Q. Huang, F.B. Wang, C.H. Yuan, Z. He, L. Rao, B. Cai, B. Chen, S. Jiang, Z. Li, J. Chen, W. Liu, F. Guo, Z. Ao, S. Chen, X.Z. Zhao, Gelatin nanoparticle-coated silicon beads for density-selective capture and release of heterogeneous circulating tumor cells with high purity, Theranostics. 8 (2018) 1624–1635. https://doi.org/10.7150/thno.23531

[99] N. Sun, X. Li, Z. Wang, R. Zhang, J. Wang, K. Wang, R. Pei, A Multiscale TiO2 Nanorod Array for Ultrasensitive Capture of Circulating Tumor Cells, ACS Appl. Mater. Interfaces. 8 (2016) 12638–12643. https://doi.org/10.1021/acsami.6b02178

[100] C. Wu, T. Schneider, M. Zeigler, J. Yu, P.G. Schiro, D.R. Burnham, J.D. McNeill, D.T. Chiu, Bioconjugation of ultrabright semiconducting polymer dots for specific cellular targeting, J. Am. Chem. Soc. 132 (2010) 15410–15417. https://doi.org/10.1021/ja107196s

[101] J. Shen, K. Li, L. Cheng, Z. Liu, S.T. Lee, J. Liu, Specific detection and simultaneously localized photothermal treatment of cancer cells using layer-by-layer assembled multifunctional nanoparticles, ACS Appl. Mater. Interfaces. 6 (2014) 6443–6452. https://doi.org/10.1021/am405924g

[102] Y. Zhang, B. Chen, M. He, B. Yang, J. Zhang, B. Hu, Immunomagnetic separation combined with inductively coupled plasma mass spectrometry for the detection of tumor cells using gold nanoparticle labeling, Anal. Chem. 86 (2014) 8082–8089. https://doi.org/10.1021/ac500964s

[103] X. Pang, C. Cui, M. Su, Y. Wang, Q. Wei, W. Tan, Construction of self-powered cytosensing device based on ZnO nanodisks@g-C3N4 quantum dots and application in the detection of CCRF-CEM cells, Nano Energy. 46 (2018) 101–109. https://doi.org/10.1016/j.nanoen.2018.01.018

[104] Y. Liu, F. Zhu, W. Dan, Y. Fu, S. Liu, Construction of carbon nanotube based nanoarchitectures for selective impedimetric detection of cancer cells in whole blood, Analyst. 139 (2014) 5086–5092. https://doi.org/10.1039/c4an00758a

[105] C.H. Wu, Y.Y. Huang, P. Chen, K. Hoshino, H. Liu, E.P. Frenkel, J.X.J. Zhang, K. V. Sokolov, Versatile immunomagnetic nanocarrier platform for capturing cancer cells, ACS Nano. 7 (2013) 8816–8823. https://doi.org/10.1021/nn403281e

[106] O. Rabin, J.M. Perez, J. Grimm, G. Wojtkiewicz, R. Weissleder, An X-ray computed tomography imaging agent based on long-circulating bismuth sulphide nanoparticles, Nat. Mater. 5 (2006) 118–122. https://doi.org/10.1038/nmat1571

[107] A. Jakhmola, N. Anton, T.F. Vandamme, Inorganic nanoparticles based contrast agents for X-ray computed tomography, Adv. Healthc. Mater. 1 (2012) 413–431. https://doi.org/10.1002/adhm.201200032

[108] J. Leike, A. Sachse, C. Ehritt, W. Krause, Biodistribution and CT-imaging characteristics of iopromide-carrying liposomes in rats, J. Liposome Res. 6 (1996) 665–680. https://doi.org/10.3109/08982109609039920

[109] Z. Gao, A.M. Kennedy, D.A. Christensen, N.Y. Rapoport, Drug-loaded nano/microbubbles for combining ultrasonography and targeted chemotherapy, Ultrasonics. 48 (2008) 260–270. https://doi.org/10.1016/j.ultras.2007.11.002

[110] J. Bzyl, W. Lederle, A. Rix, C. Grouls, I. Tardy, S. Pochon, M. Siepmann, T. Penzkofer, M. Schneider, F. Kiessling, M. Palmowski, Molecular and functional ultrasound imaging in differently aggressive breast cancer xenografts using two novel ultrasound contrast agents (BR55 and BR38), Eur. Radiol. 21 (2011) 1988–1995. https://doi.org/10.1007/s00330-011-2138-y

[111] M.A. Pysz, K. Foygel, J. Rosenberg, S.S. Gambhir, M. Schneider, J.K. Willmann, Antiangiogenic cancer therapy: Monitoring with molecular US and a clinically translatable contrast agent (BR55), Radiology. 256 (2010) 519–527. https://doi.org/10.1148/radiol.10091858

[112] M.C. Cassidy, H.R. Chan, B.D. Ross, P.K. Bhattacharya, C.M. Marcus, In vivo magnetic resonance imaging of hyperpolarized silicon particles, Nat. Nanotechnol. 8 (2013) 363–368. https://doi.org/10.1038/nnano.2013.65

[113] C. Zhang, M. Jugold, E.C. Woenne, T. Lammers, B. Morgenstern, M.M. Mueller, H. Zentgraf, M. Bock, M. Eisenhut, W. Semmler, F. Kiessling, Specific targeting of tumor angiogenesis by RGD-conjugated ultrasmall superparamagnetic iron oxide particles using a clinical 1.5-T magnetic resonance scanner, Cancer Res. 67 (2007) 1555–1562. https://doi.org/10.1158/0008-5472.CAN-06-1668

[114] P.M. Winter, K. Cai, J. Chen, C.R. Adair, G.E. Kiefer, P.S. Athey, P.J. Gaffney, C.E. Buff, J.D. Robertson, S.D. Caruthers, S.A. Wickline, G.M. Lanza, Targeted PARACEST nanoparticle contrast agent for the detection of fibrin, Magn. Reson. Med. 56 (2006) 1384–1388. https://doi.org/10.1002/mrm.21093

[115] A. Preda, M. Van Vliet, G.P. Krestin, R.C. Brasch, C.F. Van Dijke, Magnetic resonance macromolecular agents for monitoring tumor microvessels and angiogenesis inhibition, Invest. Radiol. 41 (2006) 325–331. https://doi.org/10.1097/01.rli.0000186565.21375.88

[116] S. Santra, D. Dutta, G.A. Walter, B.M. Moudgil, Fluorescent nanoparticle probes for cancer imaging, Technol. Cancer Res. Treat. 4 (2005) 593–602. https://doi.org/10.1177/153303460500400603

[117] K. Kim, J.H. Kim, H. Park, Y.S. Kim, K. Park, H. Nam, S. Lee, J.H. Park, R.W. Park, I.S. Kim, K. Choi, S.Y. Kim, K. Park, I.C. Kwon, Tumor-homing multifunctional nanoparticles for cancer theragnosis: Simultaneous diagnosis, drug delivery, and therapeutic monitoring, J. Control. Release. 146 (2010) 219–227. https://doi.org/10.1016/j.jconrel.2010.04.004

[118] X. Gao, Y. Cui, R.M. Levenson, L.W.K. Chung, S. Nie, In vivo cancer targeting and imaging with semiconductor quantum dots, Nat. Biotechnol. 22 (2004) 969–976. https://doi.org/10.1038/nbt994

[119] X. Cai, W. Li, C.H. Kim, Y. Yuan, L. V. Wang, Y. Xia, In vivo quantitative evaluation of the transport kinetics of gold nanocages in a lymphatic system by noninvasive photoacoustic tomography, ACS Nano. 5 (2011) 9658–9667. https://doi.org/10.1021/nn203124x

[120] A. De La Zerda, C. Zavaleta, S. Keren, S. Vaithilingam, S. Bodapati, Z. Liu, J. Levi, B.R. Smith, T.J. Ma, O. Oralkan, Z. Cheng, X. Chen, H. Dai, B.T. Khuri-Yakub, S.S. Gambhir, Carbon nanotubes as photoacoustic molecular imaging agents in living mice, Nat. Nanotechnol. 3 (2008) 557–562. https://doi.org/10.1038/nnano.2008.231

[121] M.D. Majmudar, J. Yoo, E.J. Keliher, J.J. Truelove, Y. Iwamoto, B. Sena, P. Dutta, A. Borodovsky, K. Fitzgerald, M.F. Di Carli, P. Libby, D.G. Anderson, F.K. Swirski, R. Weissleder, M. Nahrendorf, Polymeric nanoparticle PET/MR imaging allows macrophage detection in atherosclerotic plaques, Circ. Res. 112 (2013) 755–761. https://doi.org/10.1161/CIRCRESAHA.111.300576

[122] H. Xie, Z.J. Wang, A. Bao, B. Goins, W.T. Phillips, In vivo PET imaging and biodistribution of radiolabeled gold nanoshells in rats with tumor xenografts, Int. J. Pharm. 395 (2010) 324–330. https://doi.org/10.1016/j.ijpharm.2010.06.005

[123] O.R. Brouwer, T. Buckle, L. Vermeeren, W.M.C. Klop, A.J.M. Balm, H.G. Van Der Poel, B.W. Van Rhijn, S. Horenblas, O.E. Nieweg, F.W.B. Van Leeuwen, R.A. Valdeś Olmos, Comparing the hybrid fluorescent-radioactive tracer indocyanine green- 99mTc-Nanocolloid with99mTc-nanocolloid for sentinel node identification: A validation study using lymphoscintigraphy and SPECT/CT, J. Nucl. Med. 53 (2012) 1034–1040. https://doi.org/10.2967/jnumed.112.103127

[124] B.E. Ocampo-García, F. de M. Ramírez, G. Ferro-Flores, L.M. De León-Rodríguez, C.L. Santos-Cuevas, E. Morales-Avila, C.A. de Murphy, M. Pedraza-López, L.A. Medina, M.A. Camacho-López, 99mTc-labelled gold nanoparticles capped with HYNIC-peptide/mannose for sentinel lymph node detection, Nucl. Med. Biol. 38 (2011) 1–11. https://doi.org/10.1016/j.nucmedbio.2010.07.007

[125] B. Kumar, R. Kumar, I. Skvortsova, V. Kumar, Mechanisms of Tubulin Binding Ligands to Target Cancer Cells: Updates on their Therapeutic Potential and Clinical Trials, Curr. Cancer Drug Targets. 17 (2017) 357–375. https://doi.org/10.2174/1568009616666160928110818

[126] Y. Xu, K. Jang, T. Yamashita, Y. Tanaka, K. Mawatari, T. Kitamori, Microchip-based cellular biochemical systems for practical applications and fundamental research: from microfluidics to nanofluidics, Anal. Bioanal. Chem. 402 (2011) 99–107. https://doi.org/10.1007/s00216-011-5296-5

[127] A. Ponomaryova, E. Rykova, N. Cherdyntseva, E. Morozkin, I. Zaporozhchenko, T. Skvortsova, A. Dobrodeev, A. Zav'yalov, S. Tuzikov, V. Vlassov, P. Laktionov, P90, Eur. J. Cancer Suppl. 13 (2015) 43–44. https://doi.org/10.1016/j.ejcsup.2015.08.077

[128] X. Chen, R. Park, M. Tohme, A.H. Shahinian, J.R. Bading, P.S. Conti, MicroPET and Autoradiographic Imaging of Breast Cancer αv-Integrin Expression Using 18F- and 64Cu-Labeled RGD Peptide, Bioconjug. Chem. 15 (2003) 41–49. https://doi.org/10.1021/bc0300403

[129] Y.-S. Borghei, M. Hosseini, M.R. Ganjali, Detection of large deletion in human BRCA1 gene in human breast carcinoma MCF-7 cells by using DNA-Silver Nanoclusters, Methods Appl. Fluoresc. 6 (2017) 15001. https://doi.org/10.1088/2050-6120/aa8988

[130] S. Hannenhalli, Faculty Opinions recommendation of Epigenetically reprogrammed methylation landscape drives the DNA self-assembly and serves as a universal cancer biomarker., Fac. Opin. – Post-Publication Peer Rev. Biomed. Lit. (2018). https://doi.org/10.3410/f.734549397.793554291

[131] N. Shehada, G. Brönstrup, K. Funka, S. Christiansen, M. Leja, H. Haick, Ultrasensitive Silicon Nanowire for Real-World Gas Sensing: Noninvasive Diagnosis of Cancer from Breath Volatolome, Nano Lett. 15 (2014) 1288–1295. https://doi.org/10.1021/nl504482t

[132] F.S. Alves, J. de A. Rodrigues Do Rego, M.L. Da Costa, L.F. Lobato Da Silva, R.A. Da Costa, J.N. Cruz, D.D.S.B. Brasil, Spectroscopic methods and in silico analyses using density functional theory to characterize and identify piperine alkaloid crystals isolated from pepper (Piper Nigrum L.), J. Biomol. Struct. Dyn. 38 (2019) 2792–2799. https://doi.org/10.1080/07391102.2019.1639547

[133] Y. Zhu, P. Chandra, Y.-B. Shim, Ultrasensitive and Selective Electrochemical Diagnosis of Breast Cancer Based on a Hydrazine–Au Nanoparticle–Aptamer Bioconjugate, Anal. Chem. 85 (2012) 1058–1064. https://doi.org/10.1021/ac302923k

[134] C.-Y. Wen, L.-L. Wu, Z.-L. Zhang, Y.-L. Liu, S.-Z. Wei, J. Hu, M. Tang, E.-Z. Sun, Y.-P. Gong, J. Yu, D.-W. Pang, Quick-Response Magnetic Nanospheres for Rapid, Efficient Capture and Sensitive Detection of Circulating Tumor Cells, ACS Nano. 8 (2013) 941–949. https://doi.org/10.1021/nn405744f

[135] Y. Liu, Hybrid BaYbF5 Nanoparticles: Novel Binary Contrast Agent for High-Resolution in Vivo X-Ray Computed Tomography Angiography, Springer Theses. (2017) 105–120. https://doi.org/10.1007/978-981-10-6168-4_5

[136] E. Kang, H.S. Min, J. Lee, M.H. Han, H.J. Ahn, I.-C. Yoon, K. Choi, K. Kim, K. Park, I.C. Kwon, Nanobubbles from Gas-Generating Polymeric Nanoparticles: Ultrasound Imaging of Living Subjects, Angew. Chemie Int. Ed. 49 (2009) 524–528. https://doi.org/10.1002/anie.200903841

[137] S. Geninatti Crich, B. Bussolati, L. Tei, C. Grange, G. Esposito, S. Lanzardo, G. Camussi, S. Aime, Magnetic Resonance Visualization of Tumor Angiogenesis by Targeting Neural Cell Adhesion Molecules with the Highly Sensitive Gadolinium-Loaded Apoferritin Probe, Cancer Res. 66 (2006) 9196–9201. https://doi.org/10.1158/0008-5472.can-06-1728

[138] F. Kiessling, B. Morgenstern, C. Zhang, Contrast Agents and Applications to Assess Tumor Angiogenesis In Vivo by Magnetic Resonance Imaging, Curr. Med. Chem. 14 (2007) 77–91. https://doi.org/10.2174/092986707779313516

[139] A. Koukourakis Sofia Koukouraki Michael I., High Intratumoral Accumulation of Stealth Liposomal Doxorubicin in Sarcomas: Rationale for Combination with Radiotherapy, Acta Oncol. (Madr). 39 (2000) 207–211. https://doi.org/10.1080/028418600430789

[140] D. Urbach, M. Lupien, M.R. Karagas, J.H. Moore, Cancer heterogeneity: Origins and implications for genetic association studies, Trends Genet. 28 (2012) 538–543

[141] Y. Liu, N. Zhang, Gadolinium loaded nanoparticles in theranostic magnetic resonance imaging, Biomaterials. 33 (2012) 5363–5375. https://doi.org/10.1016/j.biomaterials.2012.03.084

Nanobiomaterials | Materials Research Forum LLC
Materials Research Foundations 145 (2023) 281-310 | https://doi.org/10.21741/9781644902370-11

Chapter 11

Nanotechnology in Environmental Clean-up

Anu Bansal[1], Rohan Samir Kumar Sachan[1], Jyostna Devgon[2], Inderpal Devgon[1], Arun Karnwal[1,*]

[1]School of Bioengineering and Biosciences, Lovely Professional University, Phagwara-144411, Punjab, India

[2]Center for Interdisciplinary & Biomedical Sciences, Adesh University, Bathinda-151009, Punjab, India

arunkarnwal@gmail.com; arun.20599@lpu.co.in

Abstract

Many technologies like adsorption, absorption, light-mediated catalysis, and various chemical reactions have been implied in the process of remediation of environmental pollutants. Still, environmental pollutants undoubtedly remain the major problem. The nanotechnologies have been explored for its use as capture and degradation of the pollutants. Nanotechnology or nanotech is simply defined as the use of particles or matters at nanoscale ranging from 1 to 100 nanometer. The reactivity of the nanotechnologies depends upon the surface area-to-volume of the nanoparticles utilized for the purpose. The most important factors for effective environmental remediation using nanoparticles are the detailed analysis of the in-situ location, kind of pollutants to be treated, and the decision of specific nanoparticles to be used is very critically important. The following chapter deals with the different kinds of nanotechnologies (Polymeric-based, semiconductor-based, pottery-based, metal-based, carbon-based nanoparticles) utilized for different environmental pollutions. The use of such nanoparticles on the remediation of environmental pollutants such as heavy metals, organic or inorganic matters, chemical herbicides, volatile compounds, aromatic compounds, etc. ubiquitously present in the environment (air, water, and soil). The implications of these nanoparticles are utilized in the environment on living organisms (plants, animals, and humans).

Keywords

Nanotechnologies, Nanoparticles, Environmental Clean-Up, Ex-Situ Nanotechnology, In-Situ Nanotechnology

Materials Research Foundations 145 (2023) 281-310 https://doi.org/10.21741/9781644902370-11

Contents

Materials Research Foundations 145 (2023) 281-310 https://doi.org/10.21741/9781644902370-11

1. Introduction

Recently, in the field of science and technology, nanotechnology has emerged as an effective technique to work at both the cellular and molecular level. It has provided novelty in research and development [1]. Nanotechnology is the specialized field that focuses on designing and developing small size particles that particular range from 1-100nm in length. It has been observed that there is a significant difference in the physical and chemical properties of the same compound in bulk size and nano-size [2]. Nanoparticle Nanoparticles (NPs) and nanostructure materials (NSMs) represent an active research area and a fully expanding technology sector in many application topics. NPs and NSMs have gained prominence in technological advancements due to their physicochemical flexibility such as melting point, wet ability, electrical and thermal performance, durability, light penetration and diffusion resulting in more advanced applications. The term nanomaterial is defined as a synthetic or synthetic substance consisting of non-linear, aggregated or aggregated particles where the outer diameter is between 1-100 nm according to the EU Commission. Nanotechnology has spread its roots in all the fields including engineering, medical, pharmaceutical, agriculture, food packing, disinfectants and cosmetics [3]. NPs are not a single molecule but it is a composition of several layers: first layer known as surface layer utilized for adsorption of small molecules, metal, emulsifiers, and polymers, the second is called the shell layer which is chemically different from the core layer and the last is the core layer that is the main part of the NPs [4]. Inferable from such outstanding attributes, these materials got monstrous enthusiasm of specialists in multidisciplinary fields. The NPs can be utilized for drug conveyance [5], synthetic and organic detecting [6], gas detecting [7-9], CO2 catching [10,11], and other related applications [12].

Among the various promising applications of nanotechnology, there are many other roles of NPs that are related to the environment. The present chapter will focus on the potential of nanotechnology concerning the environment, focusing on the benefits that this technology offers together with the implications of its use towards environment clean-up. Finally, the chapter also discusses the future development of nanotechnology in detail.

2. Types of nanomaterial

Nanoparticles are categorized into various classifications depending on their physical and chemical properties. Some of them are explained below:

2.1 Carbon-based NPs

Two significant classifications of carbon-based nanoparticles are fullerenes and carbon nanotubes (CNTs). Nanomaterial comprising in fullerenes class is made of globular empty pen, for example, allotropic types of carbon [13]. They have made significant business enthusiasm because of their electrical conductivity, high quality, structure, electron proclivity, and flexibility. In these classes, carbon units are arranged in a pentagonal and hexagonal organization where each carbon exists in sp2 hybridization state. The structure of CNTs is described as extended, round and breadth is 1-2nm. Based on CNTs

Materials Research Foundations 145 (2023) 281-310 https://doi.org/10.21741/9781644902370-11

measurement telicity they are projected as metallic or semiconducting. They are named as single-walled (SWNTs), twofold walled (DWNTs), or multi-walled carbon nanotubes (MWNTs) according to the number of graphite sheets moved in NPs arrangement. In current presented studies, it has demonstrated that CNTs are combined through synthetic fume testimony strategy that possesses efficient qualities and thereby used in industrial applications, for instance, these are used as filters [14,15], the gas adsorbent in vehicles, and industrial smokestack [16].

2.2 Metal NPs

Metal NPs are those NPs that are made of the metals or salts of metals such as copper (Cu), silver (Ag), and gold (Au). Because of notably restricted surface plasmon reverberation (LSPR) property, these NPs have novel opto-electrical potential. These NPs showed the wide ingestion band in the region of the electromagnetic sun-based range [17]. The characteristics of metallic NPs are highly reliant on their size and shape. Metal nano-particles showed a wide range of applications due to its significant optical activity, for example, gold NPs are used in SEM analysis to improve the stream of electron and helps the user to obtain excellent SEM pictures.

2.3 Ceramics NPs

An inorganic nonmetallic solid comprises the ceramic NPs that can be found in undefined, polycrystalline, thick, permeable, or empty structures [18]. This is the reason, these NPs play a vital role in catalysis, photocatalysis, photodegradation of colors, and imaging applications [19].

2.4 Semiconductor NPs

Nanoparticles act as a semiconductor when they have properties among metals and nonmetals [20,21]. NPs that acts as a semiconductor has wide bandgaps and this characteristic led to the development of critical modification in their properties with bandgap position [22]. Due to their appropriate bandgap and band-edge positions NPs act as a semiconductor in water splitting applications. Therefore, they behave as a significant substance used in photocatalysis, photograph optics, and electronic gadget [23].

2.5 Polymeric NPs

These NPs are naturally originated and also called a polymer nanoparticle (PNP) [24]. They generally occur in two forms either nano-spheres or nano-capsular. In nano-spheres overall mass is solid and other molecules are adsorbed at the outer surface of the sphere whereas in the nano-capsular model solid mass is encapsulated [25]. The studies had shown the wide applications of polymeric nanoparticles (NPs).

2.6 Lipid-based NPs

The presence of lipid molecule in these NPs extends its use in the field of biomedical applications such as drug carriers and delivery [26] and RNA release in cancer therapy

Materials Research Foundations 145 (2023) 281-310 https://doi.org/10.21741/9781644902370-11

[27]. Therefore, a specialized field has been created that focuses on the lipid NPs designing and modifications to improve its application which is known as Lipid nanotechnology [28].

Likewise, polymeric NPs, NPs made up of lipids are also spherical (diameter ranges from 10 to 1000 nm). The lipid-based nanoparticles are matrix particles whose solid core comprises lipid and matrix consist of lipophilic molecules. Other molecules such as surfactants or emulsifiers are adsorbed at the boundary of these NPs surface [29].

3. Nanoparticle characteristics

Along with the designing and development of nanoparticles the characterization of nanoparticles is conducted simultaneously. Different techniques are used for the characterization of nanoparticles including x-ray diffraction (XRD), SEM, TEM, and particle size analysis.

3.1 Morphology characterizations

Characterization of nanoparticle morphology is quite important as it influences most of the physical and chemical properties of nanoparticles. The techniques which are utilized to study the morphology of NPs are polarized optical microscopy (POM), SEM, and TEM.

Scanned electron microscope (SEM) is the technique that is used to NPs morphology as well as its dispersion in the matrix system. Various studies showed that the dispersion of SWNTs in the polymer matrix poly (butylene) terephthalate (PBT) and nylon-6 were studied using SEM technique [14,15]. In another study, SEM was used to study the morphological features of zinc oxide (ZnO) modified metal-organic frameworks (MOFs) and it was observed that MOF showed different morphology as compared to ZnO NP at different reaction conditions [30].

Transmitted electron microscope (TEM) is the technique that is used to analyze a large amount of material from very low to higher magnification. The different morphologies of gold (Au) nanoparticles are studied using TEM [31-33]. TEM also gives information about the quadrupolar hollow shell structure of Co3O4 NPs [34].

3.2 Structural characterizations

The structural characterization of NPs is important to know the composition and nature of bonds present in it. The techniques which are used widely to study NPs structure characteristics are XRD, energy dispersive X-ray (EDX), XPS, IR, Raman, BET, and zieta size analyzer.

The crystal property of NPs can be identified by using XRD [9,35,36]. It also provides information about the phase (single or multiphase) of NPs Emery et al., 2016). The elemental composition of NPs is determined by the EDX technique, for instance, EDX showed the occurrence of carbon (C), and oxygen (O) as contribution element in the elemental composition of graphene heterostructure NPs [7]. The ratio of elements and their nature of bond in nanoparticles can be found by using the XPS technique. XPS measures the composition of NPs by measuring depth electron transfer study via CeO2 supported platinum NPs. XPS determined that per ten Pt atoms, only one electron is elated from the NPs to CeO2 support [37].

FT-IR is used to study the vibrational characterization of NPs as it confirms the functionalization by displaying the signature vibrational peaks of carboxylated C–O 2033 cm−1, respectively in addition to a broader O–H peak at 3280 cm−1 [38].

3.3 Particle size and surface area characterization

There are various techniques such as SEM, XRD, TEM, AFM, and DLS that are used by most of the researchers to determine the size of the nano-particles. Out of all the mentioned techniques, the best result about the size of NPs is given by SEM, TEM, XRD, and AFM [39]. If the concentration of NPs is very low then DLS can be used to estimate the size of NPs for instance silica NPs size was measured by using DLS [40].

3.4 Optical characterizations

Optical properties are important in determining photo-catalytic applications. Therefore, it is required to know the absorption and reflectance value of NPs to gain knowledge about the basic mechanism of NPs for photo-catalytic application [41]. The optical instruments used to study the optical properties of NPs are UV, visible, photo-luminescence, and the ellipsometer. As bandgap is important to determine the photo-activity and conductivity, optical characterization helps to determine the bandgap of NPs [42,43]. Optical characterization is also used to analyze the shift in the absorption of NPs (doping, composite, or heterostructure) absorption or emission capacity of the materials and their effect on the overall excitation time of photoexciton [44]. The strong red shift was observed in the case of nanocomposite as compared to pristine MMT and LaFeO3 NPs. In addition to UV, PL also considers a valuable technique to study the optical properties of the photoactive NPs and other nanomaterials [45-48].

4. Environment nanotechnology

The branch of nanotechnology working in the environment sector is called e-nano or environmental nanotechnology which improves the overall effectiveness of remediation methods to detect and decontaminate harmful biological sources affecting the environment through the application of nanoscale particles. Several studies and commercial reports presented so far have highlighted the benefits of nanotechnology for environmental clean-up. Nowadays environmental pollution has become one of the most important concerned topics worldwide that need to be resolved on an urgent basis. Researches have reported traces of various pollutants in water bodies including the presence of pesticides, oil, heavy metals that decrease the oxygen level of water bodies and thereby affecting marine life [49]. Kale in his studies has reported an elevated amount of heavy metals namely aluminum, cadmium, chromium, iron, zinc, nickel, and lead in groundwater [50]. There is a need for biodegradable materials that can efficiently remove or accumulate

Materials Research Foundations 145 (2023) 281-310 https://doi.org/10.21741/9781644902370-11

environmental contaminants and convert to harmless byproducts. The principles of nanoparticles have enabled its utilization in the bioremediation of environmental contaminants. Among the different nanoparticles, inorganic; carbon-based; polymeric-based nanoparticles have been successfully applied for the remediation of the environmental contaminants. The various application of nanoparticles in the environment has been mentioned in Fig. **1**. Also, engineered nanoparticles especially modified surface structures with active compounds can be a great advantage to manufacture target specific materials. Various nanoparticles have been utilized for the remediation of the environmental contaminants and some are listed in Table 1.

Table 1: Various nanoparticles used for the bioremediation of various environmental contaminants

Nanoparticles applied	Against environmental contaminants	References
Iron nanoparticles	Chlorinated compounds, heavy metals	[51]
Palladium nanoparticles	Nitroarenes	[52]
$Dy_xMnFe_xO_4$ nanoparticles over mesoporous silica	Crystal violet	[53]
Copper oxide nanoparticles	Rhodamine B dye	[54]
Nickel nanoparticles	Chromium (VI)	[55]
Gold nanoparticles	Congo red and methylene blue dyes	[19]
Silver nanoparticles	Effluent Dyes	[56]

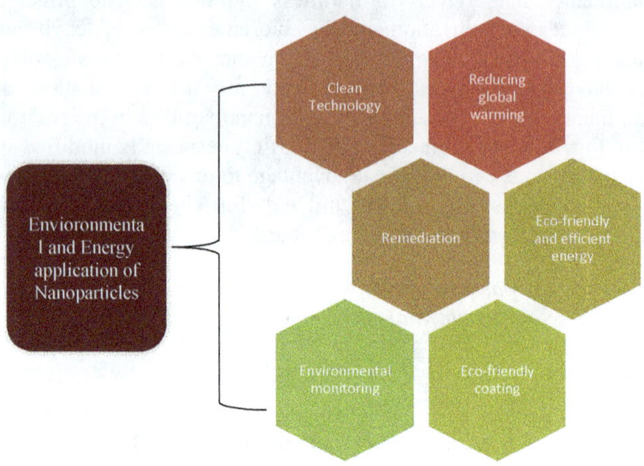

Figure 1: Various application of nanoparticles seen in the environment

Stating about various aspects of nanoparticles used for the remediation of environmental pollutants, the nanoparticles also play a keen role in soil and water remediation. Tons of organic and inorganic wastes have been dumped over the last few decades from industries. These have a deleterious effect on soil quality. The leaching of such contaminants into the groundwater has a great concern toward the well-being of the ecosystem. Various in situ propaganda of nanoparticles are utilized for more efficient that covers larger surface area along with the successive higher activity. For instance, nanoscale ZVI (Zerovalent Iron) has been used against trichloroethylene, polychlorinated biphenyls, and nitrobenzenes [57].

4.1 *Ex-situ* nanotechnology

Self-assembled monolayers on mesoporous supports (SAMMS) is one of the noticeable examples in the field of environment nanotechnology that have been demonstrated as environment contaminants adsorbent in various studies. SAMMS is having a very high surface area of approximately ~1000 meter square per gram. SAMMS showed activity against pollutants by behaving as an adsorbent for the contaminates like mercury (Hg), chromate (Cr), arsenate (As), pertechnetate (TcO −), and selenite (O3Se-2). Dendrimer which is a dendritic polymer (nano-material) is enhanced ultrafiltration used to remove Cu(II) from water and Pb(II) from the soil. These NPs adsorbents can be recovered along with the contaminated toxic material which they adsorb and therefore can be applied as ex-

Materials Research Foundations 145 (2023) 281-310 https://doi.org/10.21741/9781644902370-11

situ. Mostly organic matter is decontaminated by degradation method and hence photo-oxidation techniques are performed by nano-material photo-catalyst (< ~10 nm) such as TiO2. It is also considered as an ex-situ methodology as the technique needs to be performed in a specifically designed reactor [58].

4.2 *In situ* nanotechnology

In situ degradation of contaminants is often preferred over the ex-situ strategy because it can be applied cost-effectively. However, the major issue that relies on the in-situ approach is that it demands the transportation of the method/technique to the contamination. To overcome this issue nano-materials are best suited as they can be injected as absorptive or reactive aquifers. It can be done in two ways: creating in situ reactive zones with immobilized nano-particles or creating reactive mobile nanoparticles that can be transferred to the contaminated zones. The common examples of NPs that can be used in the in- situ approaches for environmental improvement are nonionic amphiphilic polyurethane or alumina-supported noble metals or nanoparticles containing nZVI [58].

Nano-remediation is currently endeavoring to give another novel and viable answer for environmental remediation by assuming a huge function in contamination avoidance, location, observing, and remediation [59]. Nano-remediation utilizing nanoscale particles or nano-materials holds huge potential to cost-viably address a portion of the difficulties of site remediation and improve the general proficiency of the remediation measures. Momentum costly, repetitive and halfway successful soil and groundwater remediation advances, for example, warm treatment, substance oxidation, and surfactant cosolvent flushing could be gradually eliminated throughout the next few decades [60,61] with the appearance of the fast, reasonable, and productive nanoremediation draws near. While critical budgetary speculation has been made to the clean up of defiled soil, water, and other natural media, the accumulated monetary weight for site tidy up is high. Nano-remediation techniques involve the use of receptive NMs, for example, nanoscale zeolites, metal oxides, carbon nanotubes and strands, bimetallic nanoparticles, and so on for the change and detoxification of poisons. Of these, nanoscale zero-valent iron (nZVI) is presently the most generally utilized [58, 62-63].

The nanoparticles can allow both synthetic decrease and catalysis to alleviate an assortment of contaminations of concern including chlorinated mixes, organochloride pesticides, polychlorinated biphenyls, hefty metal particles, and inorganic anions [64]. In the majority of contaminated sources, nanoscale zero-valent iron (nZVI's) are likely to be transformed from Fe0 to Fe2+ and then may undergo further oxidative transformation to Fe3+. nZVI's are considered an attractive option to transform and detoxify pollutants due to its low standard reduction potential, and optimum quantum size properties [65]. In a study, it was found that iron nanoparticles were found to perform various applications including removal of heavy metals namely arsenic and chromium, pesticides, chlorinated solvents, and organic compounds transformation [66]. As this field offers a lot of advantages therefore environmental nanotechnology has been used for treating environmental issues including soil and water remediation, oil spill decontamination, wastewater treatments, and water purification [66]. Metallic nanoparticles are used to remove metal contaminants from the water bodies [57,67,68]. Besides this carbon nanotubes (CNTs) were considered as an alternative option for adsorption of pollutants from water bodies. CNTs absorb the pollutants by attaching a functional group that increases the affinity of pollutants towards carbon nanotubes [69,70]. In addition, CNTs have also been used for the removal of heavy metals like Cr3+, Pb2+ and Zn2+ as well as metalloids such as arsenic compounds, organics biological impurities, volatile organic.

5. Role of nanoparticle in environment clean up

Nanoparticle-based advances discover applications in a few natural territories as referenced in Fig. **2**. The uses of nanoparticles in ecological advances are on the whole alluded to as "environment nanotechnology." Because nanoparticles have upgraded basic, attractive, electrical, and optical properties, they have a huge potential to supplant existing materials. As a result of their boss properties, lower amounts will be utilized that won't just lower costs yet additionally leave a decreased natural impression. Nanoparticles are discovering applications in more up to date detecting advances to identify toxins at lower fixations with more noteworthy selectivity and exactness. Due to inborn size points of interest, they can be utilized to test more hard to-arrive at areas, for example, in the subsurface climate or complex designed frameworks. Nano-materials likewise discover extraordinary pertinence in existing cycles to decrease or forestall emanations or to change effluents over to helpful results. Nanoparticles can likewise be utilized to tidy up foreign substances that are available in the climate on account of harmful past practices. The role of nanotechnology in environment cleanup categorizes in a different manner (i) producing environment-friendly products (ii) environment remediation contaminated with toxic material (iii) sensors for environment contaminants [62].

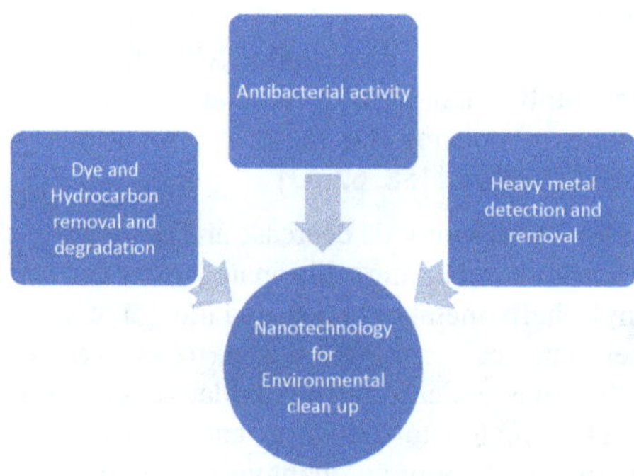

Figure 2: Application of nanotechnology in environmental cleanup

 Materials Research Forum LLC
https://doi.org/10.21741/9781644902370-11

The presence of heavy metals such as mercury, lead, thallium, cadmium, and arsenic in natural water bodies had affected the health of an individual consuming the water for various purposes. Hence, it is required to remove the heavy metals efficiently and NPs of iron oxide are considered as an effective sorbent material for the toxic soft material [71]. Nano-materials are utilized for photo-degradation. In one study NiO/ZnO NPs modified silica was used for photo-degradation. The high surface area of NPs due to a very small size (<10 nm), facilitated the efficient photo-degradation reaction [72,73].

6. Nanotechnology and pollution control

Pollution is the main concern related to environmental toxicity that is caused by excessive resources production and consumption. NPs are having the potential of controlling and preventing air and water pollution effectively.

6.1 Air pollution

Nanotechnology methodology like nano-catalysts is used for air pollution clean-up. The most common type of nanotechnologies used for the cleaning of toxic gases that causes air pollution are CNTs (carbon nanotubes) and gold particles adsorption. The CNTs have proven to be a good absorbent towards the efficient removal of organic and inorganic air pollutants like nitrous oxide, ammonia gases, carbon dioxide, and dioxin [74]. The nano-catalysts having increased surface area is used to catalyze the gaseous chemical reaction that transforms the toxic vapors released from vehicles and industries into non-toxic gases. In addition to this, the nano-fibers catalyst made of manganese oxide is also used to decontaminate the industrial smokestack by removing volatile organic compounds from it. Another approach uses nanostructured membranes that have pores small enough to separate methane or carbon dioxide from exhaust [75]. More research is going on carbon nanotubes (CNT) that are used to deceive the emission of greenhouse gases released by coal mining and power generation. This new technology both processes and separates large volumes of gas effectively, unlike conventional membranes that can only do one or the other effectively. Apart from the commonly known toxic gases, various volatile and polyaromatic compounds have also caused air pollution. The combination of manganese oxide and gold nanoparticles have been used against such pollutants because the manganese oxide-based nanoparticles have a wide surface area that allows much more effective adsorption of volatile compounds and hence increase the rate of degradation. The release of isopropyl alcohol (IPA) that is used in the semiconductor manufacturing industries has also gained top priority as unchecked IPA vapors are released in the air. The nanoparticles with adsorbent of a combination of nitric acid and sodium hypochlorite have been used for the entrapment of IPA vapors and its successful removal from the air [74].

6.2 Water pollution

Likewise air pollution clean-up, chemical reactions are preceded in water bodies also where toxic compounds can be transformed into non-toxic compounds. For instance, NPs have the potential to transform trichloroethene, toxic contamination found in industrial

Materials Research Foundations 145 (2023) 281-310 https://doi.org/10.21741/9781644902370-11

wastewater into harmful compounds. The use of nanotechnology for water clean-up is an easy, efficient, and cheap process as it requires the insertion of NPs into polluted water instead of pumping out water for treatment. For example use of nano-sized fibers as electrodes that follows the deionization method for water treatment. By and large, the water filtration framework utilizes the semi-penetrable film for water decontamination, diminishing the pore size of the layer to the nano-size will expand the exchange of particles in a more particular way. Nowadays, nano-size membranes are added in the water filtration technique that filters out the viruses also. The polymer nanoparticles have also been used for the treatment of water bodies. The property of polymer nanoparticles used here is similar to that of surfactant micelle that means the polymer nanoparticles have amphipathic property. Each molecule in the nanoparticles has two components: hydrophobic and hydrophilic part. One such example is amphiphilic polyurethane (APU) which has been used for the successful removal of phenanthrene compounds from the aquifer [76].

Additionally, resins, an organic polymer with nano-sized pores on their surface are used to separate, purify, and decontaminate the toxic compounds by the ion-exchange method. Mostly this methodology is used for water softening and water purification like in water sample toxic heavy metals is replaced by non-toxic metals like sodium (Na) or potassium (K). Though ion exchange resins have a vital role in the environment cleanup it can easily be damaged by iron (Fe), organic matter, bacteria, and chlorine (Cl). Ongoing advancements of nano-wires made of potassium manganese oxide can tidy up oil and other natural toxins while making oil recuperation conceivable. These nanowires structure a work that retains up to multiple times its weight in hydrophobic fluids while dismissing water with its water repulsing covering. Since the potassium manganese oxide is entirely steady even at high temperatures, the oil can be bubbled off the nanowires and both the oil and the nanowires would then be able to be reused. Also, nanoparticles in the form of nanofibers and nanobodies have been widely used for water purification purposes.

Nanotechnology also provides an alternative for the disinfection of water rather than chemical-based disinfection. The attachment of bioactive to nanoparticles has proven to kill the germs like bacteria and viruses in the water. The mode of action achieved by the use of such nanoparticles is through the production of reactive oxygen species in the bacteria that brings out the killing. The antimicrobial peptides, chitosan, carboxy fullerenes, CNTs, ZnO (zinc oxide), and silver nanoparticles are some of the common types used for the purpose. Its application does not affect the quality of the water nor it affects humans. The construction of such nanoparticles has been low-cost [74].

6.3 Soil pollution

Soil pollution is mainly due to the presence of contamination like heavy metals, pathogens, organic waste, and many more. Such pollutants have gained an immense concern throughout the globe due to the hazardous leaching out of harmful chemicals into the soil. Thus, such presence of contaminants possesses various harmful effects on human health including the environment. Due to this, there is an increasing demand for such technologies with more efficient, low-cost, and yet environmental friendly to detect such contaminants

Materials Research Foundations 145 (2023) 281-310
https://doi.org/10.21741/9781644902370-11

or pollutants. Not only detection is necessary, but also the treatment of such kind needs to be searched i.e. converting the harmful and toxic wastes into harmless and neutral byproducts. Nanotechnologies have been fundamentally used for the purpose and still been researched upon for its use in different sizes and shapes of concern. However, a certain amendment needs to be followed for its use during the clearing of pollutants or contaminants from the soil. These are 1) the nanoparticles used must be delivered to the contaminated zones, 2) the delivered nanoparticles must be confined to a particular area of the polluted zone and immobilized over there. To maintain the integrity and stability of the nanoparticles in the soil various researchers have come up with the idea of using certain polymers as stabilizers (Table **2**). The commonly used stabilizers are starch and carboxymethyl cellulose. The various nanoparticles used in controlling soil pollution are phosphate compounds-based nanoparticles, where the compounds are the best examples for the immobilization of the heavy metals in the soil. Next is the ZVI nanoparticles, these are also used for the immobilization of the heavy metals from the soil. There is a drawback for the use of conventional type ZVI is limited to ex-situ applications. However, through the use of stabilizers, the most common carboxymethyl cellulose has provided ZVI nanoparticles great stability to be used in *in-situ* applications [77].

Table 2: Summary of the nanoparticles used to treat pollutants present in the ecosystem

Nanoparticles	Pollutants treated	Source of the pollutants	References
Calcium-based nanoparticles	Carbon dioxide	Air	[78]
Titanium oxide coated calcium carbonate nanoparticles	Carbon dioxide	Air	[79]
Carbon nanotubes	Carbon dioxide	Air	[80]
Nickel nanoparticles	Methane	Air	[81]
Titanate nanotubes	Nitrous oxide and Nitric oxide	Air	[82]
Starch stabilized magnetite nanoparticles	Arsenate: As(V)	Soil	[83]
Carboxymethyl cellulose stabilized apatite nanoparticles	Lead	Soil	[84]
Carboxymethly cellulose stabilized zerovalent iron nanoparticles	Chromium: Cr(VI)	Soil	[85]

Materials Research Foundations 145 (2023) 281-310 https://doi.org/10.21741/9781644902370-11

Single magnetite nanoparticles	Aromatic hydrocarbons	Water	[86]
Lanthanum inserted copper chromite spinel nanoparticles	Rhodamine B and Methyl orange dyes	Water	[87]
Chitosan stabilize iron/nickel bimetallic nanoparticles	Triphenylmethane dyes	Water	[88]
Rice straw-based hydrated titanium oxide nanoparticles	Copper: Cu(II)	Water	[89]

7. Remediation methods of nps for environment cleanup

Environmental remediation has encouraged the advancement of highly effectual photo-catalysts that can contribute to detoxifying contaminants or pollutants [72]. Several advantages are associated with photo-catalyst as an agent for environmental clean-up. Photo-catalysts are capable of converting contaminants into non-toxic byproducts; the mechanism of conversion is a direct method. Photo-catalysts are of great importance as it requires oxygen as an oxidant, utilized solar energy, and self-regeneration potential. Photo-catalysts can degrade organic toxicants into CO2 and dilute mineral acids at low cost; therefore, environment clean-up through photo-catalyst is more suitable and efficient.

Environment nano-technology has the potential to improve the technologies that are presently used in environment contaminant detection, sensing, remediation, and pollution removal. Many of the applications that are related to environment nanotechnology such as nano-sensors and nano-scale are commercially available. Both nano-sensors and nano-scale coatings are replaced by a thicker and more wasteful polymer coating that prevents corrosion. Nano-sensors are used for detecting aquatic toxins whereas nano-scale biopolymers are used for improving decontamination and the recyclization of heavy metals from environmental sources. Nanoparticles made up of metals have the capacity to degrade toxic organic compounds at room temperature and therefore are considered as novel photocatalyst for environmental remediation.

7.1 Nanoparticles as adsorbent

The NPs including nano-sized metal oxides and nano-size clays remove organic and inorganic pollutants by adsorbing them on its surface [90]. Carbon nanotubes (CNTs) has the potential to adsorb a wide range of toxic organic compounds including dioxin, polynuclear aromatic hydrocarbons (PAHs), DDT and its metabolites, PBDEs, chlorobenzenes, and chlorophenols, trihalomethanes, bisphenol A and nonylphenol, phthalate esters, dyes, pesticides, and herbicides, for example, sulfur derivatives, atrazine, and dicamba that are present in an aquatic environment. Additionally, CNTs

Materials Research Foundations 145 (2023) 281-310
https://doi.org/10.21741/9781644902370-11

(hydroxylated) behave as a good adsorbent for metals like copper, nickel, cadmium, and lead. However, organometallic compounds are better to adsorb by pristine multi-walled CNTs. Modified CNTs cross-linked with nano-porous polymer showed high absorption capacity as compared to traditional CNTs for organic compounds namely p- nitro-phenol and trichloroethylene.

7.2 TiO2 and ZnO nanoparticles as semiconductors

Another nano-size material that is used for the removal of organic and inorganic pollutants is TiO2 nanoparticles which act as a semi-conductor [91,92]. These nano-sized semiconductors are used widely in various applications such as self-cleaning glass, disinfectant tiles, and filters for air purification [93]. TiO2 electrodes have the potential to determine the chemical oxygen demand (COD) of water [94]. Therefore, TiO2 is used as sensors for monitoring water contamination. Immobile TiO2 nanomaterials can be used to detoxify water and air using solar energy. The modification and engineering of nano-particles increase the interaction of NPs with organic, inorganic, and biological contaminants such as heavy metals, organochlorine pesticide, arsenic (As), and phosphates (PO_4^{3-}) in water, induced by ultraviolet (UV) light [95].

TiO2 (Titanium dioxide) leads to environment remediation by following two reactions which are oxidation and reduction. The stimulation of UV excites the TiO2 and develops the electron-hole pair that possesses sufficient oxidizing potential required to oxidized water pollutants. In addition to this combination of UV and TiO2 develops the anti-bacterial activity that can be used to degrade pathogens from the wastewater. Therefore, this approach provides the best-suited method to clean wastewater from harmful compounds and bacteria simultaneously. Apart from TiO2, there are many other nanoparticle semi-conductors such as zinc oxide (ZnO), zinc sulfide (ZnS), ferric oxide (F2O3), and cadmium sulfide (CdS) used in environmental clean-up. However, TiO2 has characteristic properties such as (i) inert (biologically and chemically) (ii) resistivity against corrosion (iii) reusable that makes it a cost-effective technique for water clean-up as compared to other oxidative nanoparticles. There is one limitation associated with TiO2 that limits its use in water treatment [94]. As treatment with TiO2 requires the presence of UV light, it can only use for the remediation of transparent wastewater. Various types of studies are going to overcome this drawback by combining the TiO2 with other material that extends the TiO2 excitation into visible light.

Unlike TiO_2, ZnO semiconductors can excite in the visible region. The excited ZnO act as hole scavengers such as phenols or iodide ions. Therefore, nanoparticles of ZnO particles are considered good candidates for use as sensors for chemical compounds. The sensor system based on ZnO nanoparticles is very sensitive as it can detect contamination sensitivity of 1ppm.

ZnO not only in the visible region but it also has the potential to degrade the toxic compounds in the presence of UV light [96]. This infers that ZnO can be used as a better approach to monitor and destroy toxic chemicals simultaneously. In one study it is demonstrated that ZnO nano-sized films have been used to sense and degrade organic compounds in water at the same time. Such a catalyst framework is valuable to actuate impurity corruption where the framework detects a focused on an atom, subsequently keeping away from the decimation of innocuous particles present in the environment.

7.3 Iron nanoparticles

Lab research has set up that nanoscale metallic iron is effective in destroying a wide variety of fundamental unfamiliar substances, for instance, chlorinated methanes, brominated methanes, trihalomethanes, chlorinated ethenes, chlorinated benzenes, other polychlorinated hydrocarbons, pesticides, and dyes [97]. The purpose behind the reaction is the disintegration of zerovalent iron in the atmosphere. Contaminants, for instance, tetrachloroethene can readily recognize the electros from iron oxidation and be diminished to ethane. In any case, nanoscale zerovalent iron (nZVI) can reduce normal unfamiliar substances just as the inorganic anions nitrate, which is lessened to soluble base perchlorate (notwithstanding chlorate or chlorite) and thereafter diminished to chloride, selenate, arsenate, arsenite, and chromate [98,99]. nZVI is also capable in wiping out separated metals from the arrangement, for example, lead (Pb) and nickel (Ni). The reaction rates for nZVI are at least 25-30 times snappier and besides sorption limit is much higher compared with granular iron. The metals are either diminished to zerovalent metals or sweetheart oxidation states, for example, Cr (III), or are surface complexed with the iron oxides that are outlined during the reaction. Some metals can manufacture the dechlorination movement of organics and besides lead to more kind things; however various metals decay the reactivity. By far maximum of the methodology have been developed using nZVI that has involved completely in groundwater and soil remediation.

7.4 Nanotechnology for hazardous waste clean-ups

Nano-scale materials showed an increased effect in the cleanup of hazardous waste due to (i) its nano-size that allows it to enter deep inside the groundwater or soil and (ii) engineered coating around the NPs that permits them to persist in groundwater a significant resource in cleanups. If possible, nano-materials could cut cleanup costs by maintaining a strategic distance from the uncommon expenses and dangers of pulling die for consumption or internment [100].

7.4.1 Nanoparticles in wastewater

The environmental organization of NPs waste could remarkably recoup constituents found in wastewater. Masters from a couple of consistent foundations around the world collaborated on the examination have focused on the lead of possibly hazardous NPs in sewage treatment plants (STP), particularly silica-shelled nanoparticles. Throughout the assessments, it was discovered that covering these nanoparticles with a business surfactant engaged their parcel from the water alongside other waste particles. The scientists express that if the nanoparticles were allowed to "settle out" thusly, by then they could be stopped

Materials Research Foundations 145 (2023) 281-310
https://doi.org/10.21741/9781644902370-11

from providing for the subsequent period of the waste treatment measure, which would help prevent mechanical incidents [100].

8. Risks associated with the use of nanoparticles

The use of NPs in industrial and household applications is increased enormously leading to the release of NPs into the environment. The excessive use of nano-material elevates the amount of them in groundwater and soil which is the main cause of environmental toxicity [101,102]. The risk of these nanoparticles towards the environment can be assessed by understanding the partition of NPs in different phases such as water and air depends based on its mobility potential, persistence, and the source magnitude [103,104]. This phase partition directs the exposure of NPs in the environment. To observe the toxic risk of NPs in the environment it is mandatory to study the concentration of NP in the environment [105]. NPs have a high surface to mass ratio due to which it causes the partitioning of absorbed contaminants in the solid/water phase. The absorption of contaminates by NPs took place either by performing co-precipitation during the NPs formation or trapped by aggregation of NPs which had contaminants adsorbed to their surface. NPs characteristics including size, composition, morphology, porosity, aggregation/disaggregation, and aggregate structure determine the absorption of pollutants on the surface of NPs. However, the luminophores are not safe in the environment and are protected from the environmental oxygen when they are doped inside the silica network [106].

As it is studied above that nZVI is one of the most important NPs among various other nano-materials but there are certain uncertainties associate with the use of nZVI for different applications. The aggregates of nZVI produce the clusters with micron size that is not suitable for the effects of nano-size materials, uncoated nZVI are not able to mobile towards contaminating zones for a long distance and might increase the risk towards the ecosystem [107]. These issues related to NPs revealed that more development and improvement is required in the environment nano-technology field. A wide range of NPs applications end up contaminating the environment system and thereby causing harmful effects on living organisms [107]. The environmental risk is caused due to dispersal, eco-toxicity, persistence, bio-accumulation, and reversibility characteristics of NPs [108,109]. Although environment nanotechnology is the best alternative to provide a beneficial impact for site remediation but research into the health and environmental effects of nanoparticles is urgently required.

8.1 Effect on ecosystem

To comprehend the danger of NPs in the climate both the dose-response impact and the exposure pathways of the NPs is important to decide how NPs enter a living being. It is suggested that focus on studies of toxic effects should not be limited towards higher organisms but also be studied in lower organisms as they are the base of the food chain [109]. It has been quantified that around every year tons of nanoparticles are used which are made from metals or metal oxides and dumped into the soil or water bodies [110,56]. Previous studies had shown that oxidation of nZVI nanoparticles generates reactive oxygen species (ROS) that affect unadulterated species of microscopic organisms including Escherichia coli, Pseudomonas fluorescens, and *Bacillus subtilis* var *niger* [111-113]. nZVI displayed the harmful effects on plants also by accumulating and assessing the phytotoxicity. nZVI toxic effects were seen in *Typha latifolia*, mixture poplars, and Typha at higher concentrations [114]. Various studies have shown adverse effects of NPs on soil microorganisms, for example, CuO and Fe3O4 at 1% and 5% w/w dry soil concentration respectively were found to cause variation in soil microbial species [115]. Conversely, Fajardo et al. [116] and Tong et al. [117] in their studies had demonstrated that even at the high dosage of nZVI (34 mg/g) and C60 (100 μg/g) showed negligible effect on cell viability and biological mechanism of microorganisms. There are many causes of the toxicities of the nanoparticles affecting the ecosystem mainly the aquatic life. For instance the use of nanocarbon tubes proven to have detrimental to aquatic life in diminishing the growth rate. Also, extensive use of the nanoparticles and their corresponding byproducts have caused a high rate of mortality rate. This proposes that the toxic effect of NPs might be firmly related to bioavailability and solubility. Henceforth, NPs delivery and its impacts on the environment ought to be observed intently, with unique consideration given to the utilization of nZVI for remediation of soil and groundwater.

8.2 Effect on human health

The risk of toxic effects on human health is related to nanotechnology incorporate inward breath or the introduction of nanoparticles at the work environment by coming in contact with contaminated air, water, or food and dermal contact [118]. The exposure of NPs leads them to penetrate inside the cell mitochondria and the nucleus [119,120] that can cause mitochondrial damage and DNA mutations that end up in cell death [119,121]. These issues may restrict the inescapable utilization of NPs for environmental clean-up. Henceforth, to make this innovation more advantageous than destructive, observing and mediation estimates should be executed immediately. The NPs can cross the blood-brain barrier due to its small size. It has been seen that these NPs interact with the proteins and enzymes, even at the molecular level, these can alter the gene expression that may result in deficient protein or enzyme production. Various in-vivo studies suggested that nanoparticles tend to possess cytotoxic effects on human cells (Fig. **3**). The combination of copper and zinc metal-based nanoparticles damages the integrity of the plasma membrane that can be achieved by the creation of the pores. The nanoparticles often form ions that can be transported into the cells via ion channels. Such ions inside the cells create DNA adducts through the formation of reactive oxygen species (oxidative stress). The formed oxidative stress reacts highly with the DNA that leads to the formation of the adduct and correspondingly affects the DNA repair mechanisms. Another effect of nanoparticles on the human cell is the destabilizing of the ribosomes and the proteins translated from it. Lastly, the mitochondria are also affected which results in the inhibition of oxidative phosphorylation pathway by producing the reactive oxygen species (ROS) [122,123]. The impact of these damages is seen in compromising the immune system and corresponding elevated cytokines. These pleiotropic cytokines are regarded as the biomarkers for the toxicity of the nanoparticles in the body.

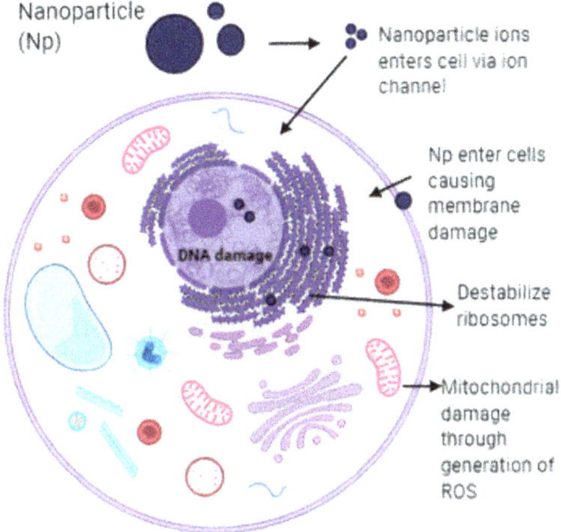

Figure 3: Impact of nanoparticles on the human cell

Concluding remarks

Many of the environmental pollutants like heavy metals, organic and inorganic matters, aromatic compounds, etc. have been successfully remediated through the use of different nanoparticles. The use of nanotechnologies for mitigating environmental pollutants has prerequisites like the use of nanoparticles in the in-situ location requires a detailed analysis of the location. Next, again a detailed analysis of the pollutant to remediate and the specific type of nanoparticle require for the purpose. Lastly, the concentration of the particles required for the efficient remediation of the pollutants.

Other nanoparticles have been checked for their efficacy under laboratory conditions and that needs to more researched for its applicability in real case scenarios. However, limited researches are dealing with what happens to the nanoparticles after successfully remediating the pollutants. Ascertain nanoparticles are potentially toxic to living organisms and there needs to for a system that would remediate those nanoparticles. Or else, the research for nanoparticles that are toxic less to the living organisms still needs to be worked on. Such challenges must be worked upon for the full potential utilization of nanotechnologies. Nevertheless, the nanoparticles will have a significant role in the coming future as alternative green means of environmental pollutant remediation.

Acknowledgements

We regard School of bioengineering and biosciences, Lovely Professional University (LPU), for infrastructure and other facilities.

References

[1] M. Faraji, Y. Yamini, M. Rezaee, Magnetic nanoparticles: Synthesis, stabilization, functionalization, characterization, and applications, 7 (2010) 1-37. https://doi.org/10.1007/BF03245856

[2] T. v. Duncan, Applications of nanotechnology in food packaging and food safety: Barrier materials, antimicrobials and sensors, J Colloid Interface Sci. 363 (2011) 1-24. https://doi.org/10.1016/j.jcis.2011.07.017

[3] a Dowling, R. Clift, N. Grobert, D. Hutton, R. Oliver, O. O'neill, J. Pethica, N. Pidgeon, J. Porritt, J. Ryan, Et Al., Nanoscience and nanotechnologies : opportunities and uncertainties, 46 (2004) 618-618. https://doi.org/10.1007/s00234-004-1255-6

[4] W.K. Shin, J. Cho, A.G. Kannan, Y.S. Lee, D.W. Kim, Cross-linked Composite Gel Polymer Electrolyte using Mesoporous Methacrylate-Functionalized SiO2 Nanoparticles for Lithium-Ion Polymer Batteries, 6 (2016) 1-10. https://doi.org/10.1038/srep26332

[5] J.E. Lee, N. Lee, T. Kim, J. Kim, T. Hyeon, Multifunctional mesoporous silica nanocomposite nanoparticles for theranostic applications, Acc Chem Res. 44 (2011) 893-902. https://doi.org/10.1021/ar2000259

[6] H. Barrak, T. Saied, P. Chevallier, G. Laroche, A. M'nif, A.H. Hamzaoui, Synthesis, characterization, and functionalization of ZnO nanoparticles by N-(trimethoxysilylpropyl) ethylenediamine triacetic acid (TMSEDTA): Investigation of the interactions between Phloroglucinol and ZnO@TMSEDTA, 12 (2019) 4340-4347. https://doi.org/10.1016/j.arabjc.2016.04.019

[7] M. Mansha, A. Qurashi, N. Ullah, F.O. Bakare, I. Khan, Z.H. Yamani, Short communication, Ceram Int. 9 (2016) 11490-11495. https://doi.org/10.1016/j.ceramint.2016.04.035

[8] I. Rawal, A. Kaur, Synthesis of mesoporous polypyrrole nanowires/nanoparticles for ammonia gas sensing application, Sens Actuators A Phys. 203 (2013) 92-102. https://doi.org/10.1016/j.sna.2013.08.023

[9] H. Ullah, I. Khan, Z.H. Yamani, A. Qurashi, Sonochemical-driven ultrafast facile synthesis of SnO2 nanoparticles: Growth mechanism structural electrical and hydrogen gas sensing properties, Ultrason Sonochem. 34 (2017) 484-490. https://doi.org/10.1016/j.ultsonch.2016.06.025

[10] M. Ganesh, P. Hemalatha, M.M. Peng, H.T. Jang, One pot synthesized Li, Zr doped porous silica nanoparticle for low temperature CO2 adsorption, 10 (2017) S1501-S1505. https://doi.org/10.1016/j.arabjc.2013.04.031

[11] P.V.R.K. Ramacharyulu, R. Muhammad, J. Praveen Kumar, G.K. Prasad, P. Mohanty, Iron phthalocyanine modified mesoporous titania nanoparticles for photocatalytic activity and CO2 capture applications, 17 (2015) 26456-26462. https://doi.org/10.1039/C5CP03576G

[12] M. Shaalan, M. Saleh, M. El-Mahdy, M. El-Matbouli, Recent progress in applications of nanoparticles in fish medicine: A review, Nanomedicine. 12 (2016) 701-710. https://doi.org/10.1016/j.nano.2015.11.005

[13] K. Saeed Ibrahim, A. Info, Carbon nanotubes-properties and applications: a review, 14 (2013) 131-144. https://doi.org/10.5714/CL.2013.14.3.131

[14] K. Saeed, I. Khan, Preparation and properties of single-walled carbon nanotubes/poly(butylene terephthalate) nanocomposites, 23 (2014) 53-58. https://doi.org/10.1007/s13726-013-0199-2

[15] K. Saeed, I. Khan, Preparation and characterization of single-walled carbon nanotube/nylon 6, 6 nanocomposites, 44 (2016) 435-444. https://doi.org/10.1080/10739149.2015.1127256

[16] J.M. Ngoy, N. Wagner, L. Riboldi, O. Bolland, A CO2 Capture Technology Using Multi-walled Carbon Nanotubes with Polyaspartamide Surfactant, Energy Procedia. 63 (2014) 2230-2248. https://doi.org/10.1016/j.egypro.2014.11.242

[17] E.C. Dreaden, A.M. Alkilany, X. Huang, C.J. Murphy, M.A. El-Sayed, The golden age: gold nanoparticles for biomedicine, Chem Soc Rev. 41 (2012) 2740-2779. https://doi.org/10.1039/C1CS15237H

[18] W. Sigmund, J. Yuh, H. Park, V. Maneeratana, G. Pyrgiotakis, A. Daga, J. Taylor, J.C. Nino, Processing and Structure Relationships in Electrospinning of Ceramic Fiber Systems, 89 (2006) 395-407. https://doi.org/10.1111/j.1551-2916.2005.00807.x

[19] S. Thomas, B.S.P. Harshita, P. Mishra, S. Talegaonkar, Ceramic Nanoparticles: Fabrication Methods and Applications in Drug Delivery, Curr Pharm Des. 21 (2015) 6165-6188. https://doi.org/10.2174/1381612821666151027153246

[20] S. Ali, I. Khan, S.A. Khan, M. Sohail, R. Ahmed, A. ur Rehman, M.S. Ansari, M.A. Morsy, Electrocatalytic performance of Ni@Pt core-shell nanoparticles supported on carbon nanotubes for methanol oxidation reaction, 795 (2017) 17-25. https://doi.org/10.1016/j.jelechem.2017.04.040

[21] I. Khan, A. Abdalla, A. Qurashi, Synthesis of hierarchical WO3 and Bi2O3/WO3 nanocomposite for solar-driven water splitting applications, Int J Hydrogen Energy. 42 (2017) 3431-3439. https://doi.org/10.1016/j.ijhydene.2016.11.105

[22] T. Hisatomi, J. Kubota, K. Domen, Recent advances in semiconductors for photocatalytic and photoelectrochemical water splitting, Chem Soc Rev. 43 (2014) 7520-7535. https://doi.org/10.1039/C3CS60378D

[23] S. Sun, C.B. Murray, D. Weller, L. Folks, A. Moser, Monodisperse FePt nanoparticles and ferromagnetic FePt nanocrystal superlattices, Science. 287 (2000) 1989-1992. https://doi.org/10.1126/science.287.5460.1989

[24] M. Mansha, I. Khan, N. Ullah, A. Qurashi, Synthesis, characterization and visible-light-driven photoelectrochemical hydrogen evolution reaction of carbazole-containing conjugated polymers, Int J Hydrogen Energy. 42 (2017) 10952-10961. https://doi.org/10.1016/j.ijhydene.2017.02.053

[25] J.P. Rao, K.E. Geckeler, Polymer nanoparticles: Preparation techniques and size-control parameters, Prog Polym Sci. 36 (2011) 887-913. https://doi.org/10.1016/j.progpolymsci.2011.01.001

[26] A. Puri, K. Loomis, B. Smith, J.H. Lee, A. Yavlovich, E. Heldman, R. Blumenthal, Lipid-based nanoparticles as pharmaceutical drug carriers: from concepts to clinic, Crit Rev Ther Drug Carrier Syst. 26 (2009) 523-580. https://doi.org/10.1615/CritRevTherDrugCarrierSyst.v26.i6.10

[27] M. Gujrati, A. Malamas, T. Shin, E. Jin, Y. Sun, Z.R. Lu, Multifunctional cationic lipid-based nanoparticles facilitate endosomal escape and reduction-triggered cytosolic siRNA release, Mol Pharm. 11 (2014) 2734-2744. https://doi.org/10.1021/mp400787s

[28] S. Mashaghi, T. Jadidi, G. Koenderink, A. Mashaghi, Lipid nanotechnology, Int J Mol Sci. 14 (2013) 4242-4282. https://doi.org/10.3390/ijms14024242

[29] M.K. Rawat, A. Jain, S. Singh, Studies on binary lipid matrix based solid lipid nanoparticles of repaglinide: in vitro and in vivo evaluation, J Pharm Sci. 100 (2011) 2366-2378. https://doi.org/10.1002/jps.22435

[30] E. Mirzadeh, K. Akhbari, Synthesis of nanomaterials with desirable morphologies from metal-organic frameworks for various applications, CrystEngComm. 18 (2016) 7410-7424. https://doi.org/10.1039/C6CE01076H

[31] N. Khlebtsov, L. Dykmana, Biodistribution and toxicity of engineered gold nanoparticles: a review of in vitro and in vivo studies, Chem Soc Rev. 40 (2011) 1647-1671. https://doi.org/10.1039/C0CS00018C

[32] N. Khlebtsov, L. Dykman, Plasmonic Nanoparticles, (2010) 37-85. https://doi.org/10.1201/9781439806296-c2

[33] N.G. Khlebtsov, L.A. Dykman, Optical properties and biomedical applications of plasmonic nanoparticles, J Quant Spectrosc Radiat Transf. 111 (2010) 1-35. https://doi.org/10.1016/j.jqsrt.2009.07.012

[34] J. Wang, N. Yang, H. Tang, Z. Dong, Q. Jin, M. Yang, D. Kisailus, H. Zhao, Z. Tang, D. Wang, Accurate Control of Multishelled Co3O4 Hollow Microspheres as

High-Performance Anode Materials in Lithium-Ion Batteries, 52 (2013) 6417-6420. https://doi.org/10.1002/anie.201301622

[35] I. Khan, S. Ali, M. Mansha, A. Qurashi, Sonochemical assisted hydrothermal synthesis of pseudo-flower shaped Bismuth vanadate (BiVO4) and their solar-driven water splitting application, Ultrason Sonochem. 36 (2017) 386-392. https://doi.org/10.1016/j.ultsonch.2016.12.014

[36] I. Khan, A.A.M. Ibrahim, M. Sohail, A. Qurashi, Sonochemical assisted synthesis of RGO/ZnO nanowire arrays for photoelectrochemical water splitting, Ultrason Sonochem. 37 (2017) 669-675. https://doi.org/10.1016/j.ultsonch.2017.02.029

[37] Y. Lykhach, S.M. Kozlov, T. Skála, A. Tovt, V. Stetsovych, N. Tsud, F. Dvořák, V. Johánek, A. Neitzel, J. Mysliveček, S. Fabris, V. Matolín, K.M. Neyman, J. Libuda, Counting electrons on supported nanoparticles, 15 (2015) 284-288. https://doi.org/10.1038/nmat4500

[38] C. Dablemont, P. Lang, C. Mangeney, J.Y. Piquemal, V. Petkov, F. Herbst, G. Viau, FTIR and XPS study of Pt nanoparticle functionalization and interaction with alumina, 24 (2008) 5832-5841. https://doi.org/10.1021/la7028643

[39] V. Kestens, G. Roebben, J. Herrmann, Å. Jämting, V. Coleman, C. Minelli, C. Clifford, P.J. de Temmerman, J. Mast, L. Junjie, F. Babick, H. Cölfen, H. Emons, Challenges in the size analysis of a silica nanoparticle mixture as candidate certified reference material, J Nanopart Res. 18 (2016). https://doi.org/10.1007/s11051-016-3474-2

[40] A. Sikora, A.G. Shard, C. Minelli, Size and ζ-Potential Measurement of Silica Nanoparticles in Serum Using Tunable Resistive Pulse Sensing, 32 (2016) 2216-2224. https://doi.org/10.1021/acs.langmuir.5b04160

[41] D.F. Swinehart, The Beer-Lambert Law, J Chem Educ. 39 (1962) 333-335. https://doi.org/10.1021/ed039p333

[42] D. Liu, C. Li, F. Zhou, T. Zhang, H. Zhang, X. Li, G. Duan, W. Cai, Y. Li, Rapid Synthesis of Monodisperse Au Nanospheres through a Laser Irradiation -Induced Shape Conversion, Self-Assembly and Their Electromagnetic Coupling SERS Enhancement, 5 (2015) 1-9. https://doi.org/10.1038/srep07686

[43] J. Liu, Y. Liu, N. Liu, Y. Han, X. Zhang, H. Huang, Y. Lifshitz, S.T. Lee, J. Zhong, Z. Kang, Metal-free efficient photocatalyst for stable visible water splitting via a two-electron pathway, Science (1979). 347 (2015) 970-974. https://doi.org/10.1126/science.aaa3145

[44] Z.B. Yu, Y.P. Xie, G. Liu, G.Q. Lu, X.L. Ma, H.M. Cheng, Self-assembled CdS/Au/ZnO heterostructure induced by surface polar charges for efficient photocatalytic hydrogen evolution, J Mater Chem A Mater. 1 (2013) 2773-2776. https://doi.org/10.1039/c3ta01476b

[45] K. Gupta, R.P. Singh, A. Pandey, A. Pandey, Photocatalytic antibacterial performance of TiO2 and Ag-doped TiO2 against S. aureus. P. aeruginosa and E. coli, 4 (2013) 345-351. https://doi.org/10.3762/bjnano.4.40

[46] Pal, M., Pal, U., Jiménez, J. M. G. Y., & Pérez-Rodríguez, F. Effects of crystallization and dopant concentration on the emission behavior of TiO2:Eu nanophosphors, Nanoscale Res Lett. 7 (2012) 1-6. https://doi.org/10.1186/1556-276X-7-1

[47] T. v. Torchynska, B. el Filali, T. v. Torchynska, B. el Filali, Emission, Defects, and Structure of ZnO Nanocrystal Films Obtained by Electrochemical Method, (2017). https://doi.org/10.5772/66335

[48] G. Lin, Q. Zhang, X. Lin, D. Zhao, R. Jia, N. Gao, Z. Zuo, X. Xu, D. Liu, Enhanced photoluminescence of gallium phosphide by surface plasmon resonances of metallic nanoparticles, RSC Adv. 5 (2015) 48275-48280. https://doi.org/10.1039/C5RA07368E

[49] Jadhav, S. D., & Jadhav, M. S. Analysis of River Water Quality with Special Reference to Nitrate Concentration of Indrayani River, Pune. Parameters (2015).

[50] S.S. Kale, A.K. Kadam, S. Kumar, N.J. Pawar, Evaluating pollution potential of leachate from landfill site, from the Pune metropolitan city and its impact on shallow basaltic aquifers, Environ Monit Assess. 162 (2010) 327-346. https://doi.org/10.1007/s10661-009-0799-7

[51] X.Q. Li, W.X. Zhang, Iron nanoparticles: The core-shell structure and unique properties for Ni(II) sequestration, 22 (2006) 4638-4642. https://doi.org/10.1021/la060057k

[52] I. Sargin, T. Baran, G. Arslan, Environmental remediation by chitosan-carbon nanotube supported palladium nanoparticles: Conversion of toxic nitroarenes into aromatic amines, degradation of dye pollutants and green synthesis of biaryls, Sep Purif Technol. 247 (2020) 116987. https://doi.org/10.1016/j.seppur.2020.116987

[53] M.M. Baig, S. Zulfiqar, M.A. Yousuf, I. Shakir, M.F.A. Aboud, M.F. Warsi, DyxMnFe2-xO4 nanoparticles decorated over mesoporous silica for environmental remediation applications, J Hazard Mater. 402 (2021) 123526. https://doi.org/10.1016/j.jhazmat.2020.123526

[54] S.A. Akintelu, A.S. Folorunso, F.A. Folorunso, A.K. Oyebamiji, Green synthesis of copper oxide nanoparticles for biomedical application and environmental remediation, Heliyon. 6 (2020) e04508. https://doi.org/10.1016/j.heliyon.2020.e04508

[55] G. Zhong, J. Huang, Z. Yao, B. Luo, K. Li, S. Xu, X. Fu, Y. Cao, Intrinsic acid resistance and high removal performance from the incorporation of nickel nanoparticles into nitrogen doped tubular carbons for environmental remediation, J Colloid Interface Sci. 566 (2020) 46-59. https://doi.org/10.1016/j.jcis.2020.01.055

[56] A. Parmar, G. Kaur, S. Kapil, V. Sharma, M.K. Choudhary, S. Sharma, Novel biogenic silver nanoparticles as invigorated catalytic and antibacterial tool: A cleaner approach towards environmental remediation and combating bacterial invasion, Mater Chem Phys. 238 (2019) 121861. https://doi.org/10.1016/j.matchemphys.2019.121861

[57] W. Yan, A.A. Herzing, X.Q. Li, C.J. Kiely, W.X. Zhang, Structural evolution of Pd-doped nanoscale zero-valent iron (nZVI) in aqueous media and implications for particle aging and reactivity, Environ Sci Technol. 44 (2010) 4288-4294. https://doi.org/10.1021/es100051q

[58] P.G. Tratnyek, R.L. Johnson, Nanotechnologies for environmental cleanup, Nano Today. 1 (2006) 44-48. https://doi.org/10.1016/S1748-0132(06)70048-2

[59] C.S.R. Rajan, Nanotechnology in Groundwater Remediation, (2011) 182-187. https://doi.org/10.7763/IJESD.2011.V2.121

[60] E. Shahsavari, G. Poi, A. Aburto-Medina, N. Haleyur, A.S. Ball, Bioremediation approaches for petroleum hydrocarbon-contaminated environments, 1 (2017) 21-41. https://doi.org/10.1007/978-3-319-55426-6_3

[61] F.E. Löffler, E.A. Edwards, Harnessing microbial activities for environmental cleanup, Curr Opin Biotechnol. 17 (2006) 274-284. https://doi.org/10.1016/j.copbio.2006.05.001

[62] K.L. Garner, A.A. Keller, Emerging patterns for engineered nanomaterials in the environment: a review of fate and toxicity studies, 16 (2014). https://doi.org/10.1007/s11051-014-2503-2

[63] Y.P. Sun, X. qin Li, J. Cao, W. xian Zhang, H.P. Wang, Characterization of zero-valent iron nanoparticles, Adv Colloid Interface Sci. 120 (2006) 47-56. https://doi.org/10.1016/j.cis.2006.03.001

[64] B. Karn, T. Kuiken, M. Otto, Nanotechnology and in Situ Remediation: A Review of the Benefits and Potential Risks, Environ Health Perspect. 117 (2009) 1813. https://doi.org/10.1289/ehp.0900793

[65] T. Tosco, M. Petrangeli Papini, C. Cruz Viggi, R. Sethi, Nanoscale zerovalent iron particles for groundwater remediation: a review, J Clean Prod. 77 (2014) 10-21. https://doi.org/10.1016/j.jclepro.2013.12.026

[66] G. Ghasemzadeh, M. Momenpour, F. Omidi, M.R. Hosseini, M. Ahani, A. Barzegari, Applications of nanomaterials in water treatment and environmental remediation, Front Environ Sci Eng. 8 (2014) 471-482. https://doi.org/10.1007/s11783-014-0654-0

[67] A. Koutsospyros, J. Pavlov, J. Fawcett, D. Strickland, B. Smolinski, W. Braida, Degradation of high energetic and insensitive munitions compounds by Fe/Cu bimetal reduction, J Hazard Mater. 219-220 (2012) 75-81. https://doi.org/10.1016/j.jhazmat.2012.03.048

[68] X. Nie, J. Liu, X. Zeng, D. Yue, Rapid degradation of hexachlorobenzene by micron Ag/Fe bimetal particles, 25 (2013) 473-478. https://doi.org/10.1016/S1001-0742(12)60088-6

[69] N. Savage, M.S. Diallo, Nanomaterials and water purification: Opportunities and challenges, 7 (2005) 331-342. https://doi.org/10.1007/s11051-005-7523-5

[70] P. Liang, Y. Liu, L. Guo, J. Zeng, H. Lu, Multiwalled carbon nanotubes as solid-phase extraction adsorbent for the preconcentration of trace metal ions and their determination by inductively coupled plasma atomic emission spectrometry, J Anal At Spectrom. 19 (2004) 1489-1492. https://doi.org/10.1039/b409619c

[71] N.C. Mueller, B. Nowack, Exposure modeling of engineered nanoparticles in the environment, Environ Sci Technol. 42 (2008) 4447-4453. https://doi.org/10.1021/es7029637

[72] E.A. Rogozea, A.R. Petcu, N.L. Olteanu, C.A. Lazar, D. Cadar, M. Mihaly, Tandem adsorption-photodegradation activity induced by light on NiO-ZnO p-n couple modified silica nanomaterials, Mater Sci Semicond Process. 57 (2017) 1-11. https://doi.org/10.1016/j.mssp.2016.10.006

[73] E.A. Rogozea, N.L. Olteanu, A.R. Petcu, C.A. Lazar, A. Meghea, M. Mihaly, Extension of optical properties of ZnO/SiO₂ materials induced by incorporation of Au or NiO nanoparticles, 56 (2016) 45-48. https://doi.org/10.1016/j.optmat.2015.12.020

[74] I.S. Yunus, Harwin, A. Kurniawan, D. Adityawarman, A. Indarto, Nanotechnologies in water and air pollution treatment, 1 (2013) 136-148. https://doi.org/10.1080/21622515.2012.733966

[75] E.F. Mohamed, Nanotechnology: Future of Environmental Air Pollution Control, 6 (2017) 429-454. https://doi.org/10.5296/emsd.v6i2.12047

[76] J.Y. Kim, S.B. Shim, J.K. Shim, Effect of amphiphilic polyurethane nanoparticles on sorption-desorption of phenanthrene in aquifer material, J Hazard Mater. 98 (2003) 145-160. https://doi.org/10.1016/S0304-3894(02)00311-4

[77] R.K. Ibrahim, M. Hayyan, M.A. AlSaadi, A. Hayyan, S. Ibrahim, Environmental application of nanotechnology: air, soil, and water, 23 (2016) 13754-13788. https://doi.org/10.1007/s11356-016-6457-z

[78] J.C. Abanades, D. Alvarez, Conversion Limits in the Reaction of CO2 with Lime, 17 (2003) 308-315. https://doi.org/10.1021/ef020152a

[79] Y. Wang, Y. Zhu, S. Wu, A new nano CaO-based CO2 adsorbent prepared using an adsorption phase technique, 218 (2013) 39-45. https://doi.org/10.1016/j.cej.2012.11.095

[80] J. Jänchen, D.T.F. Möhlmann, H. Stach, Water and carbon dioxide sorption properties of natural zeolites and clay minerals at martian surface temperature and

pressure conditions, Stud Surf Sci Catal. 170 (2007) 2116-2121. https://doi.org/10.1016/S0167-2991(07)81108-6

[81] H.Y. Wang, A.C. Lua, Development of metallic nickel nanoparticle catalyst for the decomposition of methane into hydrogen and carbon nanofibers, 116 (2012) 26765-26775. https://doi.org/10.1021/jp306519t

[82] N.H. Nguyen, H. Bai, Effect of washing pH on the properties of titanate nanotubes and its activity for photocatalytic oxidation of NO and NO2, Appl Surf Sci. 355 (2015) 672-680. https://doi.org/10.1016/j.apsusc.2015.07.118

[83] Q. Liang, D. Zhao, Immobilization of arsenate in a sandy loam soil using starch-stabilized magnetite nanoparticles, J Hazard Mater. 271 (2014) 16-23. https://doi.org/10.1016/j.jhazmat.2014.01.055

[84] R. Liu, D. Zhao, Synthesis and characterization of a new class of stabilized apatite nanoparticles and applying the particles to in situ Pb immobilization in a fire-range soil, Chemosphere. 91 (2013) 594-601. https://doi.org/10.1016/j.chemosphere.2012.12.034

[85] Y. Wang, Z. Fang, B. Liang, E.P. Tsang, Remediation of hexavalent chromium contaminated soil by stabilized nanoscale zero-valent iron prepared from steel pickling waste liquor, 247 (2014) 283-290. https://doi.org/10.1016/j.cej.2014.03.011

[86] Z. Sheikholeslami, D.Y. Kebria, F. Qaderi, Application of γ-Fe2O3 nanoparticles for pollution removal from water with visible light, J Mol Liq. 299 (2020) 112118. https://doi.org/10.1016/j.molliq.2019.112118

[87] H. Ramezanalizadeh, R. Peymanfar, N. Khodamoradipoor, Design and development of a novel lanthanum inserted CuCr2O4 nanoparticles photocatalyst for the efficient removal of water pollutions, Optik (Stuttg). 180 (2019) 113-124. https://doi.org/10.1016/j.ijleo.2018.11.067

[88] P. Anju Rose Puthukkara, T. Sunil Jose, S. Dinoop lal, Chitosan stabilized Fe/Ni bimetallic nanoparticles for the removal of cationic and anionic triphenylmethane dyes from water, Environ Nanotechnol Monit Manag. 14 (2020) 100295. https://doi.org/10.1016/j.enmm.2020.100295

[89] Y. Chen, H. Shi, H. Guo, C. Ling, X. Yuan, P. Li, Hydrated titanium oxide nanoparticles supported on natural rice straw for Cu (II) removal from water, Environ Technol Innov. 20 (2020) 101143. https://doi.org/10.1016/j.eti.2020.101143

[90] L.P. Burkhard, Estimating dissolved organic carbon partition coefficients for nonionic organic chemicals, Environ Sci Technol. 34 (2000) 4663-4668. https://doi.org/10.1021/es001269l

[91] S.O. Obare, G.J. Meyer, Nanostructured Materials for Environmental Remediation of Organic Contaminants in Water, 39 (2011) 2549-2582. https://doi.org/10.1081/ESE-200027010

[92] M.R. Hoffmann, S.T. Martin, W. Choi, D.W. Bahnemann, Environmental Applications of Semiconductor Photocatalysis, Chem Rev. 95 (1995) 69-96. https://doi.org/10.1021/cr00033a004

[93] A. Fujishima, T.N. Rao, D.A. Tryk, Titanium dioxide photocatalysis, 1 (2000) 1-21. https://doi.org/10.1016/S1389-5567(00)00002-2

[94] Y.C. Kim, K.H. Lee, S. Sasaki, K. Hashimoto, K. Ikebukuro, I. Karube, Photocatalytic sensor for chemical oxygen demand determination based on oxygen electrode, Anal Chem. 72 (2000) 3379-3382. https://doi.org/10.1021/ac9911342

[95] D. Bahnemann, Photocatalytic water treatment: solar energy applications, 77 (2004) 445-459. https://doi.org/10.1016/j.solener.2004.03.031

[96] P. v. Kamat, D. Meisel, Nanoscience opportunities in environmental remediation, 6 (2003) 999-1007. https://doi.org/10.1016/j.crci.2003.06.005

[97] W.X. Zhang, Nanoscale iron particles for environmental remediation: An overview, 5 (2003) 323-332.

[98] S.M. Ponder, J.G. Darab, T.E. Mallouk, Remediation of Cr(VI) and Pb(II) Aqueous Solutions Using Supported, Nanoscale Zero-valent Iron, Environ Sci Technol. 34 (2000) 2564-2569. https://doi.org/10.1021/es9911420

[99] S. Mitra, A. Sarkar, S. Sen, Removal of chromium from industrial effluents using nanotechnology: a review, 2 (2017) 1-14. https://doi.org/10.1007/s41204-017-0022-y

[100] B. Pandey, M. Fulekar, Nanotechnology: remediation technologies to clean up the environmental pollutants., (2012).

[101] M. Golobič, A. Jemec, D. Drobne, T. Romih, K. Kasemets, A. Kahru, Upon exposure to cu nanoparticles, accumulation of copper in the isopod porcellio scaber is due to the dissolved Cu ions inside the digestive tract, Environ Sci Technol. 46 (2012) 12112-12119. https://doi.org/10.1021/es3022182

[102] T. Masciangioli, W.X. Zhang, Peer Reviewed: Environmental Technologies at the Nanoscale, Environ Sci Technol. 37 (2003). https://doi.org/10.1021/es0323998

[103] Biotechnology and Nanotechnology Risk Assessment: Minding and Managing the Potential Threats Around Us | NHBS Academic & Professional Books, (n.d.).

[104] J. Zhuang, R.W. Gentry, Environmental application and risks of nanotechnology: A balanced view, 1079 (2011) 41-67. https://doi.org/10.1021/bk-2011-1079.ch003

[105] L. Reijnders, Cleaner nanotechnology and hazard reduction of manufactured nanoparticles, J Clean Prod. 14 (2006) 124-133. https://doi.org/10.1016/j.jclepro.2005.03.018

[106] S. Santra, P. Zhang, K. Wang, R. Tapec, W. Tan, Conjugation of biomolecules with luminophore-doped silica nanoparticles for photostable biomarkers., Anal Chem. (2001). https://doi.org/10.1021/ac010406+

[107] E. Oberdörster, S. Zhu, T.M. Blickley, P. McClellan-Green, M.L. Haasch, Ecotoxicology of carbon-based engineered nanoparticles: Effects of fullerene (C60) on aquatic organisms, Carbon N Y. 44 (2006) 1112-1120. https://doi.org/10.1016/j.carbon.2005.11.008

[108] K.D. Grieger, A. Fjordbøge, N.B. Hartmann, E. Eriksson, P.L. Bjerg, A. Baun, Environmental benefits and risks of zero-valent iron nanoparticles (nZVI) for in situ remediation: risk mitigation or trade-off?, J Contam Hydrol. 118 (2010) 165-183. https://doi.org/10.1016/j.jconhyd.2010.07.011

[109] Stockholm Convention on Persistent Organic Pollutants - United States Department of State, (n.d.).

[110] S.M. Taghavi, M. Momenpour, M. Azarian, M. Ahmadian, F. Souri, S.A. Taghavi, M. Sadeghain, M. Karchani, Effects of Nanoparticles on the Environment and Outdoor Workplaces, Electron Physician. 5 (2013) 706.

[111] M. Auffan, W. Achouak, J. Rose, M.A. Roncato, C. Chanéac, D.T. Waite, A. Masion, J.C. Woicik, M.R. Wiesner, J.Y. Bottero, Relation between the redox state of iron-based nanoparticles and their cytotoxicity toward Escherichia coli, Environ Sci Technol. 42 (2008) 6730-6735. https://doi.org/10.1021/es800086f

[112] C. Lee, Y.K. Jee, I.L. Won, K.L. Nelson, J. Yoon, D.L. Sedlak, Bactericidal effect of zero-valent iron nanoparticles on Escherichia coli, Environ Sci Technol. 42 (2008) 4927-4933. https://doi.org/10.1021/es800408u

[113] M. Diao, M. Yao, Use of zero-valent iron nanoparticles in inactivating microbes, Water Res. 43 (2009) 5243-5251. https://doi.org/10.1016/j.watres.2009.08.051

[114] X. Ma, A. Gurung, Y. Deng, Phytotoxicity and uptake of nanoscale zero-valent iron (nZVI) by two plant species, Sci Total Environ. 443 (2013) 844-849. https://doi.org/10.1016/j.scitotenv.2012.11.073

[115] T. Ben-Moshe, S. Frenk, I. Dror, D. Minz, B. Berkowitz, Effects of metal oxide nanoparticles on soil properties, Chemosphere. 90 (2013) 640-646. https://doi.org/10.1016/j.chemosphere.2012.09.018

[116] C. Fajardo, L.T. Ortíz, M.L. Rodríguez-Membibre, M. Nande, M.C. Lobo, M. Martin, Assessing the impact of zero-valent iron (ZVI) nanotechnology on soil microbial structure and functionality: a molecular approach, Chemosphere. 86 (2012) 802-808. https://doi.org/10.1016/j.chemosphere.2011.11.041

[117] Z. Tong, M. Bischoff, L. Nies, B. Applegate, R.F. Turco, Impact of Fullerene (C60) on a Soil Microbial Community, Environ Sci Technol. 41 (2007) 2985-2991. https://doi.org/10.1021/es061953l

[118] J. Wang, J.D. Gerlach, N. Savage, G.P. Cobb, Necessity and approach to integrated nanomaterial legislation and governance, 442 (2013) 56-62. https://doi.org/10.1016/j.scitotenv.2012.09.073

Materials Research Foundations 145 (2023) 281-310 https://doi.org/10.21741/9781644902370-11

[119] M. Geiser, B. Rothen-Rutishauser, N. Kapp, S. Schürch, W. Kreyling, H. Schulz, M. Semmler, V. Im Hof, J. Heyder, P. Gehr, Ultrafine Particles Cross Cellular Membranes by Nonphagocytic Mechanisms in Lungs and in Cultured Cells, Environ Health Perspect. 113 (2005) 1555. https://doi.org/10.1289/ehp.8006

[120] A.E. Porter, M. Gass, K. Muller, J.N. Skepper, P. Midgley, M. Welland, Visualizing the Uptake of C60 to the Cytoplasm and Nucleus of Human Monocyte-Derived Macrophage Cells Using Energy-Filtered Transmission Electron Microscopy and Electron Tomography, Environ Sci Technol. 41 (2007) 3012-3017. https://doi.org/10.1021/es062541f

[121] A. Nel, T. Xia, L. Mädler, N. Li, Toxic potential of materials at the nanolevel, Science. 311 (2006) 622-627. https://doi.org/10.1126/science.1114397

[122] G. Bhabra, A. Sood, B. Fisher, L. Cartwright, M. Saunders, W.H. Evans, A. Surprenant, G. Lopez-Castejon, S. Mann, S.A. Davis, L.A. Hails, E. Ingham, P. Verkade, J. Lane, K. Heesom, R. Newson, C.P. Case, Nanoparticles can cause DNA damage across a cellular barrier, Nat Nanotechnol. 4 (2009) 876-883. https://doi.org/10.1038/nnano.2009.313

[123] H.L. Karlsson, P. Cronholm, Y. Hedberg, M. Tornberg, L. de Battice, S. Svedhem, I.O. Wallinder, Cell membrane damage and protein interaction induced by copper containing nanoparticles-Importance of the metal release process, Toxicology. 313 (2013) 59-69. https://doi.org/10.1016/j.tox.2013.07.012

 https://doi.org/10.21741/9781644902370-12

Chapter 12

Ethics of Nanomedicine

Ineke Malsch
[1]Malsch TechnoValuation, PO Box 455, 3500 AL Utrecht, The Netherlands
malschtechnovaluation@xs4all.nl

Abstract

Nanomedicine is the application of nanomaterials and nanotechnologies in healthcare. Nanomedicine is a key enabler of e.g., mRNA COVID-19 vaccines, and enables digital twins, organ on chip and wearables. Introducing nanomaterials in the body in pharmaceuticals or implants raises nanosafety issues. Ethical impacts of nanomedicine are e.g., related to freedom, equality, data protection, and biosecurity. Researchers should contribute to Responsible Research and Innovation together with governments and other stakeholders. The principles of inclusiveness, anticipation, openness and responsiveness are leading.

Keywords

Nanomedicine, Ethics, Responsible Research and Innovation, Nanosafety

Contents

1. Introduction

According to the European Technology Platform on Nanomedicine: “Nanomedicine is the application of nanotechnology to achieve innovation in healthcare. It uses the properties developed by a material at its nanometric scale (10^{-9}) which often differ in terms of physics, chemistry or biology from the same material at a bigger scale.” [1]

The European Commission recently updated its definition of nanomaterial for regulatory purposes: “'Nanomaterial' means a natural, incidental or manufactured material consisting of solid particles that are present, either on their own or as identifiable constituent particles in aggregates or agglomerates, and where 50 % or more of these particles in the number-based size distribution fulfil at least one of the following conditions: (a) one or more external dimensions of the particle are in the size range 1 nm to 100 nm; (b) the particle has an elongated shape, such as a rod, fibre or tube, where two external dimensions are smaller than 1 nm and the other dimension is larger than 100 nm; (c) the particle has a plate-like shape, where one external dimension is smaller than 1 nm and the other dimensions are larger than 100 nm. In the determination of the particle number-based size distribution, particles with at least two orthogonal external dimensions larger than 100 μm need not be considered. However, a material with a specific surface area by volume of < 6 m^2/cm^3 shall not be considered a nanomaterial.”

In this chapter, the concept of Responsible Research and Innovation will be briefly introduced and an overview will be given of ethical issues in nanomedicine for cancer therapy. How scientists could contribute to responsible governance of each ethical issue will be indicated.

2. Screening ethical impacts of nanomedicine

Which ethical issues does nanomedicine raise? A screening of expected ethical impacts of nanomedicine was performed, guided by the online Ethical Impact Assessment tools developed and tested in RiskGONE[1] [2,3]. These are based on the CEN Workshop Agreement 17145-2:2017 (E) [4]). The Ethical Impact Assessment starts with a self-assessment guided by a checklist of nine general ethical categories: health, privacy,

[1] Accessible via: http://www.enaloscloud.novamechanics.com/riskgone/EIA/

Materials Research Foundations 145 (2023) 311-325 https://doi.org/10.21741/9781644902370-12

liberties, equality, common good, environment, sustainability, dual military use, and misuse. In addition to potential ethical risks, nanomedicine also offers expected ethical benefits. The strength of expected positive and negative ethical impacts should be estimated on a five point scale: 0=no, 1=minor, 2=moderate, 3=medium, 4=high, 5=severe. A rough estimate of the likely ethical risks and benefits of nanomedicine is depicted in figure 1.

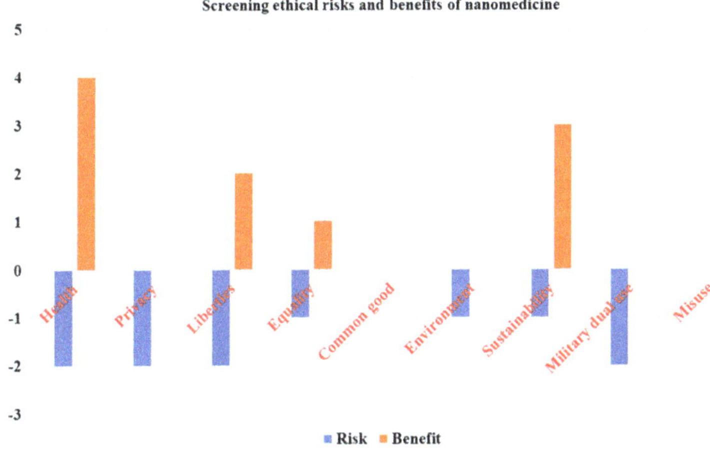

Figure 1. screening ethical risks and benefits of nanomedicine.

Related to health, all applications of nanomedicine, including mRNA COVID-19 vaccines, digital twins, organ on chip and wearables, raise general biomedical ethics issues such as the need for informed consent of patients [5], beneficence and non-maleficence [5]. The use of some nanomaterials inside the body in pharmaceuticals or implants, or in disposables, such as antimicrobial bandages, introduces uncertain nanosafety risks for the patient as well as for health professionals and relatives of the patient. These risks are additional to the health and safety risks of the used chemical substances in general. Nanosafety data is increasingly collected and curated, and some nanomaterials are shown to be more risky than others. In medical applications, expected benefits for health must outweigh the expected risks. Widespread use of applications of nanomaterials and nanoelectronics in diagnostics and lifestyle apps may contribute to influencing the healthcare system raising potential risks.

On the other hand, the use of nanomaterials in healthcare is intended to contribute to benefits for the health of patients treated with the nanomedicine. Widespread use of applications of nanomaterials and nanoelectronics in diagnostics and lifestyle apps may contribute to influencing the healthcare system, which could also offer benefits.

Furthermore, the use of nanomaterials in diagnostics and health monitoring devices raises privacy and data protection issues. In combination with other emerging technologies, these risks could be more severe than of conventional technologies. No significant positive impacts on privacy and data protection of nanomedicine are expected.

Related to liberties, freedom may be at stake in some applications, for example, when groups of professionals or citizens are obliged to be vaccinated with nanovaccines, or when healthy people are coerced into taking preventive medicine or using diagnostics. In contrast, nanomaterials enabling wearable health monitoring devices may positively contribute to the freedom of patients who can stay home rather than being hospitalised.

Related to equality, nanomedicine may worsen global inequalities in access to healthcare, e.g. when encapsulation of pharmaceuticals leads to extension of the patent period. On the other hand, nanomedicine may also positively influence global inequalities in access to healthcare, for instance if miniaturisation of diagnostics makes it more affordable.

Using nanomaterials in nanomedicine raises uncertain environmental risks if they are released into the environment. These risks are additional to the health and safety risks of the used chemical substances in general. Nanosafety data is increasingly collected and curated, and some nanomaterials are shown to be more risky than others.

Related to sustainability, some negative impacts are expected on SDG3 (good health and wellbeing, 6 (clean water and sanitation), 14 (life on land) and 15 (life under water). The impacts of nanomaterials are not significantly higher than of current pharmaceuticals and medical devices. Vice versa, more significant positive impacts are expected, mainly on SDG3.

Finally, some dual use nanomedicine research, such as targeted drug delivery crossing the blood-brain barrier raises biosecurity issues.

3. Responsible Research and Innovation

The term ‘responsibility’ is not clearly defined but is commonly used when someone voluntarily responds to a perceived need by taking care of someone or something. It is also used as another word for accountability, when someone is held responsible by others for taking care of something. In line with social contract theory [6], governments take responsibility for protecting the rights and security of their citizens and their territory, and citizens hold their governments responsible for these tasks. In daily life, citizens have individual responsibilities for taking care of themselves and others, and for complying with regulations. Science and technology escape this traditional division of responsibilities because they may introduce uncertain and unforeseen new risks which are not governed by existing legislation. Therefore, science and technology call for new forms of collective responsibilities, with role responsibilities allocated to governments, scientists, industry, civil society organisations and citizens [7].

Materials Research Foundations 145 (2023) 311-325 https://doi.org/10.21741/9781644902370-12

Different approaches could be envisaged to organising responsibility for science and technology. In the traditional division of labour, "*Science takes the credit for penicillin, while society takes the blame for the bomb*" [8]. Responsible Research and Innovation implies that scientists and all stakeholders should imagine and discuss ethical and societal aspects of new technologies and, together, change course early. This is less easy than it looks, because of the Collingridge Dilemma [9], summarised as follows: In the early stages of innovation, the technology is flexible, but potential future impacts are unclear. In late stages of innovation, the impacts are known, but by then, the technology is entrenched and inflexible. Responsible Research and Innovation (RRI) implies collective responsibility for the impacts of technology on society. Governments keep the main authority for protecting their citizens against potential risks and societal impacts of emerging technologies. They are also stimulating innovation by different means including funding R&D. To complement governmental responsibilities, governments demand co-responsibility for societal impacts from:

- The scientific community, research organisations and individual scientists
- Large industrial companies and small and medium enterprises engaged in R&D and innovation
- Civil society organisations, including trade unions, environmental and consumer movements.

The concept RRI comprises guidelines for taking this collective responsibility. Several definitions are used, but all stress the need for the four core values: i) inclusiveness, ii) anticipation, iii) openness and iv) responsiveness [10]. Inclusiveness means that all people have the right to participate in science and reap the benefits. Anticipation calls for reflecting on potential future impacts on society and the environment already during the research phase. Openness requires the publication of results of scientific research and communication about it to all people in understandable terms. Responsiveness implies that scientists should engage in public dialogue with citizens and stakeholders and take their views into account in research strategies.

RRI is a horizontal requirement in EU funded research, introduced under the Horizon 2020 programme, stimulating researchers to take more responsibility for ethical and societal impacts of research [11]. The European Commission distinguishes six keys: I) Public engagement means that projects funded by the EU must include (two-way) communication with stakeholders and citizens about the results. II) Gender equality calls for equal participation of men and women in management and research, and for addressing gender-specific issues in the research. III) Science education implies that training and education of young scientists and engaging with schools or science museums are needed. IV) Open access and open science calls for publishing results in open access publications, and offering open access of research data, etc. V) Ethics means researchers should explore and address ethical issues of the research. VI) Governance requires contributing to responsible governance and policy making of science and technology (this is mainly addressed to policy makers).

Materials Research Foundations 145 (2023) 311-325 https://doi.org/10.21741/9781644902370-12

Preceding the insertion of RRI principles as horizontal requirements in Horizon 2020, the European Commission adopted in 2008 the Recommendation on a voluntary Code of Conduct for Responsible Nanosciences and Nanotechnologies (N&N) Research. "*The Code of Conduct invites all stakeholders to act responsibly and cooperate with each other, ... to ensure that N&N research is undertaken in the Community in a safe, ethical and effective framework, supporting sustainable economic, social and environmental development.*" Scientists and other stakeholders in nanoresearch are asked to respect these seven principles: meaning, sustainability, precaution, inclusiveness, excellence, innovation, and accountability [12]. The recommendation is still referred to in the ethics guidelines to applicants of EU Horizon Europe funding.

4. Nanosafety

Introducing some nanomaterials in the body may cause diseases, even if they are administered in the form of therapeutic drugs. In line with the biomedical ethics principles of beneficence and non-maleficence, unintended health risks should be less than the intended curative benefits of nanomedicine. Toxicity varies greatly between different kinds and particle sizes of nanomaterials. To help reduce the uncertainty of possible impacts of nanomaterials on health, safety and the environment, nanomaterials safety data has increasingly been collected in projects, mainly funded by the European Union and national governments in the last twenty-odd years and curated in online databases. The European Observatory on Nanomaterials (EUON) is hosted by the European Chemicals Agency and brings together all information on nanomaterials safety and regulation, covering nanomedicine [13]. It offers a one-stop-shop to nanomaterials data collected in several public databases [14]. The collaboration of researchers participating in risk assessment and life cycle analysis research on nanomaterials to contribute sound nanosafety data is a prerequisite to responsible risk governance of nanomaterials. The EU includes contractual obligations to comply with RRI principles openness and open science, in EU funded research in Horizon 2020. Scientists should be aware of and respect these and other regulations governing nanosafety data.

Nanomaterials R&D increasingly incorporates safe-by-design: Integrate risk assessment in the research process from basic research onwards. More recently, the concept is broadened to safe and sustainable by design. This is defined as: "*A pre-market approach to chemicals that focuses on providing a function (or service), while avoiding volumes and chemical properties that may be harmful to human health or the environment, in particular groups of chemicals likely to be (eco) toxic, persistent, bio-accumulative or mobile. Overall sustainability should be ensured by minimizing the environmental footprint of chemicals in particular on climate change, resource use, ecosystems and biodiversity from a life cycle perspective*" [15,16].

From an ethical perspective, the RRI principle 'anticipation' governs nanosafety issues of nanomedicine. This calls upon researchers to reflect on potential future impacts on human health, society and the environment already during the research phase. More specifically,

Materials Research Foundations 145 (2023) 311-325
https://doi.org/10.21741/9781644902370-12

the EU nanocode calls for applying the precautionary principle: "anticipating potential environmental, health and safety impacts of N&N outcomes and taking due precautions, proportional to the level of protection, while encouraging progress for the benefit of society and the environment" [12].

5. Drug delivery

A wide variety of nanomaterials are used in targeted drug delivery, including liposomes, lipid nanoparticles, polymeric nanoparticles, polymeric micelles, dendrimers and nanogels [17]. Following successful application of nanomedicine in COVID-19 mRNA vaccines, nanomaterials are increasingly applied in regulating immunity [18].

5.1 The right to benefit from science

This application of nanomedicine raises issues related to global inequalities in access to healthcare, as illustrated by the COVID-19 vaccination status published by the World Health Organisation. On 7 July 2022, over 12 billion vaccination doses were administered and over 4 billion people were fully vaccinated, but in several, mainly African countries, still less than 20% of the population were vaccinated [18]. This is an apparent violation of the UN covenant on socio-economic rights: "*Article 12 - 1. The States Parties to the present Covenant recognize the right of everyone to the enjoyment of the highest attainable standard of physical and mental health. 2. The steps to be taken by the States Parties to the present Covenant to achieve the full realization of this right shall include those necessary for: ... (c) The prevention, treatment and control of epidemic, endemic, occupational and other diseases; ...*" [19]

From the perspective of RRI, governments and other involved stakeholders are urged to foster 'inclusiveness', respecting the basic human right of all people to participate in science and reap the benefits. However, the corresponding article in the UN covenant on socio-economic rights calls for balancing distinct rights of citizens in general on the one hand and researchers and other stakeholders involved in scientific research on the other: "*Article 15 - 1. The States Parties to the present Covenant recognize the right of everyone: ... (b) To enjoy the benefits of scientific progress and its applications; (c) To benefit from the protection of the moral and material interests resulting from any scientific, ... production of which he is the author.*" [19] The RRI-key 'Open science' of the European Commission gives more practical guidance for researchers, urging them to publish in open access publications, and offer open access of research data [13]. Related to this, patenting nanomedicine raises ethical dilemmas: "*[Nanomedicine could blur] the balance of interests whereby diagnosis, therapy and research should be available to patients without patents being a hindrance... There are risks of overly broad patents being granted that may hinder their therapeutic availability... [there is a] need for research into the manner in which the patent system can properly balance the need to reward innovation and ensure availability.*" [20]. Nanodrug delivery is sometimes used to extend the patent protection of blockbuster drugs delaying cheaper generic drug manufacturing.

Materials Research Foundations 145 (2023) 311-325 https://doi.org/10.21741/9781644902370-12

5.2 Communicating nanoscience

Public trust in COVID-19 mRNA vaccines is influenced by communication on research. To illustrate the importance of good science communication, by end of May 2021, Eurobarometer polled opinions of 26,106 Europeans of 15 years and over on their attitudes to COVID-19 vaccination. "*Key reasons for not getting vaccinated are the belief that COVID-19 vaccines have not yet been sufficiently tested and worries about the side effects of the COVID-19 vaccines, with 85% and 82%, respectively, answering these reasons are 'important'. Respondents would be keener to get vaccinated if 'more people have already been vaccinated, they see that it works and there are no major side effects' and if there is 'full clarity on how vaccines are being developed, tested and authorised' (mentioned by 30% and 26%, respectively). Overall, the benefits of vaccination against COVID-19 are recognised: 76% agree that all in all the benefits of COVID-19 vaccines outweigh possible risks. However, there are concerns about the safety of COVID-19 vaccines: half of respondents agree that COVID-19 vaccines are being developed, tested and authorised too quickly to be safe*" [21]. The responsibility of researchers is in line with the value of 'openness', meaning that scientists should publish results of scientific research and communicate about it to all people in understandable terms. In EU-funded research, scientists are encouraged to include (two-way) communication with stakeholders and citizens in research projects.

5.3 Ethical dilemmas of nanodrug delivery

Some nanomaterials can be used to cross the blood-brain barrier introducing ethical dilemmas, where an action simultaneously causes good and bad effects. Expected benefits include more effective and safer treatment of brain tumours. However, the brain also becomes more vulnerable to accidental biosafety and human-made biosecurity issues. Some nanomaterials can raise dual use biosecurity issues, most notably in interdisciplinary research on converging technologies (Nano, Bio, Info, Cogno). Nanoscience research of concern targets molecular manipulation of virulence factors or directed traversal of the blood-brain barrier by nanoparticles. Preventing misuse is a collective responsibility of governments, scientists, industry and civil society. Researchers should in any case be aware of biosecurity regulations and measures in their laboratory [22].

In addition, nanomaterials can be used in theranostics [23], such as drug delivery systems incorporating sensors and medicine circulating in the bloodstream release the drug when detecting disease biomarkers. This raises an ethical dilemma. While theranostics will allow the cure to be administered without delay upon detection of the specific disease biomarkers, the physician or patient cannot intervene. If the drug is released erroneously, who is accountable for accidental release? How can "*researchers and research organisations ... remain accountable for the ... human health impacts that their N&N research may impose on present ... generations*" [12]? What should researchers do? While pharmaceutical companies and regulators carry most responsibility for any negative consequences of theranostics products after they are finally available on the market, the RRI-principle 'anticipation' suggests that scientists already during the research phase should reflect on

Materials Research Foundations 145 (2023) 311-325 https://doi.org/10.21741/9781644902370-12

potential future impacts on society and the environment and raise discussion timely to avoid foreseeable negative impacts.

6. Personalizing therapies and diagnostics

Nanomaterials and nanotechnologies are enabling technologies for development of personalized and precision medicine, as well as organ-on-chips. These applications give rise to some ethical issues and dilemmas.

6.1 Personalized and precision nanomedicine

Nanomedicine enables personalized medicine, because it allows adapting a drug to a specific cohort of patients [24]. However, the introduction of personalised medicine raises dilemmas. For example, early (nano)diagnostics could be used to reveal personal genetic sensitivities for e.g., cancer or diabetes, allowing early treatment of the disease before the onset of symptoms, or prescribing preventive measures such as diets or changes in lifestyle. Related to this, nanomedicine is expected to enable precision medicine, promising benefits to patients and citizens. For example, "*intelligent nanoparticle design can improve efficacy in general delivery applications while enabling tailored designs for precision applications*" [25].

Some ethical issues must be addressed before introducing personalised or precision nanomedicine in healthcare practices. To begin with, the diagnostic devices collect personal data from the patient. Where and how is this data stored? Is the data storage properly secured against unlawful access? Who should be given access to the test results and for which purposes? In addition, the widespread adoption of early diagnostics may infringe on the autonomy of the person. For example, healthy people may feel obliged to test because of social pressure. Furthermore, personal freedom may be limited by prescribed diets or lifestyle changes. While governments and industry carry most responsibility for ensuring that final personalised nanomedicine products do not infringe on privacy and autonomy of citizens, the value 'responsiveness' calls upon scientists to engage in early public dialogue with citizens and stakeholders and take their views into account in research strategies, while the personalised nanomedicine is still under development.

In addition to these effects on patients themselves, the introduction of personalised or precision nanomedicine increases existing questions of distributive justice in healthcare systems. For example, increased funding for precision medicine for some patients may be balanced by decreased funding addressing health inequalities for others. While researchers do not have the main responsibility for avoiding such consequences, they should at least ensure representativeness of data cohorts allowing fair benefits sharing [20]. This way, they can live up to the value of 'inclusiveness', and respect the universal human right of all people to participate in science and reap the benefits.

Materials Research Foundations 145 (2023) 311-325
https://doi.org/10.21741/9781644902370-12

6.2 Organ-on-chip

"An organ-on-chip is a fit-for-purpose microfluidic device, containing living engineered organ substructures in a controlled microenvironment, that recapitulates one or more aspects of the organ's dynamics, functionality and (patho)physiological response in vivo under real-time monitoring" [26]. Currently, many different human organs and tissues are incorporated in such chips (for example, beating human heart cells), and some even experiment with integrating a complete set of individual organs into a human-on-chip system [27]. Applications include preclinical drug development, and toxicity testing of chemicals. Foreseen benefits include the possibility to replace animal testing with human organs-on-chips. In the longer term, cells from identifiable patients may be used to grow organs-on-chips for personalised diagnostics or in the development of orphan drugs. Like other diagnostics, diagnostic uses of organ-on-chip raises common bioethics issues, briefly mentioned earlier. However, common biomedical privacy and informed consent procedures may have to be revised if cells and tissues from identifiable patients are used to grow organs-on-chips, to manage issues related to personalised medicine. This is especially sensitive if the organ-on-chip is used to diagnose orphan diseases with only small numbers of patients whose privacy is difficult to protect.

In addition, communication about organ-on-chip technology must be sensitive to public perception issues. Scientists should be transparent about the challenges of the real scientific research work in the laboratory, and be careful how they discuss big ambitions (e.g., 'human-on-chip') which may take a long time to materialise. In line with the value 'responsiveness', researchers should engage in public dialogue with citizens and stakeholders and take their views into account in research strategies.

7. Wearables

Nanomaterials and nanotechnologies enable miniaturisation of biosensors and diagnostic devices, contributing to the increasingly widespread use of wearable nanodiagnostics. Some such wearables are explicitly marketed as medical devices subject to regulations, such as the EU medical devices regulation. However, other devices, such as mobile body vital sign tracking devices, are launched as lifestyle gadgets subject to less stringent regulations. In 2015, the European Group on Ethics (EGE) identified gaps in EU regulation on safety of digital health products [20]. Subsequently, the EU updated its medical devices Regulation (EU) 2017/745: "*Certain groups of products for which a manufacturer claims only an aesthetic or another non-medical purpose, but which are similar to medical devices in terms of functioning and risks profile should be covered by this Regulation.*" [28] In addition, the EGE recommended that scientists, industry and other stakeholders must comply better with existing legislation and standards. In line with the value 'responsiveness', researchers should engage in public dialogue with citizens and take their views into account in research strategies for developing innovative health monitoring wearables for both medical and non-medical purposes.

Materials Research Foundations 145 (2023) 311-325 https://doi.org/10.21741/9781644902370-12

8. Artificial Intelligence and big data

Converging with digital twins, Artificial Intelligence (AI) and big data, nanomedicine fosters increasing citizen participation in healthcare. On the one hand, personal health data collected through medical and non-medical wearables is considered a valuable resource for discovering disease pathways and for informing the development of new therapies. Privacy and data protection principles call for restricting access to these sensitive personal data to persons with a need to know them, including patients and medical professionals treating these patients. Secondary use of these data for research purposes is subject to informed consent and rules for anonymization or pseudonymization. The use of AI in analysing big health data could lead to disclosure of the identity of individual patients or other infringements on biomedical ethical principles. When personal data is used to develop new pharmaceuticals or therapies which are sold by commercial companies, questions of ownership of the data and what would constitute a fair share in the benefits emerge. On the other hand, converging (nano)technologies enable miniaturisation of diagnostics devices, bringing self-tests within reach of consumers and patients, who may not be properly trained to interpret the test results. This influences trust between the patient and the doctor, and puts pressure on the organisation of the healthcare system. The EGE (2015) is concerned that this convergence may shift the balance towards 'All for Health' rather than 'Health for All' (SDG 3) [20]. This means that citizens may be coerced increasingly to share their health data or submit to early testing before symptoms of diseases become noticeable. It is not clear whether these trends in healthcare technologies and markets will genuinely contribute to improved public health and universal access to healthcare.

Koen Bruynseels and colleagues discussed ethical implications of digital twins in healthcare [29]. As in engineering, digital twins are increasingly used in healthcare and virtual self models are continuously updated with increasingly refined personal health data of individuals. This enables three fundamental changes of what is considered normal health: The normal state can be defined in increasingly high resolution, can be increasingly personalised, and the personal health status becomes increasingly transparent. This again raises unprecedented privacy and data protection issues, and may drive the demand for preventive therapies before the unset of symptoms. It may also contribute to blurring the boundary between humans and machines, shift the boundary between disease and health, and could contribute to a shift from therapy to human enhancement [29].

Two conflicting RRI-values are at stake in these current trends in converging technologies for healthcare: inclusiveness and anticipation. Inclusiveness means that all people have the right to participate in science and reap the benefits. This appears to welcome the general trends in digitalisation of health, because affordable and easy to use diagnostics are expected to become within reach of larger segments of the population, and preventive measures can be taken earlier, before people fall ill. On the other hand, anticipation calls for reflecting on potential future impacts on society and the environment already during the research phase. From this perspective, researchers are advised to consider philosophical analyses of longer term and wide ranging impacts of digitalisation of healthcare [20,29] and engage in public dialogue about these potential consequences.

Materials Research Foundations 145 (2023) 311-325 https://doi.org/10.21741/9781644902370-12

9. Human enhancement

As briefly touched upon in section 7, nanomedicine may not only be used in healthcare, but also to enhance the physical or mental capabilities of healthy people. More traditional examples of sensitive trends in nanomedicine are nanotechnologies enabling neuro-prosthetics, or connecting human brains online via the internet. This raises many ethical issues, such as the following. Healthy individuals using nanomedicine to improve specific physical or mental traits, may accidentally expose themselves to unexpected health risks. As Paracelsus already warned: only the dose determines the toxicity of a substance [35]. When for instance nanopharmaceuticals are taken curatively, the benefit of fighting the disease generally outweighs the health risks of the drugs. When the same substances are taken by a healthy person, the risk-benefit balance shifts towards the risks. In addition to the impacts on the individual, use of human enhancement technologies by elite groups has implications for the organization of society, and may topple the balance between freedom and solidarity. On a more abstract level, human enhancement also raises discussion of religious and cultural aspects (e.g., 'playing God') [30].

Guided by the RRI-value 'anticipation' researchers should reflect on potential future impacts on society already during the research phase. The RRI-value 'responsiveness' calls for engaging in public dialogue with citizens and stakeholders and take their views into account in research strategies.

Conclusions

Nanomedicine can be applied in drug delivery, diagnostics and pharmaceutical research. Converging with other emerging technologies, nanomedicine enables digital twins, organ on chip and wearables monitoring vital signs. Introducing nanomaterials in the body raises nanosafety issues. Nanomedicine raises ethical issues related to freedom, equality, data protection and biosecurity, and may impact on the healthcare system. Nanomedicine may enable human enhancement, raising ethical and religious concerns. Researchers should contribute to Responsible Research and Innovation through different case-dependent strategies. The principles inclusiveness, anticipation, openness and responsiveness are leading.

Acknowledgements

This chapter is based on a lecture on *Ethics of Nanomedicine* held during the course on ethics of biomedical research organised by the VISION project. *The preparation of the lecture was supported by the EU H2020 RiskGONE project, GA no. 814425.*

References

[1] Nanomedicine European Technology Platform, What is nanomedicine? | ETPN, S/F. (n.d.). https://etp-nanomedicine.eu/about-nanomedicine/what-is-nanomedicine/ (accessed November 27, 2022)

Materials Research Foundations 145 (2023) 311-325
https://doi.org/10.21741/9781644902370-12

[2] I. Malsch, P. Isigonis, M. Dusinska, E.A. Bouman, Embedding Ethical Impact Assessment in Nanosafety Decision Support, Small. 16 (2020) 2002901. https://doi.org/10.1002/smll.202002901

[3] Draft guidelines on Identification of regulatory and ethical risk thresholds, (n.d.)

[4] Ethics assessment for research and innovation-Part 2: Ethical impact assessment framework, (2017)

[5] T. Beauchamp, The Principle of Beneficence in Applied Ethics, (2008). http://plato.stanford.edu/entries/principle-beneficence/ (accessed November 27, 2022)

[6] C. Friend, Social Contract Theory | Internet Encyclopedia of Philosophy, Internet Encycl. Philos. (2016). http://www.iep.utm.edu/soc-cont/ (accessed November 27, 2022)

[7] G. Sliwoski, S. Kothiwale, J. Meiler, E.W. Lowe, Computational methods in drug discovery, Pharmacol. Rev. 66 (2014) 334–395. https://doi.org/10.1124/pr.112.007336

[8] R.L. Burrows, A.W.H. Adkins, Merit and Responsibility: A Study in Greek Values, Clarendon Press, 1960. https://doi.org/10.2307/4344370

[9] The social control of technology (1980 edition) | Open Library, (n.d.). https://openlibrary.org/books/OL14443859M/The_social_control_of_technology (accessed November 27, 2022)

[10] R. Owen, P. Macnaghten, J. Stilgoe, Responsible research and innovation: From science in society to science for society, with society, Sci. Public Policy. 39 (2012) 751–760. https://doi.org/10.1093/scipol/scs093

[11] E. Commission, D.-G. for R. and Innovation, Responsible research and innovation : Europe's ability to respond to societal challenges, Publications Office, 2012. https://doi.org/doi/10.2777/11739

[12] COMMISSION RECOMMENDATION of 07/02/2008 on a code of conduct for responsible nanosciences and nanotechnologies research, (2008). https://ec.europa.eu/research/participants/data/ref/fp7/89918/nanocode-recommendation_en.pdf (accessed December 5, 2022)

[13] ECHA, EUON European Observatory for Nanomaterials, Euon.Echa.Europa.Eu. (n.d.). https://euon.echa.europa.eu/cosmetics1 (accessed November 27, 2022)

[14] Cosmetics - European Observatory for Nanomaterials, (n.d.). https://euon.echa.europa.eu/cosmetics1 (accessed December 5, 2022)

[15] C. Ganzleben, A. Kazmierczak, Leaving no one behind - Understanding environmental inequality in Europe, Environ. Heal. A Glob. Access Sci. Source. 19 (2020). https://doi.org/10.1186/s12940-020-00600-2

[16] J.R.C. European Commission, Safe and sustainable by design chemicals and materials - Publications Office of the EU, (n.d.). https://op.europa.eu/en/publication-

detail/-/publication/567e3b0f-a66a-11ec-83e1-01aa75ed71a1/language-en (accessed November 27, 2022)

[17] V. Buocikova, I. Rios-Mondragon, E. Pilalis, A. Chatziioannou, S. Miklikova, M. Mego, K. Pajuste, M. Rucins, N. El Yamani, E.M. Longhin, A. Sobolev, M. Freixanet, V. Puntes, A. Plotniece, M. Dusinska, M.R. Cimpan, A. Gabelova, B. Smolkova, Epigenetics in breast cancer therapy—New strategies and future nanomedicine perspectives, Cancers (Basel). 12 (2020) 1–32. https://doi.org/10.3390/cancers12123622

[18] M.M.T. van Leent, B. Priem, D.P. Schrijver, A. de Dreu, S.R.J. Hofstraat, R. Zwolsman, T.J. Beldman, M.G. Netea, W.J.M. Mulder, Regulating trained immunity with nanomedicine, Nat. Rev. Mater. 7 (2022) 465–481. https://doi.org/10.1038/s41578-021-00413-w

[19] S. Hertel, L. Minkler, International Covenant on Economic, Social, and Cultural Rights, Econ. Rights. (2009) 385–394. https://doi.org/10.1017/cbo9780511511257.019

[20] EGE Opinions, (n.d.). https://research-and-innovation.ec.europa.eu/knowledge-publications-tools-and-data/publications/all-publications/ege-opinions_en#files (accessed November 27, 2022)

[21] Attitudes on vaccination against Covid-19 - junho 2021 - - Eurobarometer survey, (n.d.). https://europa.eu/eurobarometer/surveys/detail/2512 (accessed November 27, 2022)

[22] K. Eggleson, Dual-use nanoresearch of concern: Recognizing threat and safeguarding the power of nanobiomedical research advances in the wake of the H5N1 controversy, Nanomedicine Nanotechnology, Biol. Med. 9 (2013) 316–321. https://doi.org/10.1016/j.nano.2012.12.001

[23] R.S. Kalash, V.K. Lakshmanan, C.S. Cho, I.K. Park, Theranostics, Biomater. Nanoarchitectonics. (2016) 197–215. https://doi.org/10.1016/B978-0-323-37127-8.00012-1

[24] C. Fornaguera, M.J. García-Celma, Personalized nanomedicine: A revolution at the nanoscale, J. Pers. Med. 7 (2017) 12. https://doi.org/10.3390/jpm7040012

[25] M.J. Mitchell, M.M. Billingsley, R.M. Haley, M.E. Wechsler, N.A. Peppas, R. Langer, Engineering precision nanoparticles for drug delivery, Nat. Rev. Drug Discov. 20 (2021) 101–124. https://doi.org/10.1038/s41573-020-0090-8

[26] M. Mastrangeli, S. Millet, C. Mummery, P. Loskill, D. Braeken, W. Eberle, M. Cipriano, L. Fernandez, M. Graef, X. Gidrol, N. Picollet-D'Hahan, B. van Meer, I. Ochoa, M. Schutte, J. Van den Eijnden-van Raaij, Organ-on-Chip in Development. ORCHID Final Report, 36 (2019) 650–668. https://h2020-orchid.eu/wp-content/uploads/2020/02/ORCHID-Summary.pdf/ (accessed November 27, 2022)

Materials Research Forum LLC
https://doi.org/10.21741/9781644902370-12

[27] Human Body-on-Chip platform enables in vitro prediction of drug behaviors in humans, (n.d.). https://wyss.harvard.edu/news/human-body-on-chip-platform-enables-in-vitro-prediction-of-drug-behaviors-in-humans/ (accessed November 27, 2022)

[28] EMA, Medical devices | European Medicines Agency, EMA.Europa.Eu. (2020). https://www.ema.europa.eu/en/human-regulatory/overview/medical-devices%0Ahttps://www.ema.europa.eu/en/human-regulatory/overview/medical-devices%0Ahttps://www.ema.europa.eu/en/human-regulatory/overview/medical-devices#companion-diagnostics-('in-vitro-diagn (accessed November 27, 2022)

[29] K. Bruynseels, F.S. de Sio, J. van den Hoven, Digital Twins in health care: Ethical implications of an emerging engineering paradigm, Front. Genet. 9 (2018) 31. https://doi.org/10.3389/fgene.2018.00031

[30] N.M. Noah, P.M. Ndangili, Current Trends of Nanobiosensors for Point-of-Care Diagnostics, J. Anal. Methods Chem. 2019 (2019). https://doi.org/10.1155/2019/2179718

Keyword Index

About the Editor

Jorddy Neves Cruz is a researcher at the Federal University of Pará and the Emilio Goeldi Museum. He has experience in multidisciplinary research in medicinal chemistry, drug design extraction of bioactive compounds, extraction of essential oils, food chemistry, and biological testing. He has published several research articles of international repute. He works as Editor in the Journal of Medicine, Molecules and Frontiers in Chemistry. He successfully served as an Editor to five books published by world-renowned publishers.

www.ingramcontent.com/pod-product-compliance
Lightning Source LLC
Chambersburg PA
CBHW071324210326
41597CB00015B/1334

* 9 7 8 1 6 4 4 9 0 2 3 6 3 *